Malcolm Wray Dunlop · Hermann Lühr
Editors

Ionospheric Multi-Spacecraft Analysis Tools

Approaches for Deriving Ionospheric Parameters

Editors
Malcolm Wray Dunlop
School of Space and Environment
Beihang University
Beijing, P.R. China

RAL Space, Rutherford Appleton Laboratory
STFC-UKRI
Chilton, UK

Hermann Lühr
GFZ German Research Centre for Geosciences
Helmholtz-Centre Potsdam
Potsdam, Brandenburg, Germany

https://doi.org/10.1007/978-3-030-26732-2

Dedication

Dr. Olaf Amm had originally proposed the ISSI Working Group "Multi Satellite Analysis Tools—Ionosphere", which was later implemented by the Directorate. Sadly, and far too early, he passed away on 16 December 2014, while in the midst of organizing this Working Group with the international team. In order to recover the started activities of the Working Group, the editors of this book took over the lead. The editors dedicate this volume to Olaf and have asked me, a close colleague of Olaf, to write this note.

Olaf, Senior Research Scientist at the Finnish Meteorological Institute, was a widely appreciated researcher, especially in the fields of ionospheric electrodynamics and magnetosphere- ionosphere coupling processes. He developed several new innovative approaches for creating regional maps of ionospheric currents, conductances and fields with adjustable space resolution from the basis of ground- or space-based magnetic and electric field measurements. The methods that Olaf developed are widely used by other scientists.

Olaf was born in Rendsburg, Germany. He started his studies in geophysics at the University of Münster in early 1990s and finalized his doctoral thesis at the Technical University of Braunschweig in 1998. Soon after the doctoral defense, Olaf moved to Finland, where he served at the Finnish Meteorological Institute, first as a post-doc researcher, and later as a senior scientist and supervisor for several doctoral students. He recurrently gave lectures on ionospheric physics and potential theory applications in space physics at the Department of Physics of the University of Helsinki and was nominated as a Docent in Space Physics in 2002.

Olaf was the principal investigator of the TomoScand network for ionospheric tomography measurements and a member of the PI-Team of the MIRACLE network of magnetometers and auroral cameras. He served also as co-investigator for the flux-gate magnetometers on the Double Star satellites. The methods that Olaf developed first for the Fennoscandian ground-based instrument networks are nowadays widely used elsewhere when data from multipoint sources are processed. During recent years he had several visiting professorship periods at the Universities of Nagoya (at STELAB) and Kyushu.

Olaf was eagerly awaiting the first measurements by the ESA Swarm mission, launched end of 2013, which would in many ways offer valuable reference material for his theoretical ideas. As preparatory work for Swarm, in 2013 he led an ESA STSE project where a novel analysis method for Swarm electric and magnetic field measurements was developed. Applications of his innovative approach to Swarm and other satellite data are reported in several chapters of this monograph.

Kirsti Kauristie
Finnish Meteorological Institute
Helsinki, Finland

Preface

The goal of this volume (First proposed by Olaf Amm; please see Dedication) is to provide a comprehensive "tool book" of analysis techniques for ionospheric multi-satellite missions. The immediate need for this book is motivated by the ESA Swarm satellite mission, but the tools that will be described are general and can be used for any future ionospheric multi-satellite mission with comparable instrumentation. The title is intentionally chosen to be similar to the first ISSI Scientific Report, SR-001 (Analysis Methods for Multi-Spacecraft Data), which was motivated by the ESA Cluster multi-satellite mission in the magnetosphere.

In the ionosphere, a different plasma environment prevails that is dominated by interactions with neutrals, and the Earth's main magnetic field clearly dominates the total magnetic field. Further, an ionospheric multi-satellite mission has different research goals than a magnetospheric one, namely in addition to the study of the immediate plasma environment and its coupling to other regions also the study of the Earth's main magnetic field and its anomalies caused by core, mantle, or crustal sources. Therefore, different tools are needed for an ionospheric multi-satellite mission as compared to a magnetospheric one, and different parameters are desired to be determined with those tools. Besides currents, electric fields and plasma convection, such parameters include: ionospheric conductances, Joule heating, neutral gas densities and neutral winds.

This book is an outcome of the Working Group set up at the International Space Science Institute (ISSI) in Bern entitled "Multi Satellite Analysis Tools—Ionosphere". It will focus on techniques that are able to derive such local plasma parameters from the immediate multi-satellite measurements and on techniques that can link these locally derived plasma parameters with observations made by other instruments in adjacent domains (including observations by other satellite missions, such as Cluster and ground-based observations), in order to determine the coupling between that domain and the ionosphere. In terms of the study of the Earth's main magnetic field, this book limits itself to tools that utilize the multi-satellite ionospheric observations in order to minimize errors in the main magnetic field modeling. It therefore does not include dedicated techniques that are designed to determine core, mantle, or crustal magnetic anomalies separately from the main geomagnetic field. We believe that this book will become a reference volume for the ESA Swarm mission, as well as for future ionospheric multi-satellite missions. A first meeting of this working group was held at ISSI on 14–16 September 2015. All the members presented their possible contributions. As a result we defined the outline structure of the book and assigned the chapters, which focus predominantly on currents and magnetic modeling.

The members of the ISSI Working Group are: Tomoko Matsuo (University of Colorado, Boulder, USA), Joachim Vogt (Jacobs University Bremen, Germany), Colin L. Waters (Newcastle University, Australia), Chris Finlay (Danish Technical University, Denmark), Robyn A. D. Fiori (Natural Resources Canada, Canada), Patrick Alken (National Centers for Environmental Information, NOAA, USA), and it is led by Malcolm Wray Dunlop (Rutherford Appleton Laboratory, UK) and Hermann Lühr (GFZ-German Research Centre for Geosciences, Germany).

Oxford, UK Malcolm Wray Dunlop
Potsdam, Germany Hermann Lühr

Contents

Chapter 1
Introduction

Malcolm Wray Dunlop and Hermann Lühr

Abstract Two volumes, ISSI Scientific Reports, SR-001: Analysis Methods for Multi-Spacecraft Data and SR-008: Multi-Spacecraft Analysis Methods revisited, were published to document the growing toolset using the multi-spacecraft dataset being collected by Cluster.

Two volumes, ISSI Scientific Reports, SR-001: Analysis Methods for Multi-Spacecraft Data and SR-008: Multi-Spacecraft Analysis Methods revisited, were published to document the growing toolset using the multi-spacecraft dataset being collected by Cluster. Cluster was the first phased, multi-spacecraft mission, currently in its 19th year of full science operations, to maintain a close configuration of four spacecraft, evolving around an orbit covering many mid- to outer magnetospheric regions. Such a configuration allowed the estimation of plasma and field gradients, as well as wave vector determinations for the first time. A range of spatial scales were accessed through a sequence of orbital manoeuvres, predominantly from meso- to large scale spacecraft separation distances. Although covering a vast array of science targets, Cluster did not cover the small (sub-ion) spatial scales and did not access the low-Earth orbit (LEO) altitudes suitable for the upper ionosphere.

Since Cluster the Magnetospheric Multi-Scale (MMS) mission has now taken four spacecraft measurements on spatial separation scales of tens of kilometers, but still in equatorial orbits reaching the outer magnetosphere and solar wind. Swarm is the first multi-spacecraft, LEO mission, comprising 3 spacecraft in polar circular orbits at altitudes of 460 km and 510 km. Two Swarm spacecraft have maintained approximately east-west separations of 1.4° in longitude, corresponding to distances of typically 50 km in the high latitude regions, while the third has drifted relative to the other two in local time. Swarm established full science operations in April 2014

M. W. Dunlop (✉)
School of Space and Environment, Beihang University, Beijing 100191, P.R. China
e-mail: malcolm.dunlop@stfc.ac.uk

RAL Space, Rutherford Appleton Laboratory, STFC-UKRI, Chilton, UK

H. Lühr
GFZ-German Research Centre for Geosciences, Potsdam, Germany

M. W. Dunlop and H. Lühr (eds.), *Ionospheric Multi-Spacecraft Analysis Tools*, ISSI Scientific Report Series 17,
https://doi.org/10.1007/978-3-030-26732-2_1

and has been operating nominally since. Together with Swarm, other spacecraft arrays have been achieved at low orbit, such as the AMPERE experiment of the Iridium spacecraft array. The Iridium array comprises 66 active satellites in a set of crossing polar orbits covering all local times. These new measurements have allowed both local gradients and global currents to be mapped routinely, and a new set of methodology has arisen as a result.

The present volume (Multi-Satellite Data Analysis: Approaches for Deriving Ionospheric Parameters) documents a set of methods, modelling and analysis suitable for the low altitude (ionospheric) regions. It describes a range of approaches, from the gradient calculation for current density, to new global modelling of the geomagnetic field, through techniques for polar cap mapping.

We have organized the book by grouping chapters with common themes and start with two chapters on Olaf Amm's Spherical Elementary Current Systems (SECS) approach (see Preface). Following this, Chaps. 4–8 deal with field-aligned current (FAC) estimates; Chaps. 9 and 10 describe approaches for combining observations from different sources for deriving continuous maps for physical quantities, and Chaps. 11 and 12 are examples of Swarm constellation applications for deriving current or field distributions. During the late stages of this Working Group, ESA's Swarm project management set up a workshop as part of its Swarm DISC (Data, Innovation and Science Cluster) activity to assess the various field-aligned current estimating methods: Methodology Inter-Comparison Exercise (MICE), and this provided some form of rating of the various approaches. We have therefore included in Chap. 8 a summary of this exercise.

Chapter 2: Introduction to Spherical Elementary Current Systems is a review of the basic SECS method, covering its various applications to the study of ionospheric current systems, identified by both ground-based and satellite observations. The chapter concentrates on the general approach, starting with a review of ionospheric electrodynamics and a definition of the elementary current systems. The 1-D SECS approach is outlined. The details of its application to the Swarm electric and magnetic field data are left to Chap. 3: Spherical Elementary Current Systems applied to Swarm data; in particular describing the two dimensional (in latitude and longitude) maps of the ionospheric FACs and horizontal currents which surround the satellite path. Similarly the electric field and conductance can be obtained when data is available.

Chapter 4: Local least squares analysis of auroral currents, probes the first methodology into the form of structures using multi-spacecraft constellation data and highlights the technique for the two and three spacecraft data of Swarm. The chapter also discusses techniques for the geometrical characterization of auroral current structures with observations under stationary or weakly time dependent conditions.

Chapter 5: Multi-spacecraft current estimates at Swarm, applies the older, established Curlometer technique, which has been previously used with the four-spacecraft constellation data of Cluster over the past two decades. The Curlometer directly estimates the current density from the curl of the magnetic field and this chapter focusses on the extension and application of the technique to the two and three spacecraft Swarm data. The chapter also reviews examples of the coordination of signals seen simultaneously between Cluster and Swarm and the application of the method to in situ estimates of current density in the Earth's ring current using other magnetospheric satellite data.

Chapter 6: Applying the dual-spacecraft approach to the Swarm constellation for deriving radial current density, reviews the standardized method, derived from the basic Curlometer concept, which is adopted by ESA for production of the Swarm Level 2 FAC products. As well as describing this method and possible errors, special emphasis is placed on the underlying assumptions and limitations on the approach, which include features associated with plasma instabilities and disturbances. The applicability of the method in different regions, depending on orbital constraints on the one hand and scale sizes on the other is discussed.

Chapter 7: Science data products for AMPERE, describes a methodology to analyze the magnetic field data measured by the global spacecraft array which is based on an orthogonal basis function expansion and associated data fitting. The AMPERE experiment uses magnetic field data from the attitude control system of the Iridium satellites and estimates data products based on the theory of magnetic fields and currents on spherical shells. The chapter discusses the application of the spherical cap harmonic basis and elementary current system methods to generate the AMPERE science data products.

Chapter 8: ESA Field Aligned Currents—Methodology Inter-Comparison Exercise, summarizes the MICE activity referred to above. The activity explored the possible evolution for the Swarm Level 2 FAC products by inter-comparing a number of different approaches on a test data base of Swarm auroral crossings. The chapter here describes the different strengths and assumptions in each method and outlines possible future activities. The known caveats on use of the methods are discussed in terms of the expected properties and scales of FACs.

Chapter 9: Spherical Cap Harmonic Analysis techniques for mapping high-latitude ionospheric plasma flow—Application to the Swarm satellite mission, introduces and describes a tool for mapping a variety of one, two, and three-dimensional parameters. The chapter outlines the theoretical basis through a discussion of the spherical cap coordinate system. The boundary conditions and basis-functions are discussed and practical considerations are summarized. The application of SCHA to the mapping of ionospheric plasma flow using a ground-based data set is also given and two-dimensional SCHA is shown applied to the mapping of Swarm ion drift measurements, as well as in conjunction with measurements from other instruments.

Chapter 10: Recent Progress on Inverse and Data Assimilation Procedure for High-Latitude Ionospheric Electrodynamics, discusses the development of this technique with an emphasis on the historical inversion of ground-based magnetometer observations. The method provides a way to obtain complete maps of high-latitude ionospheric electrodynamics; overcoming the limitations of a given geospace monitoring system. The chapter outlines recent technical progress, which is motivated by recent increase in availability of regular monitoring of high-latitude electrodynamics by space-borne instruments. The method description includes state variable representation by polar-cap spherical harmonics, where coefficients are estimated in the Bayesian inferential framework. Applications to SuperDARN plasma drift data,

AMPERE measurements and DMSP magnetic field and auroral particle precipitation data are covered.

Chapter 11: Estimating currents and electric fields at low-latitudes from satellite magnetic measurements, presents techniques developed for processing magnetic measurements of the equatorial electrojet (EEJ) current to extract information about the low-latitude currents and their driving electric fields. The chapter presents a multiple line current approach to recover the EEJ current density distribution and emphasises the issues relating to the cleanliness of the satellite data and the minimization of the magnetic fields arising from other internal and external sources. The electric field determination uses a combination of physical modelling and fitting of the EEJ current strengths measures by the Swarm satellites. Such methods, which attain a global knowledge of the spatial structure of the low latitude currents, give insight into the atmospheric tidal harmonics present at ionospheric altitudes.

Chapter 12: Modelling the internal geomagnetic field using data from multiple satellites and field gradients–Applications to the Swarm satellite mission, reviews how models of the main magnetic field are constructed from multiple satellites, such as Swarm. The focus is on how to take advantage of estimated field gradients, both along-track and across track. The chapter summarises recent results from the Swarm mission dealing with the core and lithospheric fields. The aim is to inform users interested in ionospheric applications. Limitations of the current generation of main field models are also discussed pointing out that further progress requires improved treatment of ionospheric current systems, particularly at polar latitudes.

These chapters cover very different methodologies, but do have overlaps in techniques, and these have been referenced within and between each chapter. We thank all the authors for the substantial amount of effort needed to put this collection of work together.

Chapter 2
Introduction to Spherical Elementary Current Systems

Heikki Vanhamäki and Liisa Juusola

Abstract This is a review of the Spherical Elementary Current System or SECS method, and its various applications to studying ionospheric current systems. In this chapter, the discussion is more general, and applications where both ground-based and/or satellite observations are used as the input data are discussed. Application of the SECS method to analyzing electric and magnetic field data provided by the Swarm satellites will be discussed in more detail in the next chapter.

2.1 Introduction

At high magnetic latitudes, the ionospheric current system basically consist of horizontal currents flowing around 100–150 km altitude, and almost vertical field-aligned currents (FAC) flowing along the geomagnetic field, thus connecting the ionospheric currents to the magnetosphere. The magnitude, spatial distribution, and temporal variations of the horizontal currents and FAC can be estimated from the magnetic field they produce. Over the years, several techniques have been developed for this task, as discussed in various Chapters of this book (see also Vanhamäki and Juusola 2018, and reference therein). The present chapter gives an overall introduction to the

The original version of this chapter was revised. Electronic Supplementary Material has been added to this chapter. The correction to this chapter is available at https://doi.org/10.1007/978-3-030-26732-2_13

Electronic supplementary material The online version of this chapter (https://doi.org/10.1007/978-3-030-26732-2_2) contains supplementary material, which is available to authorized users.

H. Vanhamäki (✉)
University of Oulu, Oulu, Finland
e-mail: Heikki.Vanhamaki@oulu.fi

L. Juusola
Finnish Meteorological Institute, Helsinki, Finland

M. W. Dunlop and H. Lühr (eds.), *Ionospheric Multi-Spacecraft Analysis Tools,* ISSI Scientific Report Series 17,
https://doi.org/10.1007/978-3-030-26732-2_2

Spherical Elementary Current System (SECS) method, while Chap. 3 deals with the specific application of the SECS method to magnetic data provided by the Swarm satellite mission.

Mathematically speaking, the elementary systems form a set of basis functions for representing two-dimensional vector fields on a spherical surface. This can, of course, be done in other ways too, e.g., by using spherical harmonic or spherical cap harmonic functions. The main difference is that the elementary systems represent the vector field in term of its divergence and curl, whereas harmonic functions are used to represent the scalar potential and stream function of the vector field. In principle, these methods should be equivalent, but in practice, each has its strengths and weaknesses. As will be seen, advantages of the SECS method include adjustable grid resolution, variable shape of the analysis region and no requirement for explicit boundary conditions.

The chapter begins with a summary of some basic electrodynamic properties of ionospheric current systems and the most commonly used approximations in Sect. 2.2. The 2D SECSs are introduced in Sect. 2.3. Their applications to analysis of two-dimensional vector fields and magnetic fields are discussed in Sects. 2.4–2.7. A one-dimensional variant of the SECS method, applicable to studies of single-satellite magnetic measurements, is discussed in Sect. 2.9. Some practical issues when applying the SECS method are discussed in Sect. 2.10. Finally, a short overview of some of the studies where the SECS method has been used is given in Sect. 2.11.

An example MATLAB code demonstrating the use of SECS in the specific task of estimating ionospheric equivalent current from ground magnetic measurements is included as supplementary material in the electronic version of the book, including data from the IMAGE (International Monitor for Auroral Geomagnetic Effects[1]) magnetometer network.

2.2 Short Review of Ionospheric Electrodynamics

A short summary of the relevant properties of ionospheric electrodynamics, especially at high magnetic latitudes (i.e., the auroral oval), is given in this section. For a more comprehensive introduction see, for example, Richmond and Thayer (2000). In the context of this chapter, ionospheric electrodynamics is described by the electric field, and the Hall and Pedersen conductivities and currents. Additionally, the magnetic perturbation created by the ionospheric currents is an important quantity in many studies. Thus, the focus is on macroscopic electric parameters, while many interesting phenomena, such as various chemical processes and particle dynamics, are ignored.

In the commonly used thin-sheet approximation (see e.g., Untiedt and Baumjohann (1993)) the ionosphere is assumed to be a thin, two-dimensional spherical shell of radius R at a constant distance from the Earth's center. The thin-sheet approxi-

[1] See http://space.fmi.fi/image/.

mation is justified by the fact that the horizontal currents flowing in the ionosphere are concentrated to a rather thin layer around 100–150 km altitude, where the Pedersen and Hall conductivities have their maxima. Thus the thickness of this layer is small compared to the horizontal length scale of typical ionospheric current systems. However, in some cases, three-dimensional modeling is required (Amm et al. 2008).

Above the ionospheric current sheet there is perfectly conducting plasma, where magnetic field lines are equipotentials, and below is the nonconductive neutral atmosphere. The electric field is assumed to be roughly constant in altitude through the thin current layer. Thus the Pedersen and Hall conductivities can be height integrated into Pedersen and Hall conductances, while the sheet current density **J** is obtained by similarly height integrating the horizontal part $\mathbf{j}_h$ of the 3D current **j**.

In summary, the main electrodynamic variables are: horizontal sheet current density **J**, field-aligned current density $j_{\|}$, horizontal electric field **E**, magnetic field **B** and height-integrated Hall and Pedersen conductances Σ_H and Σ_P. These variables are related through Maxwell's equations, Ohm's law, and current continuity:

$$(\nabla \times \mathbf{E})_r = -\frac{\partial B_r}{\partial t} \tag{2.1}$$

$$\nabla \times \mathbf{B} = \mu_0 \mathbf{j} = \mu_0 \mathbf{J}\delta(r - R) - \mu_0\, j_{\|}\, \hat{\mathbf{e}}_r \tag{2.2}$$

$$\mathbf{J} = \Sigma_P \mathbf{E} - \Sigma_H \hat{\mathbf{e}}_r \times \mathbf{E} \tag{2.3}$$

$$j_{\|} = \nabla \cdot \mathbf{J}. \tag{2.4}$$

In the last equation, the FAC density $j_{\|}$ just above the ionospheric current sheet is obtained by integrating the continuity equation $\nabla \cdot \mathbf{j} = 0 \Leftrightarrow \partial_z j_z = -\nabla_h \cdot \mathbf{j}_h$ through the current sheet.

Equations (2.1)–(2.4) employ the frequently used assumption of a radial magnetic field, so that $\hat{\mathbf{e}}_{\|} = \mathbf{B}/|\mathbf{B}| = -\hat{\mathbf{e}}_r$ at the northern hemisphere. Due to the thin-sheet approximation, only the radial component is needed in Eq. (2.1). According to Untiedt and Baumjohann (1993) and Amm (1998), the effect of the tilted field lines is negligible for inclination angles $\chi > 75°$, which covers the auroral zone. At lower latitudes the inclination of the magnetic field could be taken into account by modifying the Hall and Pedersen conductances in Eq. (2.3) (see e.g., Brekke 1997, Chap. 7.12) and by calculating the FAC as $j_{\|} = \nabla \cdot \mathbf{J}/\sin\chi$.

In a thin-sheet ionosphere the electric field **E** and horizontal current **J** are two-dimensional vector fields, each of which can be represented by two potentials

$$\mathbf{E} = -\nabla\phi_E - \hat{\mathbf{e}}_r \times \nabla\psi_E \tag{2.5}$$

$$\mathbf{J} = -\nabla\phi_J - \hat{\mathbf{e}}_r \times \nabla\psi_J. \tag{2.6}$$

The function ϕ_E is the usual electrostatic potential and ψ_E is related to the rotational inductive part of the electric field (see e.g., Yoshikawa and Itonaga 1996; Sciffer et al. 2004). It is usually assumed that $\nabla\psi_E = 0$, but this does not hold in some situations (e.g., Vanhamäki and Amm 2011, and references therein). The current potential ϕ_J is connected to FAC through Eq. (2.4), while ψ_J represents a rotational current that is closed within the ionospheric current sheet. The latter part is also related to so called ionospheric equivalent current and ground magnetic disturbance, as discussed in Sect. 2.7.

2.3 Elementary Current Systems

In Sect. 2.2, the electric field and current were described in terms of potentials. This kind of representation is very common in many fields of physics, and can be applied by expanding the potential in terms of some basis functions, such as Fourier series, spherical harmonics or spherical cap harmonics (see, for example, Backus 1986, and Chap. 9 in this Book).

However, the fields can equally well be represented in terms of their sources and rotations, that is by their divergence and curl. This approach is used in the elementary system method. It is based on Helmholtz's theorem, which states that any well-behaved (e.g., continuously differentiable) vector field is uniquely composed of a sum of curl-free (CF) and divergence-free (DF) parts.

Elementary current systems, as applied to ionospheric current systems, were introduced by Amm (1997). Although for historical reasons the name refers to currents, they can be used to represent any two-dimensional vector field. Basically, they represent a localized curl or divergence of the vector field. Such elementary systems can be defined either in spherical or Cartesian geometry, and they are called SECS and CECS, respectively. In this chapter, the spherical variant is used.

In accordance with Helmholtz's theorem, there are two different types of elementary systems: one is DF and the other CF. The spherical elementary systems, shown in Fig. 2.1, are defined in such a way that the CF system has a Dirac δ-function divergence and the DF system a δ-function curl at its pole, with uniform and oppositely directed sources elsewhere. It is easy to show (Amm 1997) that the vector fields

$$\mathbf{V}^{CF}(\mathbf{r}') = \frac{S^{CF}}{4\pi R}\cot\left(\frac{\theta'}{2}\right)\hat{\mathbf{e}}_{\theta'} \tag{2.7}$$

$$\mathbf{V}^{DF}(\mathbf{r}') = \frac{S^{DF}}{4\pi R}\cot\left(\frac{\theta'}{2}\right)\hat{\mathbf{e}}_{\phi'}. \tag{2.8}$$

have the desired properties of

$$\nabla\cdot\mathbf{V}^{CF} = S^{CF}\left(\delta(\theta',\phi') - \frac{1}{4\pi R^2}\right) \tag{2.9}$$

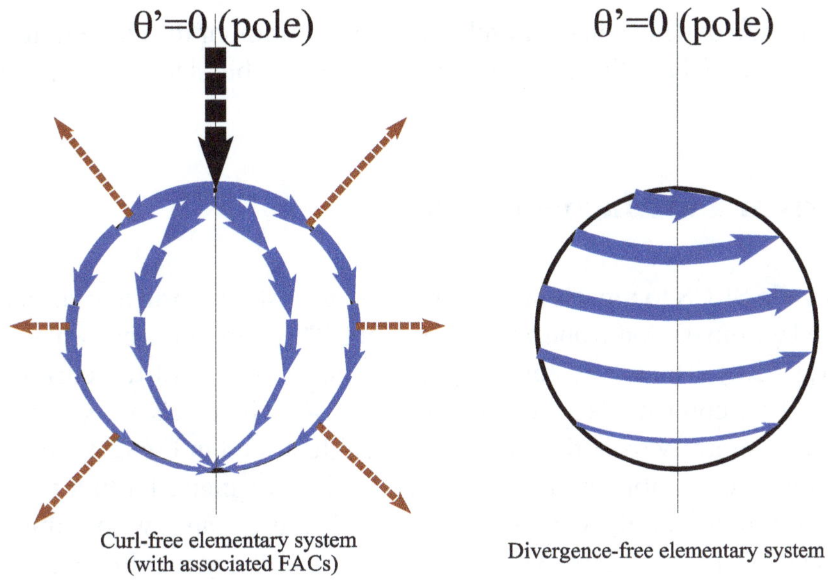

Fig. 2.1 Two-dimensional curl-free (CF) and divergence-free (DF) Spherical Elementary Current Systems (SECS). The CF SECS is shown with associated radial FAC. Adapted from Amm and Viljanen (1999)

$$\left(\nabla \times \mathbf{V}^{CF}\right)_r = 0 \tag{2.10}$$

$$\nabla \cdot \mathbf{V}^{DF} = 0, \tag{2.11}$$

$$\left(\nabla \times \mathbf{V}^{DF}\right)_r = S^{DF}\left(\delta(\theta', \phi') - \frac{1}{4\pi R^2}\right). \tag{2.12}$$

Here, S^{CF} and S^{DF} are the scaling factors of the elementary systems, while R is the radius of the sphere (e.g., ionosphere) where elementary systems are placed. The above formulas are given in a spherical coordinate system (r, θ', ϕ'), with unit vectors $(\hat{\mathbf{e}}_r, \hat{\mathbf{e}}_{\theta'}, \hat{\mathbf{e}}_{\phi'})$, oriented so that center of the elementary systems is at $\theta' = 0$. This coordinate system is used in the definition of the elementary system, as the expressions take the most simple form there. In the actual analysis, the elementary systems are rotated to a more suitable coordinate system, such as the geographical or geomagnetic system, as discussed in Sect. 2.5.

Using the theory of Green's functions it can be shown (e.g., Vanhamäki and Amm 2011) that the CF and DF SECS form a complete set of basis functions for representing two-dimensional vector fields on a sphere. An individual CF SECS with its pole located at $(R, \theta^{el}, \phi^{el})$ represents a source or sink of a vector field at that point, while a DF SECS represents rotational vector field around that point. Thus, by placing a sufficient number of CF and DF SECS at different locations at the ionosphere, one can construct any two-dimensional vector field from its sources and curls, in accordance with Helmholtz's theorem. In principle, the spatial resolution of the representation depends on the number and distribution of the elementary systems. However, in practical applications the amount of available data is a limiting factor.

2.4 Current and Magnetic Field

When using the SECS to represent currents, the DF systems form a rotational current that is closed within the ionospheric current sheet. This part of the current is described by ψ_J in Eq. (2.6). The CF systems represent the same part of the current as ϕ_J in Eq. (2.6), and are connected to the FAC via Eq. (2.4). The FACs are assumed to flow radially toward or away from the ionosphere, as illustrated in Fig. 2.1. As mentioned before, this is a reasonable assumption only at high magnetic latitudes. In addition to the δ-function at its pole, each CF SECS is also associated with a uniform FAC distributed all around the globe. However, in practice, the actual FACs are described by the δ-functions. The reasons is that if the analysis area is large enough, the sum of the SECS's scaling factors (i.e., sum or integral of the upward and downward FACs) is expected to be close to zero, so that the uniform FACs of the CF SECS will almost cancel each other.

When observing ionospheric current systems, the measured quantity is almost always the magnetic field produced by the currents. In order to use the SECS in these studies, the magnetic fields produced by the currents in individual CF and DF SECS need to be calculated, including the FAC in the case of CF SECS.

Amm and Viljanen (1999) did this calculation for the DF systems, by straightforward (although somewhat tedious) evaluation of the vector potential from the Biot–Savart law. The result is that the magnetic field has only r- and θ'-components, given by

$$B_r^{DF}(r,\theta',\phi') = \frac{\mu_0 S^{DF}}{4\pi r} \begin{cases} \frac{1}{\sqrt{1+s^2-2s\cos\theta'}} - 1, r < R, \\ \frac{s}{\sqrt{1+s^2-2s\cos\theta'}} - s, r > R. \end{cases} \tag{2.13}$$

$$B_{\theta'}^{DF}(r,\theta',\phi') = \frac{-\mu_0 S^{DF}}{4\pi r \sin\theta'} \begin{cases} \frac{s-\cos\theta'}{\sqrt{1+s^2-2s\cos\theta'}} + \cos\theta', r < R, \\ \frac{1-s\cos\theta'}{\sqrt{1+s^2-2s\cos\theta'}} - 1, \quad r > R \end{cases} \tag{2.14}$$

where $s = min(r, R)/max(r, R)$.

The magnetic field of the CF system, with associated FAC, is most easily calculated using Ampere's circuit law, following the same reasoning as in Appendix A of Juusola et al. (2006). The important thing is to first convince oneself that, due to symmetries, the magnetic field must have the form $\mathbf{B}^{CF} = B_{\phi'}(r,\theta')\,\hat{\mathbf{e}}_{\phi'}$. After that it is easy to evaluate the circuit law and obtain the field as

$$\mathbf{B}^{CF}(r,\theta',\phi') = \frac{-\mu_0 S^{CF}}{4\pi r} \begin{cases} 0, \quad r < R, \\ \cot\left(\frac{\theta'}{2}\right)\hat{\mathbf{e}}_{\phi'}, r > R. \end{cases} \tag{2.15}$$

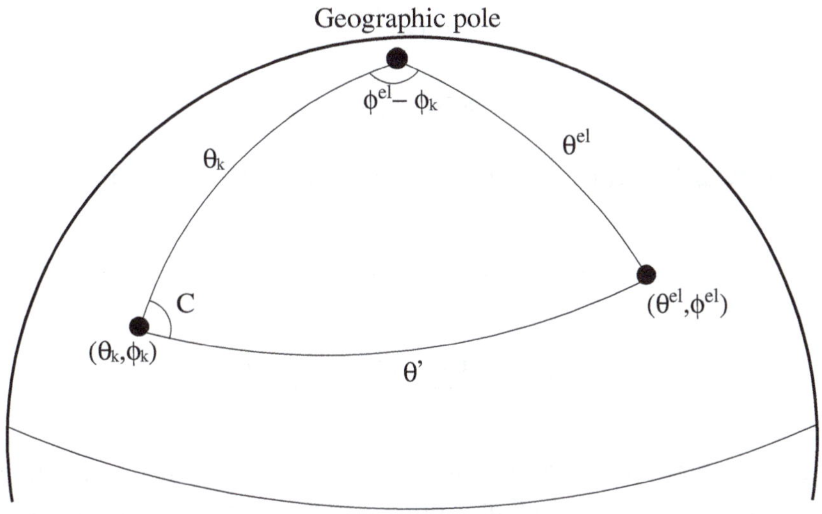

Fig. 2.2 Geometry of the coordinate transformation. Elementary system is located at $(\theta^{el}, \phi^{el}$ and the result is evaluated at (θ_k, ϕ_k). θ' is the colatitude of the point (θ_k, ϕ_k) in the coordinate system centered at the elementary system. Adapted from Vanhamäki et al. (2003)

It is left as an exercise to the reader to check that Eqs. (2.13)–(2.15) give the correct magnetic field. This is most easily done by verifying that (1) divergence of $\mathbf{B}^{CF}$ and $\mathbf{B}^{DF}$ is zero, (2) the discontinuity at the ionospheric current sheet $(r = R)$ gives the horizontal current in Eqs. (2.7) and (2.8), (3) elsewhere the curl of $\mathbf{B}^{DF}$ is zero and (4) the curl of $\mathbf{B}^{CF}$ gives the correct FAC above the ionosphere.

2.5 Coordinate Transformations

The fields of individual CF and DF SECS in Eqs. (2.7) and (2.8) and (2.13)–(2.15) are given in a coordinate system that is centered at the SECS pole. Typically, the analysis is done in the geographical or geomagnetic coordinate system, which is now the unprimed system. Assume that measurements at locations (r_k, θ_k, ϕ_k), $k = 1 \ldots K$, are available, and place the SECS at various locations (R, θ^n, ϕ^n), $n = 1 \ldots N$, in the ionosphere. In order to use the SECS, the colatitude θ' and unit vectors $(\hat{\mathbf{e}}_{\theta'}, \hat{\mathbf{e}}_{\phi'})$ need to be transformed from the SECS-centered coordinate system to the geographical or geomagnetic system. The radial coordinate and unit vector require no transformation, as they are the same in both systems.

This is a straightforward rotation of the coordinate system, but for completeness sake one possible method is presented here. The geometry of the situation is illustrated in Fig. 2.2. According to spherical trigonometry the colatitude θ' is given by

$$\cos\theta' = \cos\theta_k \cos\theta^{el} + \sin\theta_k \sin\theta^{el} \cos(\phi^{el} - \phi_k). \tag{2.16}$$

From Fig. 2.2 the unit vectors can be expressed as

$$\hat{\mathbf{e}}_{\theta'} = \hat{\mathbf{e}}_\theta \cos C - \hat{\mathbf{e}}_\phi \sin C, \tag{2.17}$$

$$\hat{\mathbf{e}}_{\phi'} = \hat{\mathbf{e}}_\theta \sin C + \hat{\mathbf{e}}_\phi \cos C. \tag{2.18}$$

It is a straightforward exercise in spherical trigonometry to show that

$$\cos C = \frac{\cos\theta^{el} - \cos\theta\cos\theta'}{\sin\theta\sin\theta'}, \tag{2.19}$$

$$\sin C = \frac{\sin\theta^{el}\sin(\phi^{el} - \phi)}{\sin\theta'}. \tag{2.20}$$

With these expressions, it is easy to calculate the current or magnetic field at geographical location (r_k, θ_k, ϕ_k) that is produced by a SECS located at geographical point $(R, \theta^{el}, \phi^{el})$.

2.6 Vector Field Analysis with SECS

In practical calculations, the elementary systems are placed at some discrete grid, and the scaling factors give the divergence and curl of the vector field in the grid cell. In some arbitrary grid cell n, the scaling factors are

$$S_n^{CF} = \int_{cell\ n} \nabla \cdot \mathbf{V}\, \mathrm{da}, \tag{2.21}$$

$$S_n^{DF} = \int_{cell\ n} (\nabla \times \mathbf{V})_r\, \mathrm{da}, \tag{2.22}$$

where da is the area element. This means that the curl and divergence distributed over the grid cell are represented by point sources at the center of the cell.

With SECS a vector field (e.g., the ionospheric horizontal current or electric field) is composed of rotational and divergent parts as

$$\mathscr{V} = \overline{\mathrm{M}}_1 \cdot \mathscr{S}^{CF} + \overline{\mathrm{M}}_2 \cdot \mathscr{S}^{DF} \tag{2.23}$$

The composite vector $\mathscr{V}$ contains the θ- and ϕ-components of the vector field $\mathbf{V}$ at the grid points $\mathbf{r}_k = (R, \theta_k, \phi_k)$,

$$\mathscr{V} = \left[V_\theta(\mathbf{r}_1),\quad V_\phi(\mathbf{r}_1),\quad V_\theta(\mathbf{r}_2),\quad \ldots\right]^T. \tag{2.24}$$

The vectors $\mathscr{S}^{CF}$ and $\mathscr{S}^{DF}$ contain the scaling factors of the CF and DF SECS, respectively, at grid points $\mathbf{r}_n^{el}$

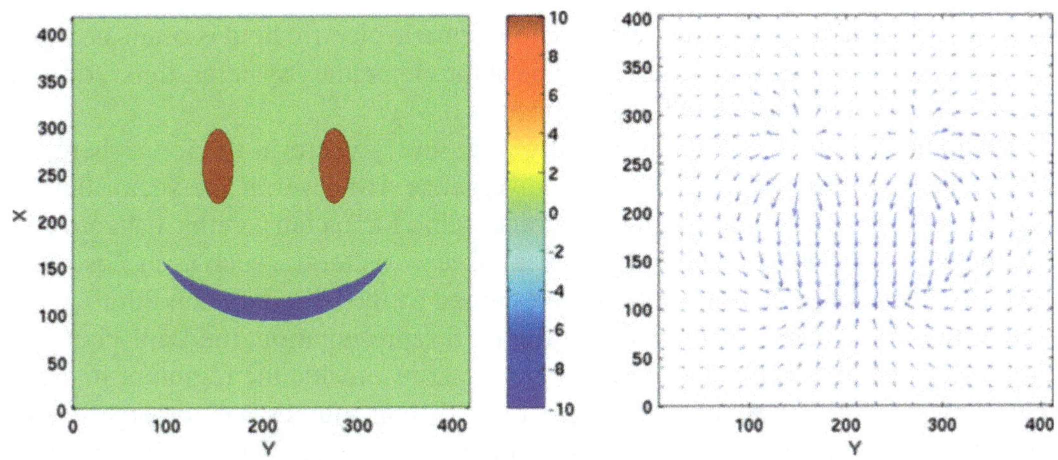

Fig. 2.3 On the left scaling factors of CF elementary systems, and on the right the corresponding vector field. Scaling factors, vector values, and lengths are in arbitrary units. From Vanhamäki (2007)

$$\mathscr{S}^{CF} = \left[S^{CF}(\mathbf{r}_1^{el}),\quad S^{CF}(\mathbf{r}_2^{el}),\quad S^{CF}(\mathbf{r}_3^{el}),\quad \ldots\right]^T, \tag{2.25}$$

$$\mathscr{S}^{DF} = \left[S^{DF}(\mathbf{r}_1^{el}),\quad S^{DF}(\mathbf{r}_2^{el}),\quad S^{DF}(\mathbf{r}_3^{el}),\quad \ldots\right]^T, \tag{2.26}$$

Here $S^{DF}(\mathbf{r}^{el})$ and $S^{CF}(\mathbf{r}^{el})$ should be interpreted as the average divergence and curl of $\mathbf{V}$ over the grid cells, as in Eqs. (2.21) and (2.22). The components of the transfer matrices $\overline{\mathrm{M}}_{1,2}$ can be calculated using Eqs. (2.7) and (2.8), as explained in detail by Vanhamäki (2011).

Figure 2.3 illustrates how an irrotational potential field can be modeled with just CF elementary systems. In this case the vector $\mathscr{S}^{DF}$ in Eq. (2.23) is zero.

A given vector field $\mathbf{V}$ could be represented with elementary systems by evaluating the integrals in Eqs. (2.21) and (2.22) over a suitable grid. However, it is often more practical to rewrite Eq. (2.23) as

$$\mathscr{V} = \overline{\mathrm{M}}_{12} \cdot \mathscr{S}^{CD}, \tag{2.27}$$

where the CF and DF parts have been combined,

$$\overline{\mathrm{M}}_{12} = \left[\overline{\mathrm{M}}_1\, \overline{\mathrm{M}}_2\right], \qquad \mathscr{S}^{CD} = \begin{bmatrix} \mathscr{S}^{CF} \\ \mathscr{S}^{DF} \end{bmatrix}. \tag{2.28}$$

Now, the equation can be inverted for the unknown scaling factors contained in the vector $\mathscr{S}^{CD}$. This inverse problem can be solved in various ways, for example, employing singular value decomposition (SVD) of the matrix $\overline{\mathrm{M}}_{12}$. The solution method and possible regularization of the inverse problem (see Sect. 2.10.3) may have some effect on the solution, especially when the matrix is under-determined (more unknowns than measurements). If it is known a priori that the vector field $\mathbf{V}$

is either curl- or divergence-free (e.g., the ionospheric electric field is often assumed curl-free), it is only necessary to use one type of the elementary systems, thus reducing the size of the inverse problem by a factor of two.

If the vector field **V** is known globally (e.g., everywhere in the ionosphere), it is completely determined by its curl and divergence. However, if the vector field is specified in only some limited region, it may contain a Laplacian part that has zero curl and divergence inside this region. In a potential representation, such as in Eqs. (2.5) and (2.6), this Laplacian part would be determined by the boundary conditions at the edge of the area where **V** is known. In the SECS representation, the Laplacian part can be included by placing some elementary systems outside the region of interest. These "external" SECS represent the effect that distant sources (i.e., divergences or curls) have inside the analysis area. Therefore, in regional studies, it is important to make the SECS grid somewhat larger than the area of interest (see Sect. 2.10.1), but it should be remembered that in the outlying areas the SECS representation is no longer unique.

This kind of vector field representation was one of the original uses of the elementary current systems. When Amm (1997) introduced the CECS and SECS ionospheric studies, he was searching for a practical way to decompose vector fields into curl-free and divergence-free parts and also to interpolate the fields in a way that would conserve their curl-free and/or divergence-free character.

2.7 Analysis of Ground Magnetic Measurements

An important application of the SECS method has been the estimation of the ionospheric current system based on the magnetic disturbance field it creates at the ground. This is a classical problem in geosciences, and many methods have been developed to tackle it, see e.g., Chapman and Bartels (1940), or Untiedt and Baumjohann (1993), Amm and Viljanen (1999) and references therein. Most of the previously used methods were based on harmonic analysis, where the magnetic field is expanded as a sum of suitable basis functions, for example, spherical harmonics. In the SECS analysis, it is the current system that is expanded in terms of elementary systems, whose amplitude is then fitted to match the measured magnetic disturbance field.

An important practical question is how to separate the disturbance field from the total magnetic field that is measured by magnetometers. Detailed discussion is beyond this review, but we mention that with ground magnetometer data this is usually done by determining some quiet-time reference level and removing it from the data. van de Kamp (2013) present one realization of this method.

The seminal work in ionospheric current studies using SECS analysis was by Amm and Viljanen (1999), who first derived analytical formulas for the magnetic field of the DF SECS and showed how the DF SECS could be used to estimate the ionospheric equivalent current from ground magnetic measurements. They also compared the SECS analysis with more traditional spherical cap harmonic analysis of the magnetic field, and demonstrated the practical advantages of the SECS method.

An important question is the relationship between the ionospheric equivalent current and the real ionospheric current. At high magnetic latitudes the curl-free part of the ionospheric horizontal current, together with associated FAC, does not produce any magnetic field below the ionosphere. Fukushima (1976) showed this by assuming uniform ionospheric conductances, but the result is valid independent of the conductance distribution (Amm 1997). The crucial assumption needed in deriving this result is that the FAC should flow radially. For strictly radial FAC, the ground magnetic disturbance from ionospheric current is produced solely by the divergence-free part, as is evident also in Eqs. (2.13)–(2.15). This is only approximately true even at the auroral zone, and breaks down completely at lower latitudes, where the magnetic field has larger inclination.

When the magnetic field lines are tilted, the FACs and associated horizontal curl-free currents make some contribution to the ground magnetic disturbance. As for example Tamao (1986) showed, this contribution can be reasonably large even at $\sim 60°$ magnetic latitude. Luckily, the ground magnetic field due to tilted FACs is typically spatially smoother than that due to divergence-free currents, as is evident in Fig. 6 by Tamao (1986). Therefore contributions from opposite FACs should readily almost cancel each other, with the remaining magnetic effect being rather small and spatially smooth. For FACs tilted in the north/south direction the ground magnetic field is mostly in the east/west direction, which should show up as north/south equivalent current (see, e.g., Fig. 12 in Untiedt and Baumjohann 1993). Taking all this into account, it should be safe to assume that at high magnetic latitudes the ionospheric equivalent current is approximately equal to the divergence-free part of the actual ionospheric current, possibly apart from a relatively small and smooth north/south directed background current. For a more thorough discussion about the concept of equivalent current see, for example, Sect. 3 in Vanhamäki and Amm (2011) and references therein.

When calculating the ionospheric equivalent current with the SECS method, the horizontal components of the ground magnetic disturbance $\mathbf{B}_G$ measured by magnetometers at locations $\mathbf{r}_n = (R_E, \theta_n, \phi_n)$ during some time instant are collected into a composite vector

$$\mathscr{B}_{G,\perp} = [B_x(\mathbf{r}_1), \quad B_y(\mathbf{r}_1), \quad B_x(\mathbf{r}_2), \quad \ldots]^T. \tag{2.29}$$

The unknown scaling factors of the DF SECS located at $\mathbf{r}_n^{el} = (R, \theta_n^{el}, \phi_n^{el})$ are collected into another vector as in Eq. (2.26). These vectors are connected by a transfer matrix $\overline{\mathrm{T}}$, so that

$$\mathscr{B}_{G,\perp} = \overline{\mathrm{T}} \cdot \mathscr{S}^{DF}. \tag{2.30}$$

The components of the transfer matrix $\overline{\mathrm{T}}$ give the magnetic field caused by each individual unit SECS at the magnetometer sites, and is therefore known and depends only on geometry. For example, $T_{2,4}$ gives the y-component (East) of $\mathbf{B}_G$ at $\mathbf{r}_1$ caused by the SECS centered at $\mathbf{r}_4^{el}$. Details of calculating the matrix $\overline{\mathrm{T}}$ and inverting Eq. (2.30) for the unknown scaling factors $\mathscr{S}^{DF}$ using truncated singular value decomposition are given by Amm and Viljanen (1999) and Pulkkinen et al. (2003b). Once the scaling factors are known, the actual ionospheric equivalent current $\mathbf{J}_{eq,ion}$ can be calculated using Eq. (2.8) for each individual DF SECS.

A matlab code included as supplementary material demonstrates the process of calculating the ionospheric equivalent current with the SECS method. Readers are encouraged to study the code and experiment with it. However, the code should not be directly applied to other magnetometer networks, as some parameters may have to be adjusted with changing geometry of the network. This is further discussed in Sect. 2.10. Quick-look plots of the equivalent currents calculated with the SECS method are provided by the Finnish Meteorological Institute.[2]

2.7.1 Separation into Internal and External Parts

In the above discussion, only the horizontal part of the ground magnetic disturbance was used, and all the elementary systems were placed at the ionosphere (radius R), thus determining the ionospheric equivalent current. However, due to geomagnetic induction, the observed ground magnetic perturbation also has internal telluric sources, especially during disturbed geomagnetic conditions (Tanskanen et al. 2001). Using all three components of the observed ground magnetic disturbance, it is possible to separate the measured field into internal and external parts, which can be represented by two layers of equivalent currents (e.g., Haines and Torta 1994).

As far as the SECS method is concerned, this kind of separation was first applied by Pulkkinen et al. (2003b). The method is very similar to the above discussion of ionospheric equivalent currents, but in this case all three magnetic field components are used and there are two layers of elementary systems, one in the ionosphere and the other inside the ground.

The measured ground magnetic disturbance $\mathbf{B}_G$ at magnetometer locations $\mathbf{r}_n = (R_E, \theta_n, \phi_n)$ are collected into a composite vector,

$$\mathscr{B}_G = [B_x(\mathbf{r}_1), \quad B_y(\mathbf{r}_1), \quad B_z(\mathbf{r}_1), \quad B_x(\mathbf{r}_2), \quad \ldots]^T. \tag{2.31}$$

The external (=ionospheric) DF SECS are located at $\mathbf{r}_n^{el,e} = (R, \theta_n^{el,e}, \phi_n^{el,e})$, while the internal DF SECS are placed at $\mathbf{r}_n^{el,i} = (R_i, \theta_n^{el,i}, \phi_n^{el,i})$. Note that in general there can be a different number of internal and external elementary systems, and they can be located at different latitudes and longitudes. The scaling factors are collected into vectors

$$\mathscr{S}^i = [S^i(\mathbf{r}_1^{el,i}), \quad S^i(\mathbf{r}_2^{el,i}), \quad S^i(\mathbf{r}_3^{el,i}), \quad \ldots]^T. \tag{2.32}$$

$$\mathscr{S}^e = [S^e(\mathbf{r}_1^{el,e}), \quad S^e(\mathbf{r}_2^{el,e}), \quad S^e(\mathbf{r}_3^{el,e}), \quad \ldots]^T. \tag{2.33}$$

These vectors are connected by transfer matrices $\overline{\mathrm{T}}_i$ and $\overline{\mathrm{T}}_e$, so that

[2]http://space.fmi.fi/MIRACLE/iono_2D.php.

$$\mathscr{B}_G = \overline{\mathrm{T}}_i \cdot \mathscr{S}^i + \overline{\mathrm{T}}_e \cdot \mathscr{S}^e. \tag{2.34}$$

These matrices can be calculated in a completely similar manner as discussed in the previous section, except that in this case, the matrices also include the vertical component of the magnetic field. For solving the unknown scaling factors, Eq. (2.34) is again written as a single matrix equation

$$\mathscr{B}_G = \overline{\mathrm{T}}_{ie} \cdot \mathscr{S}^{ie}, \tag{2.35}$$

similar to Eq. (2.27). Note that the ordering of the magnetic measurements and scaling factors in Eqs. (2.31)–(2.33) is not important, as long as the matrix $\overline{\mathrm{T}}_{ie}$ connects the elementary systems and measurements in the same order as used in the vectors.

The equivalent currents can mimic the magnetic field produced by all the currents that are located behind them, as seen from the ground surface. That is, external and internal equivalent currents represent currents that are located either above the ionospheric layer or below the internal layer, respectively. As the induced telluric currents can flow at any depth, and also very close to the surface especially in the highly conductive oceans, in principle it would be best to place the internal equivalent current just below the ground surface. However, that may lead to numerical problems, as the finite grid spacing and the singular nature of the SECS mean that one SECS pole placed close to a magnetometer station would make an unrealistically large contribution to the measurement. As a reasonable compromise between numerical stability and inclusion of near-surface currents, Pulkkinen et al. (2003b) placed the internal current layer at 30 km depth.

Making the separation into internal and external equivalent currents is in principle more accurate than calculating only the external current from Eq. (2.30). However, the separation has also some drawbacks in practice. First of all, the inverse problem becomes less stable compared to the external-only calculation, as the number of observations increase from two to three components per station, but typically the number of SECS is doubled. Furthermore, even though the separation is in principle unique when done globally (and with perfect data coverage), Thébault et al. (2006) demonstrated that in local studies, the internal and external sources mix to some degree. In practical applications, the limited amount of input data lead to further ambiguities in the solution. Thébault et al. (2006) considered spherical cap harmonic analysis of the ground magnetic field, but similar problems are expected also in SECS analysis, although this has not been studied in detail. For these reasons, the telluric contributions have been neglected in many SECS studies, which can be expected to lead to some overestimation of the ionospheric equivalent currents, especially during disturbed conditions when time variations are rapid.

2.8 Analysis of Satellite Magnetic Measurements

In the above discussion, the focus was on calculating the ionospheric equivalent current from ground magnetic measurements, which has arguably been the most successful and widely used application of the SECS method. The main limitation is that only the equivalent current, not the whole ionospheric current containing the CF, DF, and FAC parts, can be calculated from ground magnetic data alone. Getting the full current would require some further assumptions about the ionospheric electric field or electric conductivity, as discussed, e.g., by Untiedt and Baumjohann (1993) or Vanhamäki and Amm (2011). This is not a shortcoming of the SECS method, but a general limitation inherent to magnetic fields and currents.

The situation is quite different when there are magnetic measurements from low-orbiting satellites, such as CHAMP (CHAllenging Minisatellite Payload, https://www.gfz-potsdam.de/champ/) or Swarm (Olsen et al. 2013). The satellites pass through the FACs, so their effect dominates the observed magnetic disturbance. The ionospheric horizontal currents, assumed to flow in a thin sheet at E-region altitude, are usually several hundred kilometers below the satellite, and therefore make a smaller contribution to the measured field. As the satellite magnetic data contains information on the FAC, and associated CF current via Eq. (2.4), as well as the DF ionospheric current, the whole current system may be estimated by fitting both CF and DF SECS to the measurements. Therefore satellite data can in principle provide the "real" current distribution.

Often data from only one satellite at any specific region or instant of time are available. Therefore assumptions about gradients perpendicular to the satellite track have to be made, or combined data from several orbits (typically several months or years) are used. Exception to this are the Swarm mission and the AMPERE (Active Magnetosphere and Planetary Electrodynamics Response Experiment Anderson et al. 2014) project. For further discussion of AMPERE, see Chap. 8. The SECS method tailored for Swarm data analysis is discussed in detail in Chap. 3, while analysis of single-satellite passes with assumption of vanishing gradients is discussed in the next section.

Juusola et al. (2014) presented a statistical analysis of CHAMP satellite's magnetic data using the SECS method. They first projected all the magnetic measurements into a regular grid in the geomagnetic coordinate system, and then averaged and binned the data with respect to solar wind conditions. The ionospheric current system, including the FAC and both CF and DF parts of the horizontal current, was determined by fitting CF and DF SECS to the gridded magnetic data. Apart from including the CF systems, the approach is very similar to the analysis of ground magnetic data.

The gridded magnetic disturbances measured by CHAMP are collected into a composite vector similar to Eq. (2.31). The CF and DF SECS are placed at selected positions in the ionosphere, and their scaling factors are collected into vectors as in Eqs. (2.21) and (2.22). In general, there can be a different number of CF and DF elementary systems, and they can be located at different latitudes and longitudes. The vectors are connected by transfer matrices $\overline{\mathrm{T}}_{cf}$ and $\overline{\mathrm{T}}_{df}$, so that

$$\mathscr{B}_C = \overline{\mathrm{T}}_{cf} \cdot \mathscr{S}^{CF} + \overline{\mathrm{T}}_{df} \cdot \mathscr{S}^{DF}. \tag{2.36}$$

These matrices can be calculated in a completely similar manner as discussed in the previous section, using Eq. (2.15) for the CF SECSs and Eqs. (2.13) and (2.14) for the DF SECSs. The fitting problem is again combined into a single matrix equation and solved for the unknown scaling factors.

It should be noted that the assumption of perfectly radial FAC used in the CF SECS will lead to some errors when analyzing satellite data. This is most clearly manifested as a slight southward shift in the ionospheric location of the FAC, which is caused by using radial instead of field-aligned mapping from the satellite altitude (typically ∼400 km) to the ionospheric E-region. However, Juusola et al. (2014) estimated that at high latitudes the error was at most 0.9°, and thus smaller than the latitude resolution of their statistical grid.

When using only satellite data, the ionospheric currents and induced telluric currents can not be separated. The reason is that both current systems are below the satellite, so they produce qualitatively similar magnetic effects. Despite the large distance between the satellite and induced telluric current, in some cases, they may have a large effect on the measured magnetic disturbance, as shown by Vanhamäki et al. (2005). The telluric currents may be approximated by placing a perfect conductor inside the Earth at a certain depth (depending on the ground conductivity), in which case the internal currents would be mirror images of the ionospheric currents. This approach was used by Olsen (1996), but in a statistical comparison of satellite- and ground-based currents Juusola et al. (2016) found it inadequate.

2.9 1D SECS

In many situations, data are only available along a single line, and not on a two-dimensional area. Typical cases are passes of a single satellite, or a (North–South) chain of magnetometers. In these cases, for single events, some additional assumptions are necessary. For example, the SECS method discussed in Sect. 2.7 is not directly applicable, as it produces reliable results only if measurements are available in a suitably large two-dimensional area.

One approach is to use a "1D assumption", where gradients of the studied parameter (current, electric field, conductances, ...) are assumed to vanish in one specific direction. This is identified as the "zero-gradient direction", while the perpendicular direction is the "1D direction" (e.g., along a magnetic meridian). For example, assume that ionospheric current depends only on latitude so that gradients in the longitudinal direction vanish. It should be noted that this zero-gradient direction need not be exactly perpendicular to the satellite path or magnetometer chain, but the angle should still be large enough so that good coverage is achieved in the 1D direction. Also, even though the analysis maybe simplified by assuming some a priori fixed 1D direction (e.g., geomagnetic meridian), there exist methods (e.g., minimum variance analysis, Sonnerup and Scheible 1998) that can be used to determine the optimum

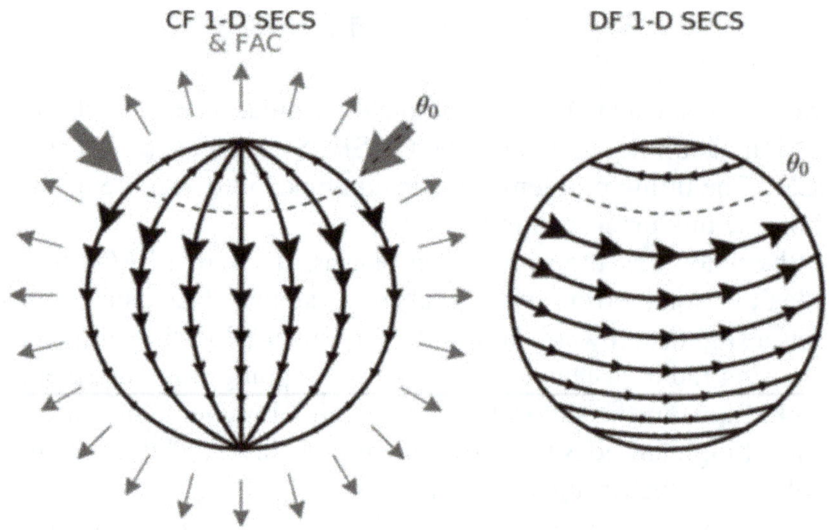

Fig. 2.4 The left panel shows a 1D CF SECS, with associated FACs. A δ-function FAC enters at colatitude θ_0, where it is connected to meridionally flowing horizontal currents. A uniform outward FAC ensures current continuity. The right panel shows a 1D DF SECS. The azimuthal horizontal current, shows a sharp shear at colatitude θ_0

direction from the data. Some method should also be used to check how good the 1D assumption is in each specific case, e.g., by estimating how small the gradients in the "zero-gradient direction" actually are, because in reality, the situation is never perfectly one-dimensional.

One-dimensional variants of the CF and DF SECS were defined by Vanhamäki et al. (2003) and Juusola et al. (2006), respectively. In order to distinguish them from the elementary systems discussed thus far, the terms 1D and 2D SECS are used here. The 1D variants can be obtained by placing the poles of the respective two-dimensional SECS around a circle at a constant latitude θ_0, essentially integrating over the position of the 2D SECS's poles (Vanhamäki et al. 2003). The resulting current systems are

$$\mathbf{J}_{1D}^{CF}(\theta, \theta_0) = \frac{S_{1D}^{CF}}{2R}\hat{\mathbf{e}}_\theta \begin{cases} -\tan(\theta/2), \theta < \theta_0 \\ \cot(\theta/2) \quad, \theta > \theta_0 \end{cases}, \tag{2.37}$$

$$\mathbf{J}_{1D}^{DF}(\theta, \theta_0) = \frac{S_{1D}^{DF}}{2R}\hat{\mathbf{e}}_\phi \begin{cases} -\tan(\theta/2), \theta < \theta_0 \\ \cot(\theta/2) \quad, \theta > \theta_0 \end{cases} \tag{2.38}$$

The 1D SECS are illustrated in Fig. 2.4 and may look deceptively similar to the 2D SECS introduced in Sect. 2.3. However, the crucial difference is that the 1D SECS are defined in the global coordinate system (often geographical or geomagnetic), where the 1D direction is in the meridional plane (all azimuthal gradients vanish). Therefore Eqs. (2.37) and (2.38) have no prime in θ or ϕ.

The 1D CF SECS has a ring of δ-function divergence at colatitude θ_0, with uniform and opposite divergence elsewhere. Similarly, the 1D DF SECS has a band of δ-

function curl, compensated by uniform curl elsewhere. This is actually an alternative way to define the 1D SECS and to derive Eqs. (2.37) and (2.38). Similar to the general 2D SECS, the CF and DF 1D SECS are basis functions for any continuously differentiable vector field on a sphere, with vanishing gradients in the azimuthal direction. By using several 1D SECS with different amplitudes and different "critical co-latitudes" θ_0, any such vector field can be constructed.

The magnetic field of the 1D DF systems (when used to represent currents) was calculated by Vanhamäki et al. (2003). With $s = min(r, R)/max(r, R)$ and defining two auxiliary functions

$$f(s) = \begin{cases} 1, r < R \\ s, r > R \end{cases}, \qquad g_l(s) = \begin{cases} 1/l & , r < R \\ -1/(l+1) \, , r > R \end{cases} \tag{2.39}$$

the components can be written more compactly as

$$B^{DF}_{r,1D}(r,\theta,\theta_0) = \frac{\mu_0 S^{DF}_{1D}}{2r} f(s) \sum_{l=1}^{\infty} s^l P_l(\cos\theta_0)\, P_l(\cos\theta) \tag{2.40}$$

$$B^{DF}_{\theta,1D}(r,\theta,\theta_0) = \frac{\mu_0 S^{DF}_{1D}}{2r} f(s) \sum_{l=1}^{\infty} s^l g_l(s)\, P_l(\cos\theta_0)\, P^1_l(\cos\theta). \tag{2.41}$$

Here, P_l and P^1_l are the unnormalized 0th- and first-order- associated Legendre polynomials. The magnetic field of the 1D CF SECS (with associated radial FAC) was calculated by Juusola et al. (2006) using Ampere's law:

$$\mathbf{B}^{CF}_{1D}(r,\theta,\theta_0) = \frac{\mu_0 S^{CF}_{1D}}{2r} \hat{\mathbf{e}}_\phi \begin{cases} \tan(\theta/2) & , r > R \wedge \theta < \theta_0 \\ -\cot(\theta/2) & , r > R \wedge \theta > \theta_0 \\ 0 & , r < R \end{cases}, \tag{2.42}$$

The 1D SECS are used in a completely analogous way to the 2D systems discussed in Sect. 2.7. For ground magnetic analysis, the B_θ-component (southward in the chosen coordinate system) can be used to calculate the ionospheric equivalent current, which in a 1D situation has only ϕ-component (as 1D current in θ-direction can not be completely divergence-free). Alternatively, both the B_r- and B_θ-components can be used for the internal/external separation. The B_ϕ-component should be much smaller than the other two, which can be used as one check of the quality of the 1D assumption.

In satellite applications, the southward current J_θ is computed from the eastward magnetic disturbance by fitting 1D CF SECS to the data. The eastward divergence-free current is associated with magnetic disturbances in the radial- and θ-directions. If only B_r is used in fitting the 1D DF SECS, the measured B_θ may be compared to the magnetic disturbance calculated from the fitted DF SECS in order to estimate how good the 1D assumption is. This line of reasoning was applied by Juusola et al. (2007), who used it to search for the best 1D direction by allowing the North Pole of the coordinate system to move. In case they did not find good enough agreement in B_θ for any North Pole location, the event was considered 2D and removed from analysis. It is better to use B_r in the fitting and B_θ in the checking rather than the other way round, as the horizontal component is more easily affected by 2D structures in the FAC. However, due to non-radial FAC, it would be even better to use field-aligned magnetic disturbance in fitting the 1D DF SECS. That can be done by taking the appropriate linear combination of the r- and θ-components.

2.10 Some Practical Considerations

In this section, several practical issues related to the application of the SECS methods are considered. Some of them are also discussed and solved in the example code that is included as supplementary material in the book. However, issues such as grid selection and regularization of the matrix inversion depend on the geometry of each situation, and must be adjusted for each magnetometer network.

2.10.1 Grid and Boundary Effects

The SECS method does not require any explicit boundary conditions, even when applied to regional studies. This is different from potential representations of the electric field or current, like Eqs. (2.5) and (2.6), which require explicit boundary information. In contrast, in the SECS analysis, there is an implicit condition that the vector fields are smooth and source-free outside the analysis area. The CF and DF SECS represent all sources of the vector field, so in regions where there are no SECSs both the curl and divergence must vanish. Of course, explicit boundary conditions may be added using virtual data points at the edges of the analysis area, requiring that the vector field has a certain value at these points.

As mentioned in Sect. 2.6, in order to minimize boundary effects caused by the implicit boundary conditions and the possible presence of a Laplacian field, the SECS grid should be somewhat larger than the area of interest. This is illustrated in Fig. 2.5, where a given vector field shown in panel (a) is divided into CF and DF parts using Eq. (2.27). The original vector field was constructed from a curl-free part with sources inside the shown area and a divergence-free part with sources outside, see panels (b–d). In this case, the data region is the area where the vector field shown in panel (a) is given. The SECS grid used in the analysis is the colored area in panels (e–d), where also the data region is shown as a black rectangle. Note that although this example was done with Cartesian Elementary Current Systems (CECS), the same principle holds for SECS.

When the given vector field in panel (a) is decomposed into DF and CF parts using elementary systems, the local curl-free part is correctly represented in terms of CF systems inside the area where the vector field was originally specified. This is seen

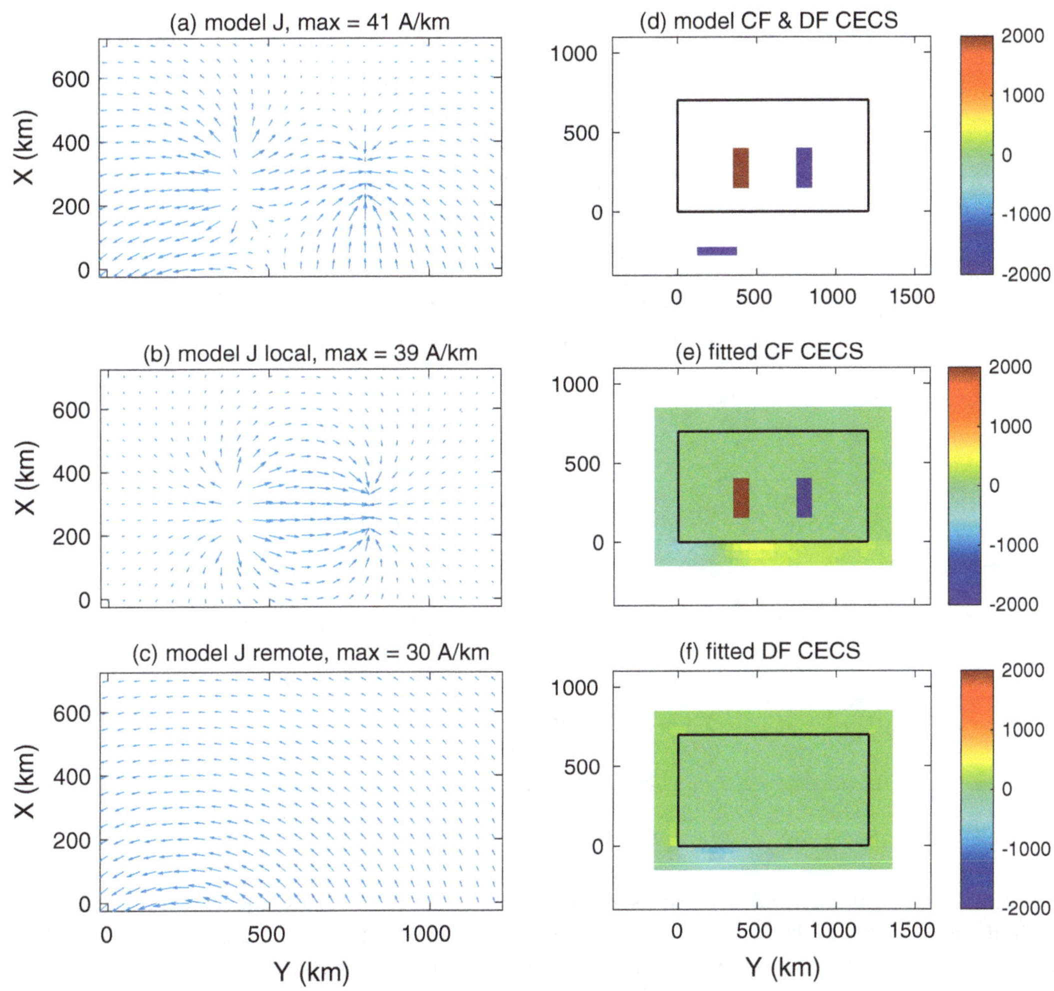

Fig. 2.5 An example of representing a given vector field (in this case current) with Cartesian Elementary Current Systems (CECS). The model current system is shown in panels (**a**, **d**). The estimated CF and DF scaling factors shown in panels (**e**) and (**f**), respectively, are obtained from fit to the current shown in panel (**a**). The black rectangle in the right side plots denoted the area shown in the vector plots. In panel **d** all the nonzero CF CECS are inside the black rectangle, while all the nonzero DF CECS are outside. The local field in panel **b** correspond to the model CF CECS in panel (**d**), while the remote field in panel **c** is given by the model DF CECS. Adapted from Vanhamäki (2007)

by comparing the estimated scaling factors shown in panels (e–f) with the model CF scaling factors that are inside the black rectangle in panel (d). The estimated CF scaling factors *inside the data region* agree very well with the model, while the estimated DF scaling factors are nearly zero. However, the remote divergence-free part gets represented in terms of both CF and DF systems located just outside the data region. This is seen by comparing the areas *outside the black rectangle* in panels (d–f): In the model there are only DF CECS outside the black rectangle, but in the fit results both CF and DF CECS have nonzero amplitudes there.

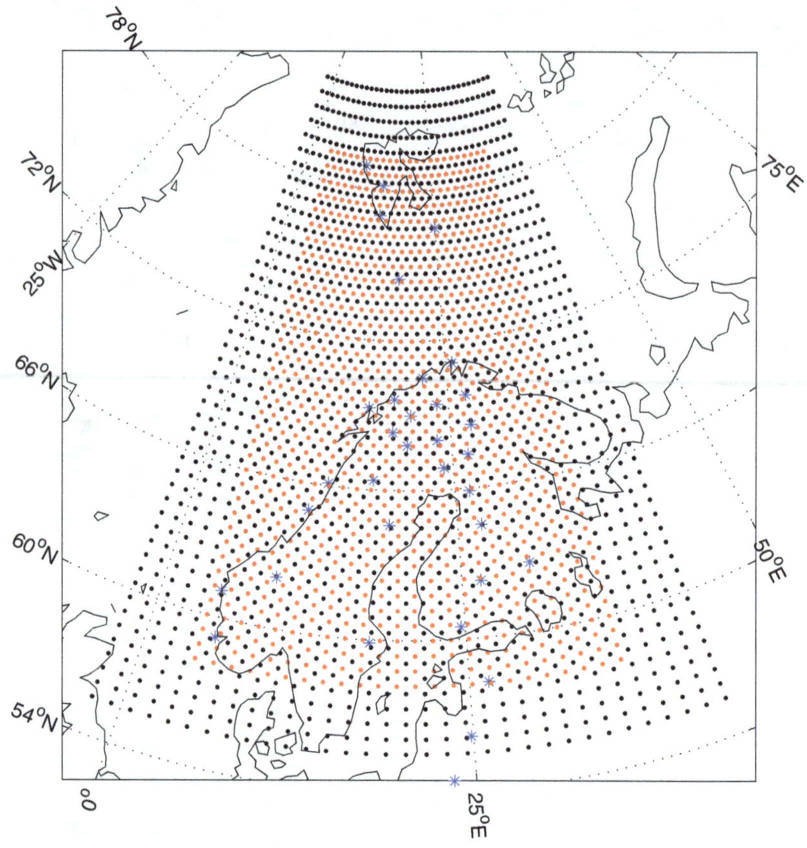

Fig. 2.6 An example of analysis grid used when calculating ionospheric equivalent currents from IMAGE magnetometer data in the example program accompanying this chapter. Black dots show the DF SECS's positions, red dots the points where the equivalent current vectors are calculated and blue stars show the magnetometer locations

This demonstrates that in general, outside the data region, the decomposition is no longer unique, and a (locally) Laplacian field may equally well be caused by either remote curls or divergences, or some combination of them. In global studies, there is no such fundamental ambiguity, because the Laplacian field must vanish or be physically unreasonable. In the case of Fig. 2.5, the Laplacian field corresponds to the remote currents shown in panel (c), caused by the DF CECS located outside the data region in panel (b).

In summary, the division of local field into CF and DF parts is in principle unique, but remote fields, whose sources are outside the data region, can not be decomposed in a unique way. This is not a limitation of the elementary system method, but similar ambiguities (related to boundary conditions) would appear also in potential representations like Eq. (2.6). Finally, it should be kept in mind that in practical applications data availability and quality are often serious limiting factors. Measurements are rarely as extensive, detailed and noise-free as the model field shown in panel (a) of Fig. 2.5.

The recommendation is to make the SECS grid somewhat larger than area of interest. Figure 2.6 illustrates typical SECS and output grids used in the calculation of equivalent currents with data from the IMAGE magnetometer network. Role of the outlying SECS is to provide an equivalent representation of distant current systems, that do not have sources (in this case curls) directly above the magnetometer network.

2.10.2 *Singularities*

The elementary systems defined in Eqs. (2.7) and (2.8) are unfortunately singular, as there is divergence at the SECS pole where $\theta' = 0$. Consequently, the magnetic field of a DF SECS given in Eqs. (2.13) and (2.14) has a singular point at $(r = R, \theta' = 0)$, while the CF SECS's field in Eq. (2.15) is singular along the line $(r \geq R, \theta' = 0)$. These singularities should be kept in mind, as they may cause numerical problems.

In many applications, it is sufficient to select the SECS grid carefully, so that there is no need to evaluate the fields (either the SECS's vector field or magnetic field) too close to singular points. This is usually the case, e.g., in the calculation of equivalent currents from ground magnetic data, discussed in Sect. 2.7 and demonstrated in the example code. If the vertical separation between the SECS layers and the magnetometers is large enough, the singularities in the magnetic field do not matter. Similarly, the resulting equivalent current vectors can be calculated at the midpoints between the SECS locations, as illustrated in Fig. 2.6.

However, in some applications it is necessary to calculate the fields near the singularities. In this case, the elementary systems in Eqs. (2.7) and (2.8) may be modified to

$$\mathbf{V}^{CF}(\mathbf{r}') = \frac{S^{CF}}{4\pi R}\hat{\mathbf{e}}_{\theta'} \begin{cases} \alpha \tan\left(\frac{\theta'}{2}\right) & \theta' < \theta_0 \\ \cot\left(\frac{\theta'}{2}\right) & \theta' \geq \theta_0 \end{cases} \tag{2.43}$$

$$\mathbf{V}^{DF}(\mathbf{r}') = \frac{S^{DF}}{4\pi R}\hat{\mathbf{e}}_{\phi'} \begin{cases} \alpha \tan\left(\frac{\theta'}{2}\right) & \theta' < \theta_0 \\ \cot\left(\frac{\theta'}{2}\right) & \theta' \geq \theta_0 \end{cases} \tag{2.44}$$

When $\alpha = \cot^2(\theta_0/2)$, the vector fields are continuous at $\theta' = \theta_0$. Moreover, the δ-function source at the elementary system's pole is now spread uniformly inside a spherical cap of width θ_0.

In a similar way the 1D SECS can be redefined so that the divergence or curl that in Eqs. (2.37) and (2.38) is a δ-function at colatitude θ_0 is uniformly spread to a spherical zone $\theta_0 - \Delta \leq \theta \leq \theta_0 + \Delta$. Inside this zone the 1D DF SECS has current density

$$\mathbf{J}_{1D}^{DF}(\theta, \theta_0, \Delta) = \frac{S_{1D}^{CF}}{2R} \frac{\cos\theta_0 \cos\Delta - (1 - \sin\theta_0 \sin\Delta)\cos\theta}{\sin\theta_0 \sin\Delta \sin\theta}\hat{\mathbf{e}}_{\phi}, \tag{2.45}$$

while outside it is the same as in Eq. (2.38). The current density of a 1D CF SECS has the same expression, but in the $\hat{\mathbf{e}}_\theta$ direction.

Assuming radial FAC, the magnetic field of the modified CF SECS can be calculated using Ampere's law, as before. In fact, for a general ionospheric curl-free current $\mathbf{J}^{cf}(\theta, \phi)$, with $(\nabla \times \mathbf{J}^{cf})_r = 0$, and radial FAC, the magnetic field is

$$\mathbf{B}(r, \theta, \phi) = \begin{cases} \frac{\mu_0 R}{r} \hat{\mathbf{e}}_r \times \mathbf{J}^{cf}(\theta, \phi), & r > R \\ 0, & r < R \end{cases} \tag{2.46}$$

This general result is evident from the magnetic field and current of CF SECS, given in Eqs. (2.15) and (2.7), respectively. Remember that the CF SECS form a complete set of basis functions for curl-free vector fields. However, the result can also be verified by checking that the magnetic field is divergence-free and gives the correct current distribution via Ampere's law. Equation (2.46) forms the basis for many analysis techniques for satellite magnetic data, including the analysis of AMPERE data, discussed further in Chap. 8.

In contrast, it is quite unlikely that the magnetic field of the modified nonsingular DF SECS could be calculated in a closed form. Possibly a series expansion could be derived using the same methods as in Vanhamäki et al. (2003). However, in the attached example code this problem is simply ignored: Equation (2.44) is used for the vector field, while the magnetic field is calculated using Eqs. (2.13) and (2.14). This is slightly inconsistent, but does not appear to affect practical applications.

The effects of the singularities can be further reduced in a rather straightforward manner by subdividing the SECS into smaller units. This is not the same as making the original SECS grid finer, as that would increase the number of scaling factors. Rather, if one SECS with scaling factor S_n is normally placed into a grid cell n, then the cell is divided into N equal parts and SECSs with amplitudes S_n/N are located into each one. This way the size of the system matrix in Eq. (2.27) or (2.30) stays the same, but the matrix elements are calculated as sums of sub-elementary systems placed at different corners of the original grid cells. This does not completely remove the singularity, but reduces it into a smaller area, which can be either handled by using the redefined nonsingular SECS in Eqs. (2.43) and (2.44), or ignored completely. For example, if the original grid cell is divided into 100 sub-cells, then removing the one sub-cell where the calculation point is located should amount to roughly 1% error.

2.10.3 Inversion Regularization

When applying the SECS method, matrix equations such as Eq. (2.27) or (2.30) need to be inverted. Often these are either under-determined (more unknown scaling factors than measurements), or otherwise ill-conditioned. In either case, direct attempt to invert the equation will lead to nonsensical results. In this kind of situation, the problem requires regularization, either by adding some constraints or assumptions about the solution.

There are several possible methods to deal with these situations, but the traditional method of choice in SECS analysis has been the Singular Value Decomposition (SVD, see, e.g., Press et al. 1992, Sect. 2.6). In SVD the system matrix, e.g., $\overline{\mathrm{T}}$ in Eq. (2.30), is decomposed into a product of three matrices

$$\overline{\mathrm{T}} = \overline{\mathrm{U}} \cdot \overline{\mathrm{S}} \cdot \overline{\mathrm{V}}^*, \tag{2.47}$$

where $\overline{\mathrm{U}}$ and $\overline{\mathrm{V}}$ are unitary matrices, $\overline{\mathrm{S}}$ is a diagonal matrix containing the singular values (nonnegative, arranged from largest to smallest) and * denotes conjugate transpose. In the case of Eq. (2.30), the rows of $\overline{\mathrm{V}}^*$ represent different, mutually orthogonal configurations of the SECS scaling factors, while the columns of $\overline{\mathrm{U}}$ give the corresponding (also mutually orthogonal) magnetic field configurations at the magnetometer stations. In some sense the corresponding singular values in $\overline{\mathrm{S}}$ indicate how distinguishable these modes are in the magnetic field. A large value $S_{n,n}$ means that the corresponding magnetic field configuration is easy to find in the data, while those with small values can be lost in the noise.

Thus SVD may be used to locate and remove the ill-conditioned parts of the system matrix, making the inversion numerically stable. In practice Eq. (2.30) is inverted as

$$\mathscr{S}^{DF} = \overline{\mathrm{V}} \cdot \overline{\sigma} \cdot \overline{\mathrm{U}}^* \cdot \mathscr{B}_{G,\perp}, \tag{2.48}$$

where $\overline{\sigma}$ is a diagonal matrix with elements

$$\sigma_{n,n} = \begin{cases} 1/S_{n,n} & \text{if } S_{n,n} > \varepsilon S_{1,1} \\ 0 & \text{otherwise} \end{cases} \tag{2.49}$$

Here, ε is a parameter that determines the cut-off point for small singular values, with respect to the largest value $S_{1,1}$.

The important question is how to choose ε. Too small a value will lead to problems with noisy data (e.g., spurious structures appearing in the solution), while too large value means that good data are rejected. Perhaps the only sure way is to test the analysis with simulated data, where the correct answer is known, and try different ε-values. In these tests, it is important to use realistic models and to add a realistic amount of noise to the simulated data. This kind of ε-optimization has been done, e.g., by Weygand et al. (2011) and Vujic and Brkic (2016).

The SVD approach seems to work well in practice, but there are also other possible ways to regularize the inversion problem. One could add extra constraints to the system matrix by demanding that the spatial gradient of the SECS scaling factors must be as small as possible. Readers are encouraged to consider and test alternatives to the SVD.

2.10.4 Tilted Field Lines

When representing ionospheric currents with SECS, the FAC is connected to the CF systems. In the present formulation, the FAC are assumed to flow radially, as shown in Fig. 2.1 and assumed in deriving Eq. (2.15). As noted in Sect. 2.4, this assumption is a reasonable approximation only at high magnetic latitudes, where inclination of the magnetic field is large. At lower latitudes, the field lines are noticeably tilted, and Eq. (2.15) becomes an increasingly worse approximation.

However, in principle, this is a problem only when analyzing satellite magnetic measurements. The CF and DF systems still form a basis for representing horizontal vector fields (including the horizontal current) at middle and low latitudes, and the ground magnetic field can still be represented in terms of equivalent currents. Unfortunately, interpretation of the ionospheric equivalent current at lower latitudes is more problematic, as it equals the divergence-free part of the real current only at high magnetic latitudes.

In satellite analysis, one can try to correct small errors caused by the radial/tilted discrepancy, e.g., by introducing a "forbidden zone" between the assumed and actual locations of the FAC at satellite altitude (see Fig. 3 in Juusola et al. 2006), and by shifting the resulting FAC and curl-free current slightly poleward by the amount the field line moves between the satellite altitude and ionospheric E-layer (e.g., Juusola et al. 2016). However, these approximate corrections are reasonably accurate only at high magnetic latitudes, where the field lines are almost vertical.

The CF systems could be improved by assuming a more realistic geometry for the FAC. At high and middle latitudes it might be sufficient to model the FAC as semi-infinite line currents that are oriented along the magnetic field. There is a closed-form analytical expression for the magnetic field of such a line current, so the 2D CF SECS could be redefined by replacing the semi-infinite radial line current at the pole (the δ-function current) with a tilted one. There is no pressing need to redefine the uniform radial FAC, as those should mostly cancel when summing several different 2D CF SECS with different amplitudes. For even lower latitudes the semi-infinite line currents should probably be replaced with FAC flowing along the actual magnetic field lines, or at least along a dipole field. For the 1D CF SECS the horizontal current may be redefined so that it is antisymmetric between the (geomagnetic) hemispheres,

$$\mathbf{V}_A(\theta, \theta_0) = \mathbf{V}^{CF}_{1D}(\theta, \theta_0) - \mathbf{V}^{CF}_{1D}(\theta, \pi - \theta_0) = \frac{V^{CF}}{R}\,\hat{\mathbf{e}}_\theta \begin{cases} -1/\sin\theta, & \theta_0 < \theta < \pi - \theta_0 \\ 0, & otherwise \end{cases} \tag{2.50}$$

Assuming that the FAC flows along dipole field lines between conjugate points, the magnetic field can be calculated using Ampere's law (Juusola et al. 2006; Deguchi 2014),

$$\mathbf{B}_A(r, \theta) = \mu_0 \left(\frac{R}{r}\right)^{3/2} \mathbf{V}(\theta_I, \theta_0) \times \hat{\mathbf{e}}_r = \frac{V^{CF}\mu_0}{R} \left(\frac{R}{r}\right)^{3/2} \hat{\mathbf{e}}_\phi \begin{cases} 1/\sin\theta_I, & \theta_0 < \theta_I < \pi - \theta_0 \\ 0, & otherwise \end{cases} \tag{2.51}$$

Here, $\theta_I = \arcsin\left(\sqrt{\frac{R}{r}}\sin\theta\right)$ is the colatitude mapped along a dipole field to the ionosphere. One could also consider other modifications, where the FAC would be terminated at the equatorial plane (Deguchi 2014). They would have the advantage that the current system is not forced to be anti-symmetrical between the hemispheres. For the 2D CF SECS, these non-radial modifications have not been investigated.

2.10.5 Equivalent Current as a Proxy for FAC

As mentioned in Sect. 2.7, the divergence-free equivalent current may be calculated using only ground magnetic data. Thus, strictly speaking, no information is available about FAC. However, it is well known that certain patterns in the equivalent current are good indicators of FAC (e.g., Untiedt and Baumjohann 1993). These estimates can be made more formal by noting that under certain conditions the curl of the equivalent current is directly proportional to the FAC (see, e.g., Amm et al. 2002).

First of all, assume that the equivalent current is equal to the divergence-free part of the actual ionospheric current. This should be valid at high magnetic latitudes, as discussed in Sect. 2.7, although distortions created by internal induced currents and magnetospheric current systems may cause small deviations. More crucially, further assume that the Hall to Pedersen conductance ratio $\alpha = \Sigma_H/\Sigma_P$ is spatially constant and that conductance gradients are perpendicular to the electric field. Under these assumptions $j_\parallel = -(\nabla \times \mathbf{J}_{eq})_r/\alpha$, which is easy to verify by comparing the curl and divergence of ionospheric Ohm's law in Eq. (2.3).

This kind of reasoning has been used from time to time (e.g., Amm et al. 2002; Juusola et al. 2009; Weygand and Wing 2016), but it should be kept in mind that this relation is only approximate and relies on assumptions that are not generally valid. Therefore $(\nabla \times \mathbf{J}_{eq})_r$ should be only considered as a proxy for FAC.

2.11 How SECS Have Been Used

As mentioned in Sect. 2.6, the elementary systems, as used in ionospheric studies, were originally introduced by Amm (1997) in order to optimally interpolate vector fields and to divide them into CF and DF parts. Since then the SECS method has found many other applications, most prominently in the analysis of satellite or ground-based magnetic measurements. The method to calculate ionospheric equivalent currents from ground-based data was developed by Amm and Viljanen (1999), as discussed in Sect. 2.7. It was extensively tested and expanded to include the internal/external separation by Pulkkinen et al. (2003a) and Pulkkinen et al. (2003b), while Vanhamäki et al. (2003) introduced the 1D variant for ground-based analysis. Since then the method has been used in numerous studies, especially with the IMAGE magnetometer network.

Other research groups have adapted the SECS method. For example McLay and Beggan (2010) applied the method to very sparse magnetometer arrays in order to interpolate the external magnetic disturbance field over large distances. Weygand et al. (2011) used the ground-based SECS method to calculate equivalent currents over North America and Greenland, by constructing an irregularly shaped grid for the elementary systems. They also carefully validated and optimized the inversion method by using simulated measurements based on a known ionospheric current model. Instead of calculating equivalent currents, Vujic and Brkic (2016) used the SECS method to construct a regional model of the crustal magnetic field using data from repeat stations and ground survey sites around the Adriatic Sea.

Satellite applications of the SECS method were developed by Juusola et al. (2006) and Juusola et al. (2014) for the 1D and 2D cases respectively, as described in Sects. 2.8 and 2.9. Juusola et al. (2007) carried out a large statistical study of the ionospheric current system by analyzing 6112 individual CHAMP passes with the 1D SECS method. To our knowledge the 2D SECS analysis of gridded and averaged CHAMP measurements by Juusola et al. (2014) was the first study where the 2D ionospheric current system, both CF and DF horizontal currents as well as FAC, was directly estimated from satellite magnetic data. Amm et al. (2015) developed a tailored SECS-based method for analyzing electric and magnetic data from the Swarm multi-satellite mission. This application is discussed in detail in Chap. 3.

Apart from magnetic data analysis, the SECSs can be used as basis functions for representing general vector fields and potentially transforming differential and integral equations into algebraic ones. This is very similar to using spherical harmonic functions in solving differential equations. For example, Vanhamäki et al. (2006) and Vanhamäki (2011) have used the elementary systems for solving ionospheric induction problems starting from Ohm's law and Maxwell's equations. Meanwhile, Vanhamäki and Amm (2007) introduced a new, local variant of the KRM (Kamide–Richmond–Matsushita) method (Kamide et al. 1981) for calculating the ionospheric electric field from ground magnetic data and estimated ionospheric conductances. In these applications, the elementary systems are used to transform the partial differential equations into matrix equations, which can be solved much more easily.

Finally, Amm et al. (2010) used the SECS method for local analysis of the ionospheric plasma convection (or electric field) measured by the SuperDARN radars. This application is very close to the original purpose of Amm (1997), as here the SECS method was used to combine and interpolate/extrapolate the radar line-of-sight velocity measurements into a divergence-free map of the plasma convection. The main advantages over the standard SuperDARN analysis (Ruohoniemi and Baker 1998) is that the SECS method can be used locally, relies only on measured data without any underlying statistical model, and does not require any explicit boundary conditions.

Acknowledgements The authors thank Ari Viljanen and Kirsti Kauristie for their continuous support in developing and applying the SECS-based methods, and checking the example code. The authors thank the International Space Science Institute in Bern, Switzerland, for supporting the ISSI Working Group: Multi-Satellite Analysis Tools Ionosphere, in which frame this study was performed. Colin Waters provided a large number of comments, which improved the text considerably. The authors thank the institutes who maintain the IMAGE magnetometer array. IMAGE magnetometer data are available at http://www.space.fmi.fi/image. The editors thank Akimasa Yoshikawa for his assistance in evaluating this chapter. This work was supported by the Academy of Finland project 314664.

References

Amm, O., H. Vanhamäki, K. Kauristie, C. Stolle, F. Christiansen, R. Haagmans, A. Masson, M.G.G.T. Taylor, R. Floberghagen, and C.P. Escoubet. 2015. A method to derive maps of ionospheric conductances, currents, and convection from the Swarm multisatellite mission. *Journal of Geophysical Research Space Physics* 120: https://doi.org/10.1002/2014JA020154.

Amm, O. 1997. Ionospheric elementary current systems in spherical coordinates and their application. *Journal Geomagnetism and Geoelectricity* 49: 947–955. https://doi.org/10.5636/jgg.49.947.

Amm, O. 1998. Method of characteristics in spherical geometry applied to a Harang-discontinuity situation. *Annals of Geophysics* 16: 413–424. https://doi.org/10.1007/s00585-998-0413-2.

Amm, O., and A. Viljanen. 1999. Ionospheric disturbance magnetic field continuation from the ground to the ionosphere using spherical elementary current systems. *Earth Planets Space* 51: 431–440. https://doi.org/10.1186/BF03352247.

Amm, O., M. Engebretson, T. Hughes, L. Newitt, A. Viljanen, and J. Watermann. 2002. A traveling convection vortex event study: Instantaneous ionospheric equivalent currents, estimation of field-aligned currents, and the role of induced currents. *Journal of Geophysical Research* 107 (A11): 1334. https://doi.org/10.1029/2002JA009472.

Amm, O., A. Aruliah, S.C. Buchert, R. Fujii, J.W. Gjerloev, A. Ieda, T. Matsuo, C. Stolle, H. Vanhamäki, and A. Yoshikawa. 2008. Towards understanding the electrodynamics of the 3-dimensional high-latitude ionosphere: Present and future. *Annals of Geophysics* 26: 3913–3932. https://doi.org/10.5194/angeo-26-3913-2008.

Amm, O., A. Grocott, M. Lester, and T.K. Yeoman. 2010. Local determination of ionospheric plasma convection from coherent scatter radar data using the SECS technique. *Journal of Geophysical Research* 115: A03304. https://doi.org/10.1029/2009JA014832.

Anderson, B.J., H. Korth, C.L. Waters, D.L. Green, V.G. Merkin, R.J. Barnes, and L.P. Dyrud. 2014. Development of large-scale Birkeland currents determined from the Active Magnetosphere and Planetary Electrodynamics Experiment. *Geophysical Research Letters* 41: 3017–3025. https://doi.org/10.1002/2014GL059941.

Backus, G. 1986. Poloidal and toroidal fields in geomagnetic field modeling. *Reviews of Geophysics* 24: 75–109. https://doi.org/10.1029/RG024i001p00075.

Brekke, A. 1997. *Physics of the upper polar atmosphere*. John Wiley & Sons, ISBN 0-471-96018-7.

Chapman, S., and J. Bartels. 1940. *Geomagnetism*, vol. II. New York: Oxford University Press.

Deguchi, R. 2014. *Unpublished master's thesis*, Department of Earth and planetary Sciences: Kyushu university.

Fukushima, N. 1976. Generalized theorem for no ground magnetic effect of vertical currents connected with Pedersen currents in the uniform-conductivity ionosphere. *Report of Ionosphere and Space Research in Japan* 30: 35–40.

Haines, G.V., and J.M. Torta. 1994. Determination of equivalent current sources from spherical cap harmonic models of geomagnetic field variations. *Geophysical Journal International* 118: 499–514. https://doi.org/10.1111/j.1365-246X.1994.tb03981.x.

Juusola, L., O. Amm, and A. Viljanen. 2006. One-dimensional spherical elementary current systems and their use for determining ionospheric currents from satellite measurement. *Earth Planets Space* 58: 667–678. https://doi.org/10.1186/BF03351964.

Juusola, L., O. Amm, K. Kauristie, and A. Viljanen. 2007. A model for estimating the relation between the Hall to Pedersen conductance ratio and ground magnetic data derived from CHAMP satellite statistics. *Annales Geophysicae* 25: 721–736. https://doi.org/10.5194/angeo-25-721-2007.

Juusola, L., R. Nakamura, O. Amm, and K. Kauristie. 2009. Conjugate ionospheric equivalent currents during bursty bulk flows. *Geophysical Journal International* 114: A04313. https://doi.org/10.1029/2008JA013908.

Juusola, L., S.E. Milan, M. Lester, A. Grocott, and S.M. Imber. 2014. Interplanetary magnetic field control of the ionospheric field-aligned current and convection distributions. *Journal of Geophysical Research: Space Physics* 119: 31303149. https://doi.org/10.1002/2013JA019455.

Juusola, L., K. Kauristie, H. Vanhamäki, A. Aikio, and M. van de Kamp. 2016b. Comparison of auroral ionospheric and field-aligned currents derived from swarm and ground magnetic field measurements. *Journal of Geophysical Research* 121: 9256–9283. https://doi.org/10.1002/2016JA022961.

Kamide, Y., A.D. Richmond, and S. Matsushita. 1981. Estimation of ionospheric electric fields, ionospheric currents and field-aligned currents from ground magnetic records. *Journal of Geophysical Research* 86: 801. https://doi.org/10.1029/JA086iA02p00801.

McLay, S.A., and C.D. Beggan. 2010. Interpolation of externally-caused magnetic fields over large sparse arrays using Spherical Elementary Current Systems. *Annales Geophysicae* 28: 1795–1805. https://doi.org/10.5194/angeo-28-1795-2010.

Olsen, N. 1996. A new tool for determining ionospheric currents from magnetic satellite data. *Geophysical Research Letters* 23: 3635–3638. https://doi.org/10.1029/96GL02896.

Olsen, N., E. Friis-Christensen, R. Floberghagen, et al. 2013. The Swarm Satellite Constellation Application and Research Facility (SCARF) and Swarm data products. *Earth Planet Space* 65: 1. https://doi.org/10.5047/eps.2013.07.001.

Press, W., S. Teukolsky, W. Vetterling and B. Flannery. 1992. *Numerical Recipes in Fortran 77*, 2nd ed. The Art of Scientific Computing. 973 p., Campridge University Press, Cambridge.

Pulkkinen, A., O. Amm, A. Viljanen, and BEAR Working Group. 2003a. Ionospheric equivalent current distributions determined with the method of spherical elementary current systems. *Journal of Geophysical Research*, 108 (A2): 1053. https://doi.org/10.1029/2001JA005085.

Pulkkinen, A., O. Amm, A. Viljanen, and BEAR Working Group. 2003b. Separation of the geomagnetic variation field on the ground into external and internal parts using the spherical elementary current system method. *Earth Planets Space* 55: 117–129. https://doi.org/10.1186/BF03351739.

Richmond, A.D. and J. P. Thayer. 2000. Ionospheric electrodynamics: a tutorial. In *Magnetospheric current systems (Geophysical Monograph Series 118)*, ed. S. Ohtani et al. (AGU, Washington D.C.), pp. 131–146.

Ruohoniemi, J.M., and K.B. Baker. 1998. Large-scale imaging of high-latitude convection with super Dual Auroral Radar Network HF radar observations. *Journal of Geophysical Research* 103: 20797. https://doi.org/10.1029/98JA01288.

Sciffer, M.D., C.L. Waters, and F.W. Menk. 2004. Propagation of ULF waves through the ionosphere: Inductive effect for oblique magnetic fields. *Annals of Geophysics* 22: 1155–1169. https://doi.org/10.5194/angeo-22-1155-2004.

Sonnerup, B. and M. Scheible. 1998. Minimum and maximum variance analysis. in analysis methods for multi-spacecraft data (ISSI Scientific Reports Series Vol. 1), ed. G. Paschmann and P. Daly (ESA/ISSI), ISBN:1608-280X, pp. 185–220.

Tamao, T. 1986. Direct contribution of oblique fieldaligned currents to ground magnetic fields. *Journal of Geophysical Research* 91 (A1): 183–189. https://doi.org/10.1029/JA091iA01p00183.

Tanskanen, E., A. Viljanen, T. Pulkkinen, R. Pirjola, L. Häkkinen, A. Pulkkinen, and O. Amm. 2001. At substorm onset 40% of AL comes from underground. *Journal of Geophysical Research* 106 (A7): 13119–13134. https://doi.org/10.1029/2000JA900135.

Thébault, E., J.J. Schott, and M. Mandea. 2006. Revised spherical cap harmonic analysis (R-SCHA): Validation and properties. *Journal of Geophysical Research* 111: B01102. https://doi.org/10.1029/2005JB003836.

Untiedt, J., and W. Baumjohann. 1993. Studies of polar current systems using the IMS Scandinavian magnetometer array. *Space Science Reviews* 63: 245–390. https://doi.org/10.1007/BF00750770.

van de Kamp, M. 2013. Harmonic quiet-day curves as magnetometer baselines for ionospheric current analyses. *Geoscientific Instrumentation, Methods and Data Systems* 2: 289–304. https://doi.org/10.5194/gi-2-289-2013.

Vanhamäki, H. 2007. Theoretical modeling of ionospheric electrodynamics including induction effects. Finnish Meteorological Institute Contributions, 66, electronic version available at http://ethesis.helsinki.fi/.

Vanhamäki, H. and L. Juusola. 2018. Review of data analysis techniques for estimating Ionospheric currents based on miracle and satellite observations, Chapter 24 in AGU monograph 235, In *Electric currents in geospace and beyond (Geophysical Monograph Series 235)*, ed. A. Keiling et al. (AGU, Washington D.C.), pp. 407–426, https://doi.org/10.1002/9781119324522.ch24.

Vanhamäki, H. 2011. Inductive ionospheric solver for magnetospheric MHD simulations. *Annals of Geophysics* 29: 97–108. https://doi.org/10.5194/angeo-29-97-2011.

Vanhamäki, H., and O. Amm. 2007. A new method to estimate ionospheric electric fields and currents using data from a local ground magnetometer network. *Annals of Geophysics* 25: 1141–1156. https://doi.org/10.5194/angeo-25-1141-2007.

Vanhamäki, H., and O. Amm. 2011. Analysis of ionospheric electrodynamic parameters on mesoscales a review of selected techniques using data from ground-based observation networks and satellites. *Annals of Geophysics* 29: 467–491. https://doi.org/10.5194/angeo-29-467-2011.

Vanhamäki, H., O. Amm, and A. Viljanen. 2003. One-dimensional upward continuation of the ground magnetic field disturbance using elementary current systems. *Earth Planets Space* 55: 613–625. https://doi.org/10.1186/BF03352468.

Vanhamäki, H., A. Viljanen, and O. Amm. 2005. Induction effects on ionospheric electric and magnetic fields. *Annals of Geophysics* 23: 1735–1746. https://doi.org/10.5194/angeo-23-1735-2005.

Vanhamäki, H., O. Amm, and A. Viljanen. 2006. New method for solving inductive electric fields in the non-uniformly conducting ionosphere. *Annals of Geophysics* 24: 2573–2582. https://doi.org/10.5194/angeo-24-2573-2006.

Vujić, E., and M. Brkić. 2016. Spherical elementary current systems method applied to geomagnetic field modeling for the adriatic. *Acta Geophysica* 64: 930. https://doi.org/10.1515/acgeo-2016-0045.

Weygand, J.M., and S. Wing. 2016. Comparison of DMSP and SECS region-1 and region-2 ionospheric current boundary. *Journal of Atmospheric and Solar-Terrestrial Physics* 143–144: 8–13. https://doi.org/10.1016/j.jastp.2016.03.002.

Weygand, J.M., O. Amm, A. Viljanen, V. Angelopoulos, D. Murr, M.J. Engebretson, H. Gleisner, and I. Mann. 2011. Application and validation of the spherical elementary currents systems technique for deriving ionospheric equivalent currents with the North American and Greenland ground magnetometer arrays. *Journal of Geophysical Research* 116: A03305. https://doi.org/10.1029/2010JA016177.

Yoshikawa, A., and M. Itonaga. 1996. Reflection of shear Alfvén waves at the ionosphere and the divergent Hall current. *Geophysical Research Letters* 23: 101–104. https://doi.org/10.1029/95GL03580.

Chapter 3
Spherical Elementary Current Systems Applied to Swarm Data

Heikki Vanhamäki, Liisa Juusola, Kirsti Kauristie, Abiyot Workayehu and Sebastian Käki

Abstract This chapter describes how the Spherical Elementary Current Systems (SECS) are applied to analyze the magnetic and electric field measurements provided by the Swarm spacecraft. The Swarm/SECS method produces two-dimensional (latitude–longitude) maps of the ionospheric horizontal and field-aligned currents around the satellite paths. If also electric field data are available, similar maps of the electric field and conductances can be obtained.

3.1 Introduction

The Spherical Elementary Current Systems (SECS) described in the previous Chap. 2 have been applied to the Swarm mission by Amm et al. (2015). Naturally, the SECS methods had been used to analyze magnetic measurements from other satellite missions before that. However, with the previous single satellite missions, one had to either use the 1D assumption of vanishing gradients in one direction (often either cross-track or zonal direction), or statistically average the data before analysis. These approaches were used for example by Juusola et al. (2007) and Juusola et al. (2014), when they estimated the ionospheric horizontal current $\mathbf{J}$ and field-aligned current $j_{\|}$ from Champ magnetic field data.

With data from the multi-satellite Swarm mission, it becomes possible to estimate $\mathbf{J}$ and $j_{\|}$ from individual passes of two or three satellites, without any simplifying assumptions apart from stationarity during the pass. The goal of the Swarm/SECS analysis method introduced by Amm et al. (2015) is to derive two-dimensional (latitude/longitude) maps of the current in a limited region around the satellite paths. As the amount and spatial coverage of the data is still quite limited (usually two, sometimes three, parallel satellite tracks), Amm et al. (2015) decided to use a hybrid

H. Vanhamäki (✉) · A. Workayehu
University of Oulu, Oulu, Finland
e-mail: Heikki.Vanhamaki@oulu.fi

L. Juusola · K. Kauristie · S. Käki
Finnish Meteorological Institute, Helsinki, Finland

M. W. Dunlop and H. Lühr (eds.), *Ionospheric Multi-Spacecraft Analysis Tools*, ISSI Scientific Report Series 17,
https://doi.org/10.1007/978-3-030-26732-2_3

1D/2D technique, where 1D SECS are fitted to the magnetic data first in order to capture the large-scale electrojet currents, and 2D SECS are only fitted to the residual, which cannot be explained by the 1D systems.

SECS can be applied to any vector field on a sphere, and a similar analysis of the plasma drift measured by Swarm can be carried out, effectively interpolating and also slightly extrapolating the measurements into a 2D map of the electric field **E**. A straightforward combination of the sheet current and electric field maps then gives the height-integrated Pedersen and Hall conductances (Σ_P and Σ_H, respectively).

The Swarm/SECS analysis method was largely developed before the Swarm spacecraft were launched in November 2014. Therefore, initial testing was performed with synthetic datasets taken from data-based models of typical ionospheric current systems, as well as from global MHD simulations. After Swarm data became available, Juusola et al. (2016b) made an extensive comparison of the currents estimated from Swarm and the ground-based MIRACLE magnetometer network. Due to the lack of good electric field data from the Swarm satellites, the electric field analysis and estimation of the Hall and Pedersen conductances has only been carried out in one limited case study by Juusola et al. (2016a).

This chapter first describes the Swarm/SECS analysis method and the various tests that were performed by Amm et al. (2015). It continues with some examples of selected event studies, and summarizes the statistical comparisons of Swarm- and MIRACLE-based currents. Finally, some ongoing work and future prospects are discussed. This includes a dataset of ionospheric current maps derived with the Swarm/SECS method that has been calculated and published by the Finnish Meteorological Institute.[1]

3.2 The Swarm/SECS Analysis Method

The SECS method, in general, and the curl-free (CF) and divergence-free (DF) basis functions are described in Chap. 2. It also includes a brief summary of ionospheric electrodynamics.

3.2.1 Current from Magnetic Field Analysis

A novel combination of the 1D and 2D SECS techniques is employed to make maximal use of the Swarm measurements. The default setting is to use data from the two lower satellites, but measurements from the upper satellite can also be included, when the three satellites are in close conjunction.

Necessary inputs are the positions ($\mathbf{r}_{sat}$) and the magnetic ($\mathbf{B}_{sat}$) measurements of the Swarm satellites. The Earth's main field, lithospheric field and magnetospheric

[1] http://space.fmi.fi/MIRACLE/Swarm_SECS/.

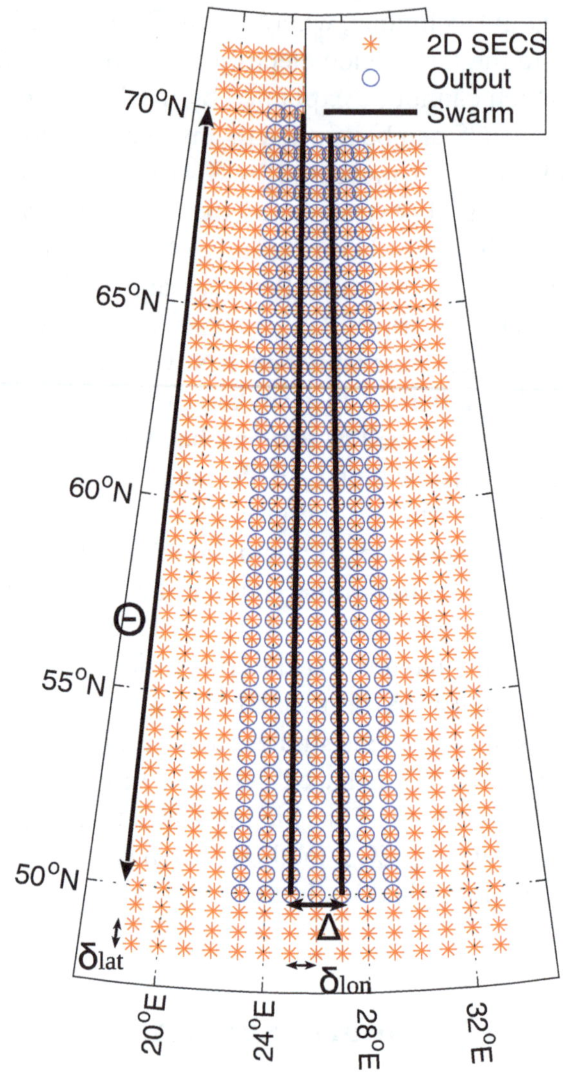

Fig. 3.1 Sketch of geometry for Swarm data analysis. Black lines show the ionospheric projection of the lower Swarm satellites orbits, blue circles the output grid, and red stars the positions of the 2D SECS poles. The 1D SECS are not shown, but they are placed at same latitudes as the 2D SECS. From Amm et al. (2015)

contributions need to be subtracted from the Swarm magnetic field data using for example the CHAOS or POMME models (Finlay et al. 2016; Maus et al. 2006, respectively). Output parameters are the height-integrated horizontal current and FAC along a strip around the ionospheric projection of the satellite tracks.

The input data can be in any spherical coordinate system (r, θ, ϕ), where θ is the colatitude and ϕ is the longitude. The most typical choices are the geographic or geomagnetic systems. The analysis grid and output grid are generated around the ionospheric footpoints of the two lower Swarm satellites. Typically altitude for the ionospheric sheet current is about 110 km. The default analysis grid has the following parameters:

- Latitudinal separation is 0.5° and longitudinal spacing is half of the satellite separation for both the 2D SECS grid and the output grid.

- The 2D SECS grid has 15 points in the longitudinal direction and it is extended 3 latitude points outside the output data area. 1D SECS are placed at the same latitudes as the 2D SECS.
- The output grid has 7 points in the longitudinal direction.

All these parameters can be changed by the user, but these were found to be a reasonable choice in the various test cases described in Sect. 3.3. An example of the default grids is shown in Fig. 3.1.

The four different current systems (1D/2D and CF/DF SECS) are fit one by one to the measured magnetic variation field. The analysis steps are as follows:

(1) Fit 1D divergence-free SECS using only the parallel magnetic component $B_{\|}$,
(2) fit 2D divergence-free SECS using the residual $B_{\|}$,
(3) fit 1D curl-free SECS using the residual eastward component B_{ϕ}, and
(4) fit 2D curl-free SECS using the residual southward and eastward components, B_{θ} and B_{ϕ}, respectively.

Ordering of the above analysis steps is a result of two factors. First, 1D SECS are used to fit the large-scale electrojet-type current systems whenever possible, as the amount of input data is limited to two satellite tracks. Second, $B_{\|}$ is mostly produced by the divergence-free ionospheric currents, whereas the perpendicular disturbances are dominated by FAC connected to the curl-free ionospheric current. Naturally, it would be possible to combine all these four steps into one large fitting problem, but keeping them separate gives better control over the analysis.

Each of the above steps results in a matrix equation between the measured field components and the unknown SECS scaling factors. For example, in the first step, it is necessary to form and solve relation

$$\mathscr{B}_{\|} = \overline{\mathrm{T}} \cdot \mathscr{S}_J^{1D,DF}, \tag{3.1}$$

where the vector $\mathscr{B}_{\|}$ contains the field-aligned magnetic disturbance components measured by the Swarm satellites at locations $\mathbf{r}_n = (r_n, \theta_n, \phi_n)$,

$$\mathscr{B}_{\|} = [B_{\|}(\mathbf{r}_1),\quad B_{\|}(\mathbf{r}_2),\quad B_{\|}(\mathbf{r}_3),\quad \ldots]^T, \tag{3.2}$$

and the vector $\mathscr{S}_J^{1D,DF}$ contains scaling factors of the 1D DF SECS located at $\mathbf{r}_k^{el} = (R_I, \theta_k^{el}, \phi_k^{el})$,

$$\mathscr{S}_J^{1D,DF} = [S(\mathbf{r}_1^{el}),\quad S(\mathbf{r}_2^{el}),\quad S(\mathbf{r}_3^{el}),\quad \ldots]^T. \tag{3.3}$$

The components of the transfer matrix $\overline{\mathrm{T}}$ give the parallel magnetic field components caused by each individual unit SECS at the measurement points. The transfer matrix depends only on the geometry, i.e., locations of the measurement points and the SECS poles. Detailed formulas for calculating the matrix elements and possible ways to invert the linear equation for the unknown scaling factors are presented in Sect. 4–8 of Chap. 2.

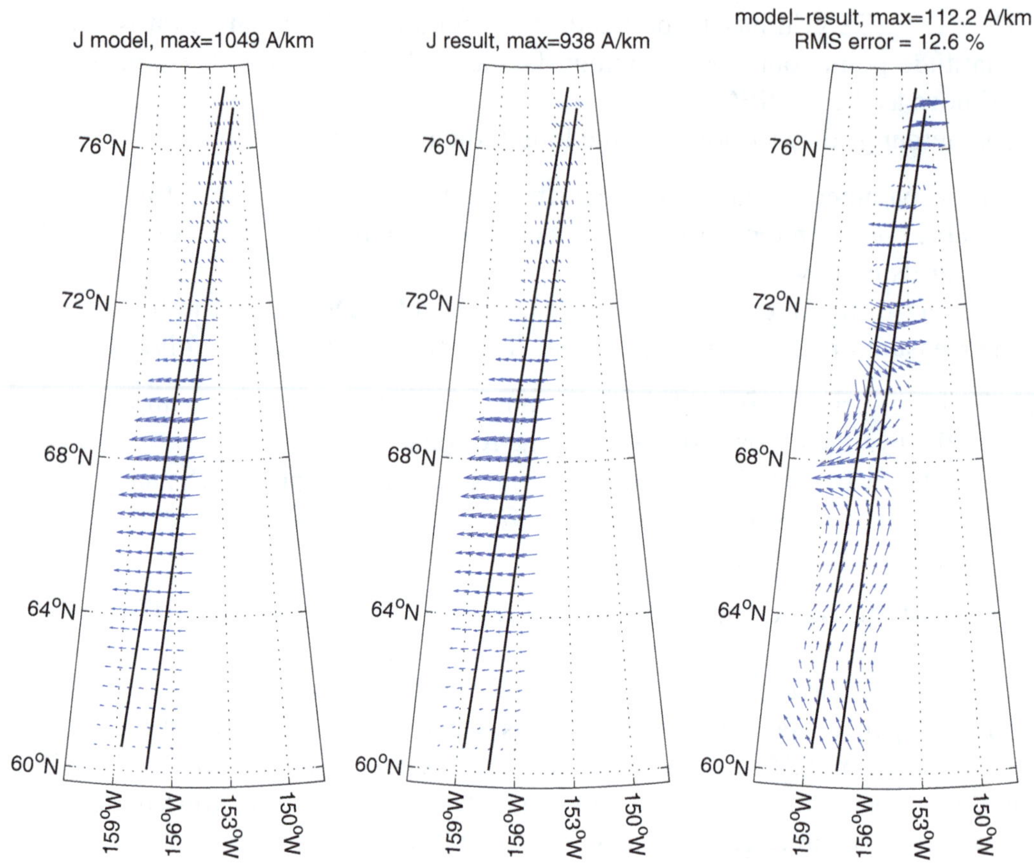

Fig. 3.2 From left to right: Ionospheric horizontal current from the MHD simulation (moderate activity), current calculated from virtual Swarm measurements with the Swarm/SECS analysis method, and the difference between the two. Note the different scales. Tracks of the two satellites used in the analysis are also shown. Corresponding results for the electric field and Hall conductance are shown in Figs. 3.3 and 3.4, respectively

In the second step, the magnetic field explained by the 1D DF SECS is removed from the measured magnetic disturbance, and the fitting is repeated using 2D DF SECS. In a similar fashion the 1D and 2D CF SECS are fitted in steps 3 and 4, respectively.

Once all the scaling factors of the different SECSs have been determined, the ionospheric horizontal current can be calculated as a sum of the individual elementary systems. An example of the output current map calculated using data from the lower pair of the Swarm satellites is shown in Fig. 3.2. This is one of the synthetic test cases further discussed in Sect. 3.3 and by Amm et al. (2015). In this test, the ionospheric current system and corresponding Swarm measurements are taken from a global magnetosphere–ionosphere simulation. The FAC can be calculated either directly from the scaling factors of the CF SECS, which describe the divergence of the current within the grid cells, or by numerically estimating $\nabla \cdot \mathbf{J}$ with finite differences.

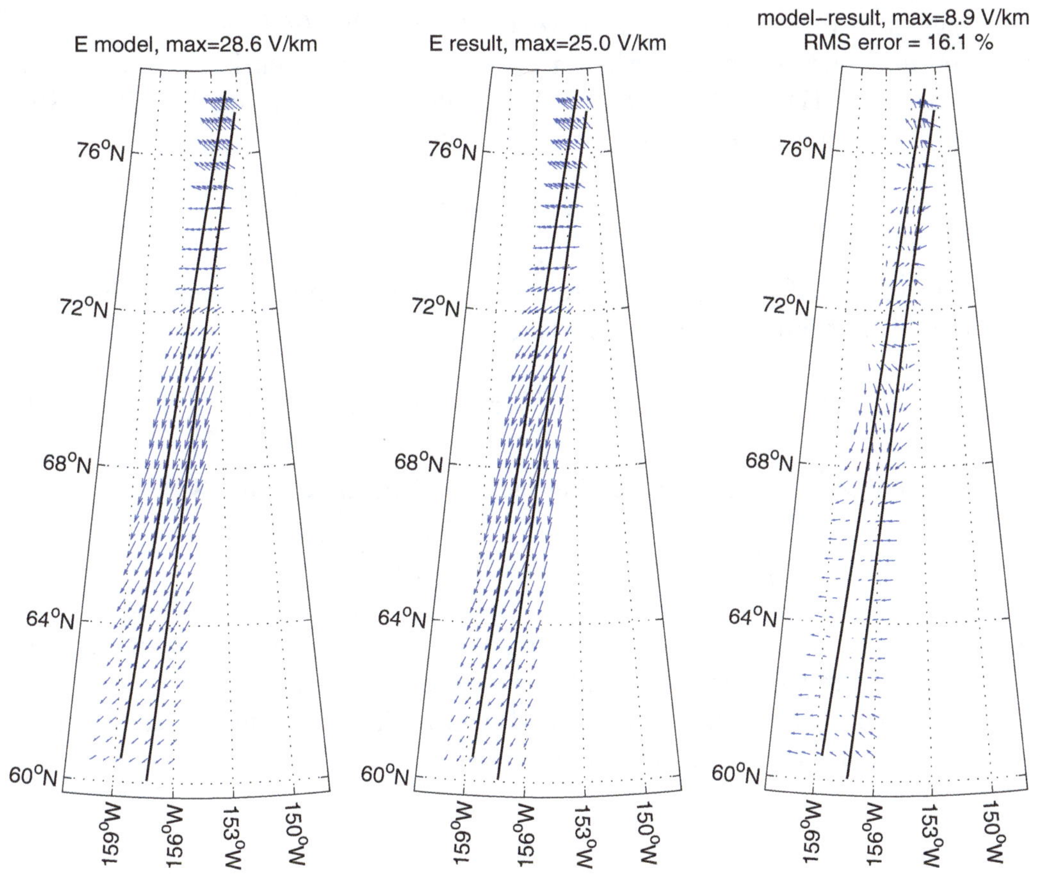

Fig. 3.3 Same as Fig. 3.2, but for the electric field analysis

3.2.2 Fitting the Electric Field with CF SECS

Electric field data from the Swarm satellites can be analyzed in a similar fashion. In this case only CF 1D and 2D elementary systems are needed, as the curl of **E** is assumed to vanish. The analysis is carried out in three steps:

(1) Map the electric field measurements down to the ionospheric current layer,
(2) fit 1D curl-free systems using the θ-component of electric field, and
(3) fit 2D curl-free systems using the residual E_θ and E_ϕ.

In this case, the Swarm data already gives the electric field along the satellite tracks, so the purpose of fitting the CF elementary systems is to interpolate and extrapolate the measurements to an extended latitude–longitude maps. This is in contrast to the magnetic field analysis, where there is a need to estimate the ionospheric currents from the measured magnetic field.

One practical way to map the electric field data from the satellite altitude down to the E-region current sheet is to use the apex coordinates, with readily available conversion library (see Laundal and Richmond 2016, and references therein). Once that has been done, the data can be fitted with the CF elementary systems. This is analogous to representing a given vector field with SECS, a topic which was discussed in Sect. 6 of Chap. 2. Also in this case there are matrix equations for the scaling factors of 1D and 2D CF SECS. For example, when fitting the 2D CF SECS in the third step, there is equation

$$\boldsymbol{\delta\mathscr{E}}_{\perp} = \overline{\mathbf{M}} \cdot \mathscr{S}_{E}^{2D,CF}, \tag{3.4}$$

where the vector $\boldsymbol{\delta\mathscr{E}}_{\perp}$ contains the residual of the downward mapped horizontal electric field after fitting the 1D CF SECS,

$$\boldsymbol{\delta\mathscr{E}}_{\perp} = [E_{\theta}^{map}(\mathbf{r}_1),\ \ E_{\phi}^{map}(\mathbf{r}_1),\ \ E_{\theta}^{map}(\mathbf{r}_2),\ \ \ldots]^T, \tag{3.5}$$

and the vector $\mathscr{S}_{E}^{2D,CF}$ contains scaling factors of the 2D CF SECS,

$$\mathscr{S}_{E}^{2D,CF} = [S(\mathbf{r}_1^{el}),\ \ S(\mathbf{r}_2^{el}),\ \ S(\mathbf{r}_3^{el}),\ \ \ldots]^T. \tag{3.6}$$

Also in this case, the transfer matrix $\overline{\mathbf{M}}$ depends only on the geometry, and it can be inverted using similar methods as with the magnetic field analysis.

An example of the output electric field map is shown in Fig. 3.3. It is from the same synthetic test case as the current map in Fig. 3.2.

3.2.3 Conductances from Ohm's Law

Once the ionospheric electric field and horizontal current have been determined, the height-integrated Pedersen and Hall conductances can be solved from Ohm's law,

$$\Sigma_P = \frac{\mathbf{J}_{\perp} \cdot \mathbf{E}_{\perp}}{|\mathbf{E}_{\perp}|^2}, \qquad \Sigma_H = \frac{\mathbf{J}_{\perp} \times \mathbf{E}_{\perp}}{|\mathbf{E}_{\perp}|^2} \tag{3.7}$$

An example of the output Hall conductance map is shown in Fig. 3.4. It has been calculated from the current and electric field shown in Figs. 3.2 and 3.3, respectively. It should be noted that the MHD model used in this test is completely self-consistent: The model electric field and conductance distribution produce the horizontal and field-aligned currents, which in turn were used to calculate the magnetic disturbance. However, in the Swarm/SECS analysis, the electric and magnetic data have been analyzed completely independently from each other, so the inevitable analysis errors accumulate in the conductance estimation. In fact, nothing guarantees that the conductances produced by the Swarm/SECS method are even positive. Nevertheless, the Swarm analysis result shown in the middle panel is in good qualitative agreement with the model, and especially between the satellite tracks the errors are small. This is quite an impressive result, when one takes into account the fact that there is measured data only along the two tracks.

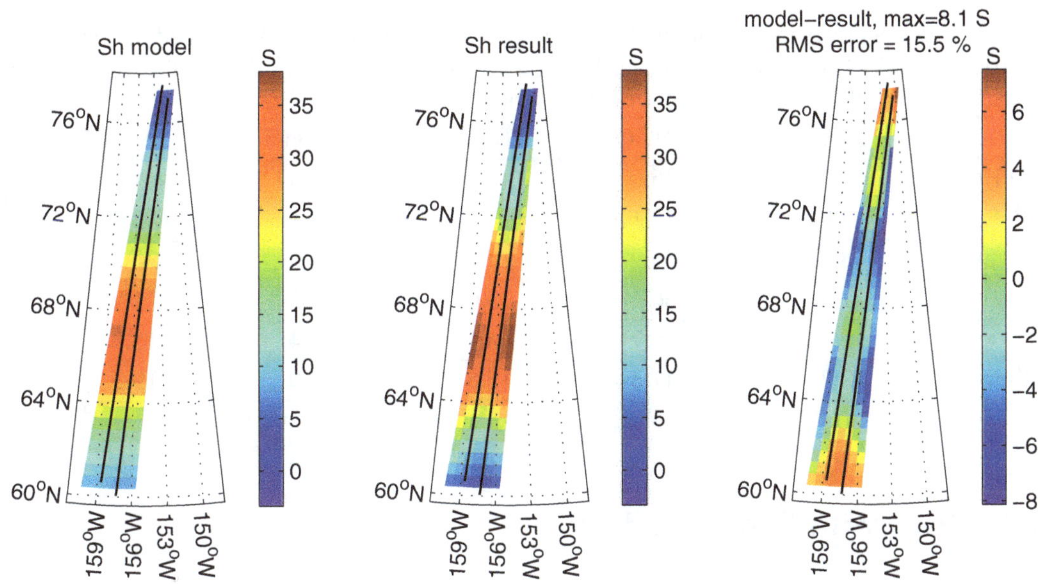

Fig. 3.4 Same as Figs. 3.2 and 3.3, but showing the Hall conductance calculated from the current and electric field

In order to improve the conductance estimation and to mitigate effects of the unexpectedly large noise in the Swarm electric field measurements, Vanhamäki and Amm (2014) developed an alternative analysis scheme that guarantees positive Hall and Pedersen conductances. In this "positive definite" approach the current map is taken as is, but the Hall and Pedersen conductances are represented with positive definite functions, such as $exp(x)$. A 2D map of the positivity parameter x is then obtained by fitting the Swarm electric field measurements, with possible additional constraints on spatial smoothness. This results in a nonlinear minimization problem that can be solved with standard techniques. However, the positive definite method has not really been applied in practice.

An alternative approach has been suggested by Marghitu et al. (2017). They also take the current map as given, but try to modify the measured electric field by linear transformations of the form

$$E'_x = aE_x + b, \qquad E'_y = cE_y + d. \tag{3.8}$$

Their goal is to find such coefficients a, b, c, d that the conductances calculated from Ohm's law match the conductances calculated from Robinson's formulas (Robinson et al. 1987). The average energy of precipitating electrons and the total energy flux needed in the Robinson's formulas are taken from the upward FAC and the conductance ratio estimated from Ohm's law. This proposed method is essentially an attempt to validate and, if necessary, recalibrate the Swarm electric field measurements.

Table 3.1 RMS errors in the solution for the horizontal current ($\mathbf{J}_\perp$), field-aligned current (FAC), electric field (**E**), Pedersen conductance (Σ_P), and Hall conductance (Σ_H) in the synthetic test cases. See Eq. (3.9) for definition of the RMS error

	$\mathbf{J}_\perp$ (%)	FAC (%)	**E** (%)	Σ_P (%)	Σ_H (%)
1D Ejet	14.8	36.0	4.9	26.6	11.6
2D Ejet	7.9	42.1	2.7	16.1	12.0
Vortex	18.7	155.9	15.1	23.0	22.8
Low	31.3	66.4	25.6	34.3	21.7
Moderate	12.6	55.7	16.1	24.4	15.5
High	12.7	47.3	15.5	37.0	19.3

3.3 Tests with Synthetic Data

Amm et al. (2015) created three simple but still realistic test cases for the Swarm/SECS analysis tool. These test cases were (1) a one-dimensional electrojet where ionospheric electric field and conductances are independent of longitude, (2) a two-dimensional electrojet whose strength varies along the jet, and (3) a fully two-dimensional current vortex. Additionally, Amm et al. (2015) took three different test cases from a self-consistent, coupled magnetosphere–ionosphere MHD simulation. These cases correspond to low, moderate and high activity in the simulated magnetosphere. More details of the test cases and the simulation setup are given by Amm et al. (2015). Moreover, Juusola et al. (2016b) considered an additional Ω-band test case constructed from direct observational data, which will be further discussed in Sect. 3.5.1.

As an overview of the standard analysis results obtained using the two lower satellites, Table 3.1 shows the RMS errors of the output parameters (horizontal current, FAC, electric field, Pedersen conductance and Hall conductance) in the three synthetic and three simulated test cases. In general, the electric field, horizontal current, and Hall conductance seem to be reproduced most reliably. The Pedersen conductance results exhibit a slightly larger error than the Hall conductance results, but in each case FAC is the most difficult parameter to reproduce accurately.

The RMS error is calculated as

$$RMSerror = 100 * \frac{\sqrt{<|model - result|^2>}}{\sqrt{<|model|^2>}}. \tag{3.9}$$

Here, <> means spatial average over the output area. The RMS errors for the horizontal current, electric field and Hall conductance given in the "Moderate" row of Table 3.1 correspond to Figs. 3.2, 3.3 and 3.4. According to the above equation, the RMS error is obtained by dividing the square root of the averaged and squared analysis error shown in the rightmost panels, by the similarly averaged model field shown in the left panels. Thus, the RMS error provides a single number characterizing the

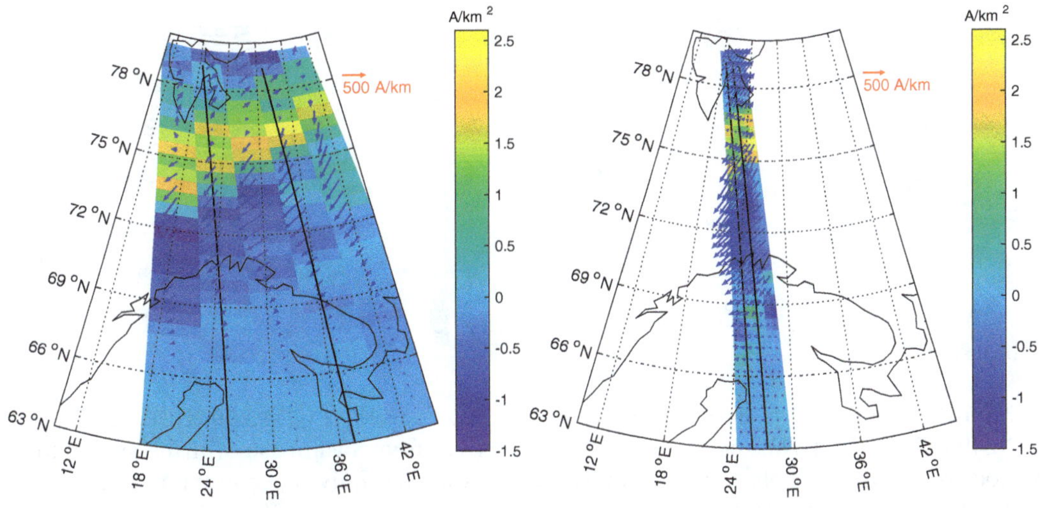

Fig. 3.5 Swarm-A, B, and C passing over the MIRACLE network, 30 July 2014, 02:08:49-02:13:48 UT. The left panel shows Swarm/SECS analysis results for the AB satellite pair, the right panel for the AC pair. From Marghitu et al. (2017)

overall relative error in the analysis results. While the RMS error is a useful parameter, it is also important to pay attention to the general structure and spatial pattern of the solution, as shown in Figs. 3.2, 3.3 and 3.4.

3.4 Examples of Event Studies

Auroral oval crossings in the early phase of the mission allow Swarm/SECS analyses both with AC and AB satellite pairs. The first example in Fig. 3.5 shows some results from such an exercise. This oval crossing took place in the dawn sector of the oval (MLT $\sim$ 0430) during a period of relatively weak global geomagnetic activity, with the Kp-index (Siebert and Meyer 1996) having value 2+. The currents in the Fennoscandian sector of the auroral oval were anyway rather strong, being larger than 500 A/km. Main features in both horizontal and field-aligned currents are similar for the AC (right panel in Fig. 3.5) and AB (left panel in Fig. 3.5) pairings. A closer look reveals some small differences in the horizontal currents along the northern parts of the Swarm-A trajectory (at latitudes poleward of 76°), where the AC pair yields currents with larger east components than in the results from the AB pair. The latter results are considered to be more reliable, because Swarm/SECS as applied to the AC pair tends to underestimate the north component of DF currents (c.f. Sect. 3.5.1).

Figure 3.6 shows the second example. It is from an event study by Juusola et al. (2016b), who applied Swarm/SECS for an overflight above the MIRACLE network[2] (Amm et al. 2001) of magnetometers and auroral cameras. The overflight

[2]see http://www.space.fmi.fi/MIRACLE/.

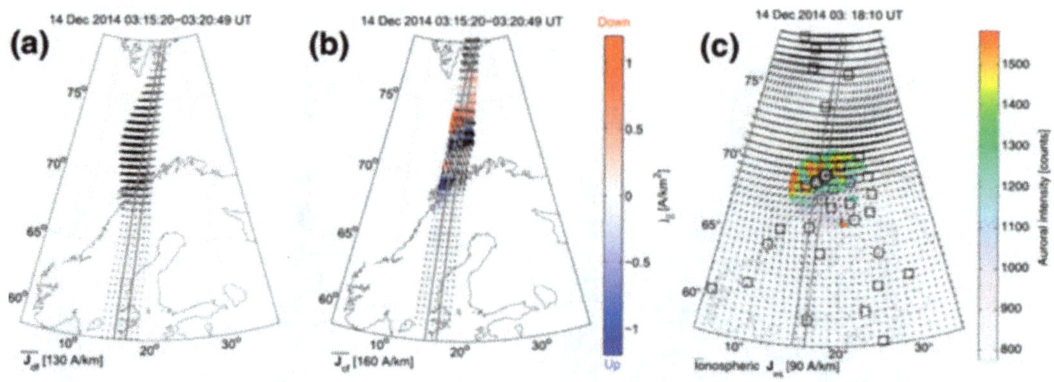

Fig. 3.6 Swarm-A and C flying over the MIRACLE network, 14 December 2014, 03:15:20-03:20:49 UT. Horizontal and field-aligned currents derived from Swarm data with the Swarm/SECS method are shown in panels **a** and **b**, while panel **c** shows the equivalent current derived from the ground-based MIRACLE magnetometers and the auroral intensity from Kilpisjärvi all-sky camera. From Juusola et al. (2016b)

took place in the dawn sector of the auroral oval (MLT $\sim$ 0530) during slightly stronger geomagnetic activity ($Kp = 3$) and pulsating auroras. The DF and CF currents by Swarm/SECS are shown in the panels (a) and (b) of Fig. 3.6, respectively. Panel (c) shows the DF currents (equivalent currents) as derived from MIRACLE data. An estimate of telluric currents (derived with a layer of SECSs at the ground surface) has been subtracted from the equivalent currents.

The satellite-based DF and ground-based equivalent current distributions are not completely identical. This is discussed in more detail in Sect. 3.5.1. In the case of Fig. 3.6 the Swarm/SECS method yields a westward DF current at magnetic latitudes $65° - 72°$, which is roughly consistent with the average oval location for $Kp = 3$ according to statistics by Juusola et al. (2009). Also, the ground- based equivalent currents are westward, but the electrojet is tilted approximately along constant geomagnetic latitude direction and it is weaker and slightly narrower than the currents estimated with the Swarm/SECS method. Concerning the strength and width of the electrojet, the picture by Swarm/SECS method is most likely more reliable than that of MIRACLE, because the electrojet is above the Arctic Sea where the coverage of ground-based magnetometers is very limited. Concerning the tilt, however, the ground-based equivalent currents yield a more reasonable result, which is consistent with the pattern of CF currents in panel (b): The bands of radial currents have the same tilt as the equivalent currents and horizontal CF currents flow in the perpendicular direction to the equivalent currents.

In the field of view of the all-sky camera (ASC) at Kilpisjärvi (magnetic latitude $\sim 66°$), the distribution of auroras have a sharp equatorward boundary, which is roughly colocated with the equatorward boundary of the electrojet (c.f. panel (c) in Fig. 3.6). Also DF and CF currents by the Swarm/SECS method are weak southward of this boundary. In this context, it is good to note that in Swarm/SECS analysis the coordinates of the Swarm magnetic field measurements have to be replaced with those corresponding to their ionospheric magnetic conjugacy points, in order to avoid obvious mismatches with spatial distributions in the auroras (for further discussion, see Juusola et al. 2016b). As electron precipitation causing auroras is also known to enhance ionospheric Hall conductances, the colocation of equivalent, and DF current and auroral equatorward boundaries is not surprising. Interpreting the distribution of CF currents and FAC in Fig. 3.6, however, is more complicated. In large scales the direction of FAC are in accordance with the standard R1/R2 pattern in the morning sector (downward current poleward of upward current), but the R2-currents are split into two bands. Swarm/SECS results show hints of a R2 band at the equatorward boundary of the electrojet, as anticipated, but another band of R2 appears poleward of the Kilpisjärvi ASC's field of view. With the available observations it is difficult to judge whether this structure is associated with an auroral structure or with a zone of enhanced electric field (Archer et al. 2017).

The third and last example is taken from Juusola et al. (2016a), who demonstrated how the Swarm/SECS method can be used to derive latitude profiles of Hall and Pedersen conductances across the auroral oval, when both the **E**- and **B**-field data are available. Figure 3.7 shows conductance estimates and the electric field in the vicinity of a post-midnight auroral arc (MLT ~ 02) with $Kp = 4-$. These results have been derived with the 1D Swarm/SECS analysis method applied to the **E**- and **B**-measurements from the Swarm-A satellite. The 1D approach assumes that longitudinal gradients and the east component of **E** are insignificant when compared to latitudinal gradients and the north component of **E**, respectively. The assumption of small longitudinal gradients in currents and conductances is supported in this case by equivalent current maps and auroral camera data from the MIRACLE network. In 1D cases, the DF and CF horizontal currents can be interpreted as Hall and Pedersen currents, respectively, and can be combined separately with electric field to produce Hall and Pedersen conductance profiles in different latitudinal resolutions. The resolutions of Hall conductance and the conductance ratio α in Fig. 3.7 are limited by that of the DF component (2°, c.f. discussion in Sect. 3.5.1). For consistency, the Pedersen conductance is shown in the same resolution, although it could be estimated with a better latitude resolution up to about 0.5°. The electric field values shown in Fig. 3.7 are somewhat higher than reported in some previous studies on auroral arcs (for example Opgenoorth et al. 1990; Aikio et al. 2002). However, the latitudinal distributions of **E**, currents and conductances follow closely the previously found pattern for morning sector arcs: Conductances are high at the arc location while **E** is high in a broader latitude range and particularly on the poleward side of the auroral arc. The westward directed Hall currents are strong in a latitudinal band, where currents in the poleward part are controlled by the strong **E** and in the equatorward part by enhanced conductances. These findings lead Juusola et al. (2016a) to suggest that this particular electrojet is in the transition zone between two different types of westward currents, conductance dominated (midnight sector) and electric field dominated (morning sector) electrojets, which are introduced in the substorm scenario by Kamide and Kobun (1996).

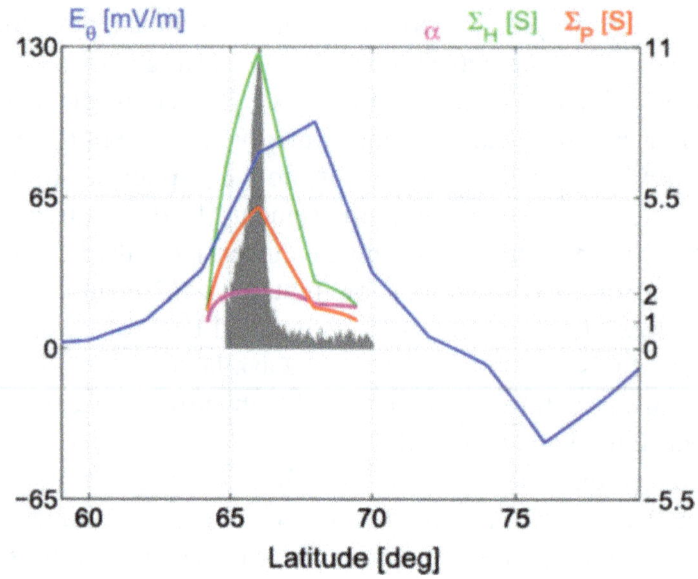

Fig. 3.7 1D fit to the electric field (blue, left axis) measured by Swarm-A on August 31, 2014 between 23:08:29 and 23:13:59 UT. The electric field has been mapped down to 110 km altitude along the Earths main field. Pedersen (Σ_P, red) and Hall (Σ_H, green) conductances and their ratio $\alpha = \Sigma_H/\Sigma_P$ (magenta, right axis), calculated from the horizontal current density and electric field. The gray shading is the auroral intensity in arbitrary units from the Sodankylä all-sky camera, projected to 110 km altitude. From Juusola et al. (2016a)

3.5 Statistical Studies

3.5.1 Swarm-MIRACLE Comparisons

It is possible to do a limited comparison between the ionospheric currents derived from Swarm magnetic field measurements using the Swarm/SECS method with those derived from ground magnetic field measurements. Such an analysis has been carried out for Swarm crossing of the MIRACLE network in 2013–2014 by Juusola et al. (2016b). A similar exercise was done previously by Ritter et al. (2004), who compared divergence-free currents estimated from CHAMP data with ground-based equivalent currents from the MIRACLE network. However, with the single CHAMP satellite only the east/west (or cross-track) electrojet current could be determined.

MIRACLE includes 38 magnetometers that form an irregular 2D network in Northern Europe between 58° and 79° geographic latitude and 5° and 36° longitude. The latitude range covers the auroral oval at most times. From the ground-based MIRACLE data, it is possible to derive 2D maps of the ionospheric equivalent current, as discussed in Sect. 7 of Chap. 2.

There are several issues that need to be taken into account when comparing currents derived from Swarm and MIRACLE. First, the comparison is limited to the DF component, as the equivalent current obtained from ground measurements should be identical to the DF part of the total current. Second, it takes Swarm approximately

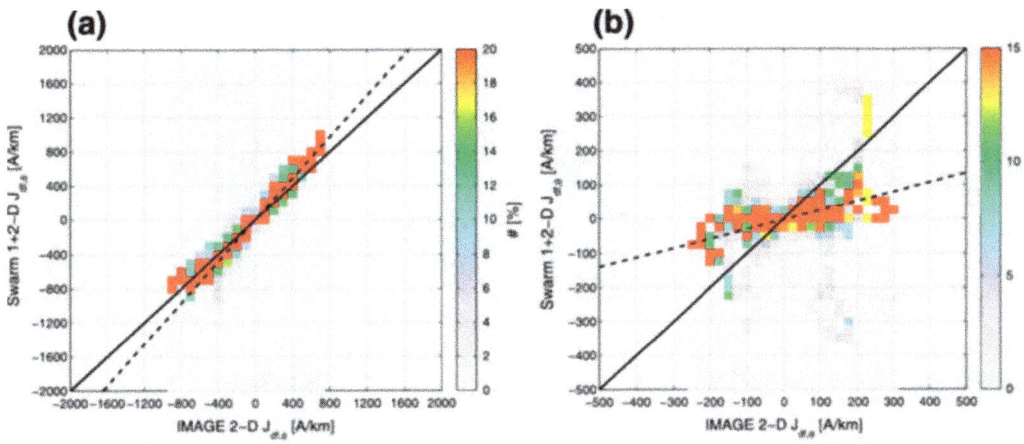

Fig. 3.8 Comparison of the DF current derived from Swarm and MIRACLE using the SECS analysis methods. Panel **a** is for the eastward component, and **b** for the northward component. From Juusola et al. (2016b)

6 min to cross the MIRACLE network, during which time the currents should remain stationary for the Swarm/SECS analysis. With MIRACLE data it is possible to analyze the instantaneous distribution every 10 s. Juusola et al. (2016b) resolved this discrepancy by averaging the MIRACLE currents over the Swarm crossing time. Third, the spatial resolutions of the DF currents derived from Swarm and the equivalent current from MIRACLE are not the same. Both the distance between the measurement and source current as well as the spatial resolution of the measurements play a role. Juusola et al. (2016b) concluded that the best spatial resolution for the MIRACLE currents is ~50 km and for the DF currents derived from Swarm ~200 km. The fourth issue in the Swarm-MIRACLE comparison has to do with the telluric currents. Juusola et al. (2016b) showed that accounting for these currents is important when analyzing ground-based magnetic data. In Swarm analysis, they can mostly be ignored, except for the most active events. For those cases, the mirror current method (Olsen 1996) works well, but it is not a good approach for quiet or only moderately active events.

Comparison has revealed that although the east component of the DF current density derived from Swarm agrees very well with that from MIRACLE (see Fig. 3.8), the south component given by Swarm is generally much weaker than that given by MIRACLE. The explanation suggested by Juusola et al. (2016b) is that the longitudinal distance between the AC satellite pair (~60 km at MIRACLE latitudes) is too small compared to the distance to the E-region current layer (~350 km) to provide sufficient gradient information in this direction, effectively rendering the measurements equivalent to those of a single satellite and imposing the 1D limitation.

Although proper comparison of the CF component derived from Swarm is not possible with ground measurements, it was noted that unlike the DF component, the CF component showed 2D features compatible with those resolved by MIRACLE (for example tilt of an electrojet). Application of the Swarm analysis to a synthetic model of an Ω-band further confirmed that while the full CF current and the zonal compo-

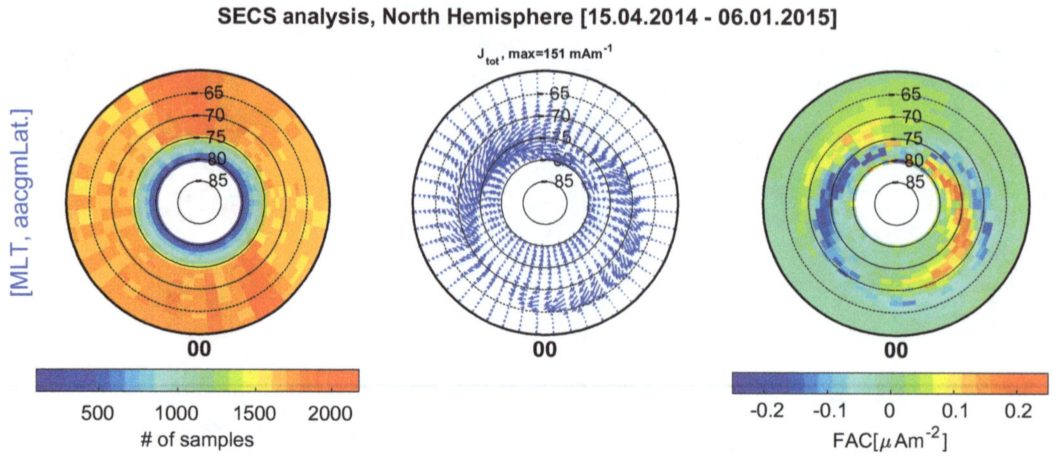

Fig. 3.9 The average pattern of the FAC and horizontal ionospheric current in the northern hemisphere as determined from 8.5 months of Swarm magnetic field data. Each pass is analyzed with the Swarm/SECS method and the results are averaged into a 1° times 0.5 MLT hour grid

nent of the DF current are well resolved, the meridional DF component is too weak. However, as the full DF component can be obtained from ground measurements, a combined SECS analysis of MIRACLE and Swarm measurements, with telluric currents taken into account, could produce the total horizontal ionospheric current density around the Swarm-AC footprints. An added benefit would an improved spatial resolution: as the CF component is directly connected to Swarm altitude through FACs, the spatial resolution of the resolved currents is only limited by the density of the measurement points. Thus, the combined analysis could increase the spatial resolution of the total current density from the ~200 km imposed by the DF part of Swarm/SECS analysis to the ~50 km of the MIRACLE analysis.

3.5.2 Global Current Systems with the Swarm/SECS Method

The Swarm mission offers a possibility to study statistical properties of the global ionospheric and field-aligned current systems without many of the simplifying assumptions employed in previous works that relied on data from single satellites. With the side-by-side AC pair, for the first time, it is possible to get reliable estimates of the current also in those regions and situations where the current geometry is far from the ideal 1D electrojet.

Figure 3.9 shows a statistical picture of the northern hemispheric FAC and horizontal currents, based on Swarm/SECS analysis of about 8.5 month of data from the AC pair. Their orbital planes precess about 2.7 h of (solar) local time in one month, giving a complete local time coverage in about 4.4 months. Thus the magnetic local time coverage shown in the left panel of Fig. 3.9 is rather uniform.

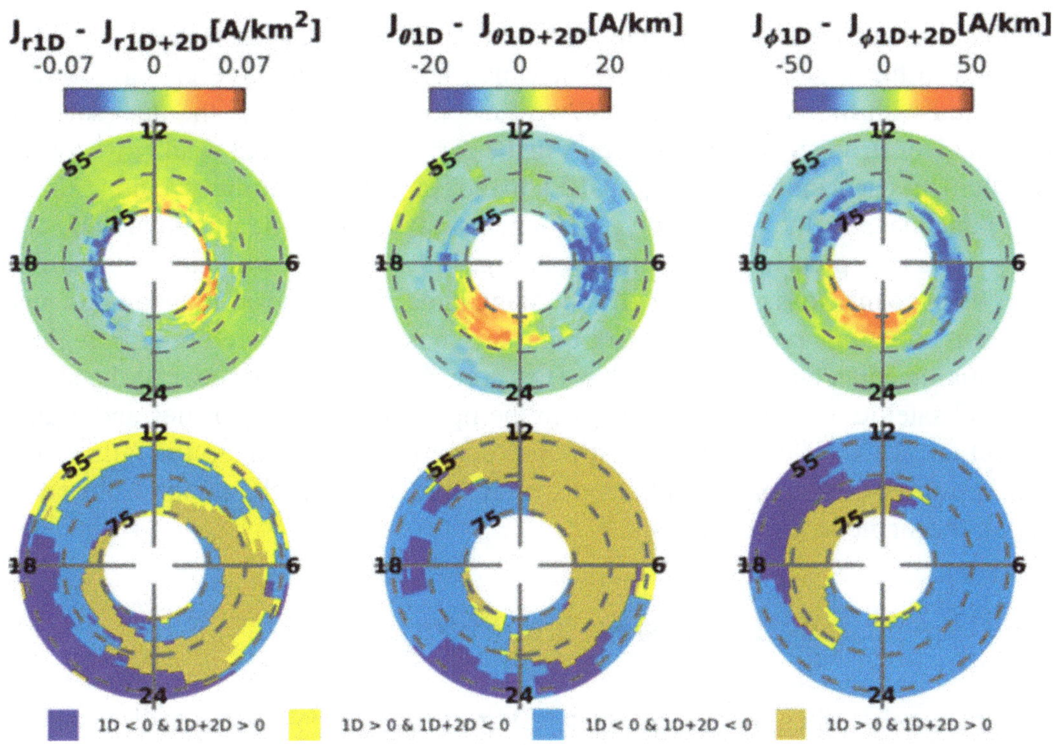

Fig. 3.10 The top row shows the difference in medians of the current densities obtained with a 1D (single satellite) and the 1D+2D Swarm/SECS (dual satellite) methods. J_θ is southward and J_ϕ is eastward. The bottom row shows in different colors the bins with matching and opposing signs between the two analysis methods

The middle and right panels of Fig. 3.9 show the median FAC and **J**, respectively. Each orbit was divided into 4 segments, consisting of ascending/descending auroral oval crossings in the northern/southern hemispheres. The passes are analyzed separately using the Swarm/SECS method, and the results were binned to a regular grid in magnetic latitude and local time (AACGM system, Shepherd 2014). The median FAC shows the familiar R1/R2 pattern, while the eastward and westward electrojets are clearly visible in the horizontal current.

As mentioned above, using the AC satellite pair, it is possible to reliably estimate currents also in those situations where single satellite methods do not work well due to complicated current geometry. Figure 3.10 shows the differences in the statistical results obtained using a single satellite method and the Swarm/SECS analysis. In both cases data from the A and C satellites was used, but in the single satellite analysis, the were analyzed using only 1D SECS, assuming vanishing gradients in the magnetic zonal direction. In order to get a feeling of relative magnitudes, the differences shown in Fig. 3.10 can be compared with the median currents shown in Fig. 3.9.

Both the 1D and Swarm/SECS methods give very similar pattern for the median currents, but the difference plots shown in the upper row of Fig. 3.10 reveal some systematic dissimilarities between the two methods. The dual satellite Swarm/SECS method gives slightly stronger R1 currents in both dawn and dusk sides, and also slightly stronger downward R2 currents around noon. These differences are also reflected in the southward current (J_θ), which is mostly curl-free closure current associated with the FAC. For the east/west electrojet current J_ϕ, the dual satellite method gives stronger eastward currents in the early afternoon, but weaker westward current in the early morning (03–07 MLT). In the Harang discontinuity region (22–24 MLT) the dual satellite method gives stronger westward current at high latitudes.

Nevertheless, plots in the bottom row of Fig. 3.10 show that while there are differences, the two methods generally give the same direction for the current components. Different directions (yellow and dark blue) are mostly encountered only in regions with small median current amplitudes. However, the differences between the single and dual satellite techniques get larger in the upper and lower quartiles and higher percentiles. This indicates that the stronger currents that dominate the tails of the distributions exhibit more complicated spatial structures, which are not captured so well in the single satellite analysis.

3.6 Conclusions, Discussion, and Future

This chapter describes how the Spherical Elementary Current Systems, discussed in detail in Chap. 2, can be used to analyze data from the multi-satellite Swarm mission. By fitting 1D/2D and CF/DF SECS to the magnetic data from the AC satellite pair, both the field-aligned and ionospheric horizontal currents can be estimated. This way the Swarm/SECS method gives a two-dimensional latitude–longitude picture of the currents around the satellite paths. In contrast to ground-based magnetic measurements, which give only the ionospheric equivalent currents, the Swarm/SECS method gives an estimate of the actual current. Data of the third Swarm satellite can also be used in the fitting.

If electric field data were available, a similar estimate of the two-dimensional distribution of the ionospheric electric field could be obtained. By combining the estimated horizontal current and electric field, the ionospheric Pedersen and Hall conductances could also be estimated.

A large number of the 2D current maps have been calculated and released to the community. The analysis has been carried out at the Finnish Meteorological Institute, and the results are available at http://space.fmi.fi/MIRACLE/Swarm_SECS/. This dataset forms an excellent basis for event studies and statistical investigations of auroral current systems, as discussed in Sects. 3.4 and 3.5 above.

As polar orbiting spacecraft, Swarm regularly pass over various ground-based instrument networks, such as the MIRACLE and other instruments located in northern Europe. These passes offer excellent opportunities for combining different datasets, both in event studies and statistical efforts. For example, in the case of currents, satellite magnetic measurements can reveal the FAC and curl-free part of the ionospheric horizontal currents, whereas ground-based measurements, due to their smaller distance from the ionospheric currents, give a better view of the divergence-free (or equivalent) currents. In fact, there is no reasons why both ground- and satellite-based magnetic data could not be used simultaneously in the Swarm/SECS analysis, which might further improve the results over magnetometer networks.

Combined with other data from the Swarm satellites and ground-based instruments, like electric field data and auroral images, such precise estimates of ionospheric horizontal currents and FAC can reveal the details of ionospheric electrodynamics and magnetosphere–ionosphere coupling.

Acknowledgements We thank the International Space Science Institute (ISSI) in Bern, Switzerland for supporting the Working Group Multi-Satellite Analysis Tools—Ionosphere from which this chapter resulted. The editors thank Ian Mann for his assistance in evaluating this chapter. Robyn Fiori provided a large number of comments, which improved the text considerably. The European Space Agency (ESA) is acknowledged for providing the Swarm data and for financially supporting the work on developing the Swarm/SECS method. This work was supported by the Academy of Finland project 314664.

References

Aikio, A.T., T. Lakkala, A. Kozlovsky, and P.J.S. Williams. 2002. Electric fields and currents of stable drifting auroral arcs in the evening sector. *Journal of Geophysical Research* 107 (A12): 1424. https://doi.org/10.1029/2001JA009172.

Amm, O., P. Janhunen, K. Kauristie, H.J. Opgenoorth, T.I. Pulkkinen, and A. Viljanen. 2001. Mesoscale ionospheric electrodynamics observed with the MIRACLE network: 1. Analysis of a pseudobreakup spiral. *Journal of Geophysical Research* 106(A11): 24675–24690. https://doi.org/10.1029/2001JA900072.

Amm, O., H. Vanhamäki, K. Kauristie, C. Stolle, F. Christiansen, R. Haagmans, A. Masson, M.G.G.T. Taylor, R. Floberghagen, and C.P. Escoubet. 2015. A method to derive maps of ionospheric conductances, currents, and convection from the Swarm multisatellite mission. *Journal of Geophysical Research* 120: 3263–3282. https://doi.org/10.1002/2014JA020154.

Archer, W.E., D.J. Knudsen, J.K. Burchill, B. Jackel, E. Donovan, M. Connors, and L. Juusola. 2017. Birkeland current boundary flows. *Journal of Geophysical Research* 122: 4617–4627. https://doi.org/10.1002/2016JA023789.

Finlay, C.C., N. Olsen, S. Kotsiaros, N. Gillet, and L. Tøffner-Clausen. 2016. Recent geomagnetic secular variation from Swarm and ground observatories as estimated in the CHAOS-6 geomagnetic field model. *Earth Planets Space* 68: 112. https://doi.org/10.1186/s40623-016-0486-1.

Juusola, L., O. Amm, K. Kauristie, and A. Viljanen. 2007. A model for estimating the relation between the Hall to Pedersen conductance ratio and ground magnetic data derived from CHAMP satellite statistics. *Annals of Geophysics* 25: 721–736. https://doi.org/10.5194/angeo-25-721-2007.

Juusola, L., K. Kauristie, O. Amm, and P. Ritter. 2009. Statistical dependence of auroral ionospheric currents on solar wind and geomagnetic parameters from 5 years of CHAMP satellite data. *Annals of Geophysics* 27: 1005–1017. https://doi.org/10.5194/angeo-27-1005-2009.

Juusola, L., S.E. Milan, M. Lester, A. Grocott, and S.M. Imber. 2014. Interplanetary magnetic field control of the ionospheric field-aligned current and convection distributions. *Journal of Geophysical Research* 119: 3130–3149. https://doi.org/10.1002/2013JA019455.

Juusola, L., W. Archer, K. Kauristie, H. Vanhamäki, and A. Aikio. 2016a. Ionospheric conductances of a morning-sector auroral arc from swarm measurements. *Geophysical Research Letters* 43: 11519–11527. https://doi.org/10.1002/2016GL070248.

Juusola, L., K. Kauristie, H. Vanhamäki, A. Aikio, and M. van de Kamp. 2016b. Comparison of auroral ionospheric and field-aligned currents derived from swarm and ground magnetic field measurements. *Journal of Geophysical Research* 121: 9256–9283. https://doi.org/10.1002/2016JA022961.

Kamide, Y., and S. Kokubun. 1996. Two-component auroral electrojet: Importance for substorm studies. *Journal of Geophysical Research* 101 (A6): 13027–13046. https://doi.org/10.1029/96JA00142.

Laundal, K.M., and A.D. Richmond. 2016. Magnetic coordinate systems. *Space Science Reviews* 206: 27–59. https://doi.org/10.1007/s11214-016-0275-y.

Marghitu, O., H. Vanhamäki, I. Madalin, L. Juusola, A. Blagau, and K. Kauristie. 2017. A tentative procedure to assess/optimize the swarm electric field data and derive the Ionospheric conductance in the auroral region. In *Paper presented at the 4th Swarm Science meeting*, Banff, 20–24 March, 2017.

Maus, S., M. Rother, C. Stolle, W. Mai, S. Choi, H. Lühr, D. Cooke, and C. Roth. 2006. Third generation of the Potsdam Magnetic Model of the Earth (POMME). *Geochemistry, Geophysics, Geosystems* 7: Q07008. https://doi.org/10.1029/2006GC001269.

Olsen, N. 1996. A new tool for determining ionospheric currents from magnetic satellite data. *Geophysical Research Letters* 23: 3635–3638. https://doi.org/10.1029/96GL02896.

Opgenoorth, H.J., I. Häggström, P.J.S. Williams, and G.O.L. Jones. 1990. Regions of strongly enhanced perpendicular electric fields adjacent to auroral arcs. *Journal of Atmospheric and Terrestrial Physics* 52: 449–458. https://doi.org/10.1016/0021-9169(90)90044-N.

Ritter, P., H. Lühr, A. Viljanen, O. Amm, A. Pulkkinen, and I. Sillanp. 2004. Ionospheric currents estimated simultaneously from CHAMP satellite and IMAGE ground based magnetic field measurements: A statisticalstudy at auroral latitudes. *Annals of Geophysics* 22: 417–430. https://doi.org/10.5194/angeo-22-417-2004.

Robinson, R.M., R.R. Vondrak, K. Miller, T. Dabbs, and D. Hardy. 1987. On calculating ionospheric conductances from the flux and energy of precipitating electrons. *Journal of Geophysical Research* 92: 2565–2569. https://doi.org/10.1029/JA092iA03p02565.

Shepherd, S.G. 2014. Altitude-adjusted corrected geomagnetic coordinates: Definition and functional approximations. *Journal of Geophysical Research* 119: 7501–7521. https://doi.org/10.1002/2014JA020264.

Siebert, M. and J. Meyer. 1996. Geomagnetic activity indices. In *The upper atmosphere*, ed. by W. Dieminger, G.K. Hartman, and R. Leitinger (Springer, Berlin Heidelberg New York), pp. 887–911. https://doi.org/10.1007/978-3-642-78717-1.

Vanhamäki, H. and O. Amm. Positive definite 2D maps of ionospheric conductances derived from Swarm electric and magnetic measurements. Paper presented at the 3rd Swarm Science meeting, Copenhagen, 19–20 June, 2014.

Chapter 4
Local Least Squares Analysis of Auroral Currents

Joachim Vogt, Adrian Blagau, Costel Bunescu and Maosheng He

Abstract Multi-spacecraft probing of geospace allows the study of physical structures on spatial scales dictated by orbital and instrumental parameters. This chapter highlights multi-point array analysis methods for constellations of two or three spacecraft such as Swarm, and also discusses multi-scale techniques for the geometrical characterisation of auroral current structures using observations of stationary or weakly time-dependent current structures along the tracks of individual satellites. Linear estimators are based on a least squares approach which is local in the sense that only few measurements around a reference point are considered for the reconstruction of geometrical and physical parameters. Local least squares estimators for field-aligned currents are compared with non-local counterparts and also local estimators based on finite differences. Uncertainties, implementation and other practical aspects are discussed. The techniques are illustrated using selected Swarm crossings of the auroral zone.

4.1 Introduction

Coupling processes in the auroral zone are to a large extent controlled by electric currents flowing parallel to the ambient magnetic field lines (e.g., Lysak 1990; Paschmann et al. 2002; Vogt 2002). Auroral field-aligned currents (FACs) connect remote plasmas in geospace and are associated with global magnetospheric dynamics such as large-scale convection and substorms. The type of electrodynamic coupling in the auroral current circuit depends on the spatial scale L of FACs. Quasi-static coupling of the equatorial magnetosphere and the auroral ionosphere is expected on large

J. Vogt (✉) · A. Blagau · C. Bunescu · M. He
Jacobs University Bremen, Bremen, Germany
e-mail: j.vogt@jacobs-university.de

A. Blagau · C. Bunescu
Institute for Space Sciences, Bucharest, Romania

M. He
Leibniz-Institut für Atmosphärenphysik, Kühlungsborn, Germany

M. W. Dunlop and H. Lühr (eds.), *Ionospheric Multi-Spacecraft Analysis Tools*, ISSI Scientific Report Series 17,
https://doi.org/10.1007/978-3-030-26732-2_4

scales $L \gtrsim \lambda_P \sim 100\,\text{km}$ (Lyons 1980; Lotko et al. 1987). Alfvénic coupling should be important on intermediate scales $L \gtrsim \lambda_A \sim 10\,\text{km}$ (Vogt and Haerendel 1998). On even smaller spatial scales in the kilometre range or below, auroral phenomena are typically very dynamic and short-lived, and not correlated with long-range magnetosphere–ionosphere coupling processes. The importance of spatial scales in FACs and their association with auroral processes and coupling regimes was confirmed by Lühr et al. (2015) using data from the initial phase of ESA's three-spacecraft Swarm mission when a range of inter-spacecraft distances was covered, see also Chap. 6 of this volume. Since April 2014, two Swarm satellites SwA and SwC orbit side-by-side in the auroral ionosphere, while the third Swarm satellite SwB moves at a higher altitude and lines up only occasionally with the lower pair to form a three-spacecraft constellation. SwA and SwC can be understood as a two-satellite array that allows the study of auroral FACs on a regular basis but restricted to the spatial scale given by their orbital separation.

To resolve electric current systems or other geospace structures by means of single-spacecraft data or multi-spacecraft observations, analysis methods must provide an adequate spatial resolution. The single-spacecraft approach where an individual satellite is assumed to move across a stationary or at most weakly time-dependent structure may resolve spatial scales given by the product of relative speed V and sampling interval Δt. Satellite constellations such as Cluster and Swarm allow to adopt a multi-point array perspective with the spatial resolution given by the inter-spacecraft distance scale Δr. For geospace phenomena and missions, $V\Delta t$ is typically smaller than Δr. Single-spacecraft methods allow the study of smaller scales but only in the along-track direction. Multi-point array techniques provide information also about the across-track variability (and possibly temporal changes) but only on larger spatial scales (e.g., Russell et al. 1983; Dunlop et al. 1988; Neubauer and Glassmeier 1990; Pinçon and Lefeuvre 1991; Chanteur and Mottez 1993; Dudok de Wit et al. 1995; Bauer et al. 2000; Balikhin et al. 2001; Dunlop et al. 2002; De Keyser et al. 2007; Vogt et al. 2008a), see also the ISSI Scientific Reports SR-1 and SR-8 (Paschmann and Daly 1998, 2008).

The category of analysis methods addressed in this report may be termed *local least squares (LS)* techniques. We are concerned mainly with multi-point array estimation of auroral FACs using vector magnetometer data from satellite missions such as Swarm, but also with multi-scale geometrical characterisation of current structures in the data of individual spacecraft. In local LS analysis, the least squares principle is applied to a confined region of interest, typically comprising only a few measurement points, equivalent to general (non-local) least squares modelling with localised basis functions but requiring less computational effort. Compared to local estimators based on finite differencing, local LS estimators turn out to be more robust with regard to non-regular (skewed or stretched) satellite constellations. To cover the multi-scale nature of auroral processes, in particular also intermediate and possibly even smaller scales, a multi-scale analysis technique developed by Bunescu et al. (2015, 2017) is included here as a localised version of the popular single-spacecraft minimum variance analysis or MVA that has its roots also in the least squares approach (Sonnerup and Scheible 1998).

Multi-spacecraft array techniques based on the local LS principle are presented in Sect. 4.2 (methodology) and Sect. 4.3 (implementation and applications). The multi-scale variant of MVA is discussed in Sect. 4.4 before we conclude in Sect. 4.5 with a summary.

4.2 Methodology of Multi-spacecraft Array Techniques

Multi-spacecraft estimation of spatial gradients, electric currents, or boundary parameters is based on a set of satellite positions and corresponding observations that are interpreted as array measurements. In the simplest and most straightforward case, each satellite of a constellation contributes one position vector to the array, and all measurements are taken at the same time. This perspective was adopted, e.g., by Dunlop et al. (1988), Chanteur (1998), Harvey (1998), as well as Vogt and Paschmann (1998) in the preparation phase of ESA's four-spacecraft mission Cluster to develop analysis methods for electric currents and spatial gradients. The corresponding three-spacecraft case of the LS approach was addressed by Vogt et al. (2009). The resulting estimators are instantaneous, thus perfectly localised in time, and also local in space at the lower resolution limit given by the inter-spacecraft separation scale that for Cluster ranged between 100 km and 10,000 km.

For the Swarm mission with only two satellites SwA and SwC close enough to be considered a multi-point array on a regular basis, the instantaneous approach was relaxed to include additional measurements shifted in time, thus generating a virtual planar four-point satellite array (Ritter and Lühr 2006; Ritter et al. 2013; Shen et al. 2012a; Vogt et al. 2013). Virtual along-track separations approximately equal to the distance between SwA and SwC of somewhat less than 100 km in the auroral zone are obtained using time shifts of about 10 s, considerably smaller than the variation time scales of FACs associated with quasi-static coupling. Localisation in space is characterised by the effective inter-spacecraft separation in the virtual quad, typically of the order of 100 km. In the context of this report, we refer to this class of methods as *local* analysis techniques.

When spatial distributions of electrodynamic variables for large parts of or even the entire auroral region are to be reconstructed from satellite crossings (e.g., He et al. 2012; Amm et al. 2015), possibly in combination with other ground-based data, the methodology is usually termed modelling rather than analysis, and should be referred to as a *regional* method (spatial extent of the order of 1000 km), applicable to structures that do not vary on time scales less than the satellite crossing time of the order of a few minutes. The most popular modelling approach is based on the least squares principle that we choose as a starting point for our discussion.

4.2.1 General Linear Least Squares

Least squares modelling can be characterised as a statistical technique to find the parameters of a model (m) that gives the best approximation of a given data set (d) of S measurements contaminated by random errors (residuals r). The measurements d_σ form the components of a data vector $\mathbf{d}$. If all parameters $a_1, a_2, \ldots, a_N$ enter the model function m linearly, then $m = \sum_\nu a_\nu f_\nu$ with basis functions f_ν. The model parameters a_ν are also called amplitudes and can be cast into a vector $\mathbf{a}$. The estimation problem can be written in the form $\mathsf{M}\mathbf{a} = \mathbf{d}$ with a matrix M and the solution $\mathbf{a} = \mathsf{M}^{\text{ils}}\mathbf{d}$ where M^{ils} is the pseudo-inverse of M in the least squares sense. For technical details of the general linear least squares approach, see Appendix A.

Regional least squares modelling of auroral field-aligned currents was carried out by He et al. (2012, 2014) who condensed ten years of CHAMP (Reigber et al. 2002) magnetic field measurements into the empirical FAC model MFACE using a set of data-adaptive basis functions called *empirical orthogonal functions (EOFs)*. In the first modelling step, the set of EOFs was constructed from the data in a coordinate frame centred on the dynamic auroral oval to capture its magnetic local time-dependent expansion and contraction during magnetospheric activity. The EOFs then serve as basis functions in an expansion of the form $j_\parallel(\Delta\beta|\mathbf{p}) = \sum_\nu a_\nu(\mathbf{p}) f_\nu(\Delta\beta)$ where the parameter vector $\mathbf{p}$ is formed by a set of predictor variables (magnetic local time, seasonal and solar wind parameters, AE), and $\Delta\beta = \beta - \beta_{\text{ACC}}$ is magnetic latitude in auroral oval coordinates, i.e. relative to the latitude of the auroral current centre β_{ACC}. In a second step, the functional dependences $\beta_{\text{ACC}} = \beta_{\text{ACC}}(\mathbf{p})$ and $a_\nu = a_\nu(\mathbf{p})$ were determined also through least squares regression. The geophysical parameters that drive the model (such as the IMF, solar wind parameters and geomagnetic indices) are obtained from NASA's OMNIWeb service. Two sample outputs are shown in Fig. 4.1. The times chosen for producing the diagrams correspond to selected Swarm auroral crossings that are used also for demonstrating multi-point FAC estimators in Sect. 4.3.2 and multi-scale MVA analysis of FACs in Sect. 4.4.1.

With regard to the discussion below of gradient and curl estimation from multi-spacecraft magnetic field measurements, it is important to note that the normal equations produce parameter estimates that are *linear in the data*: we may write

$$\mathbf{a} = \sum_\sigma \mathbf{q}_\sigma d_\sigma \tag{4.1}$$

where the vectors $\mathbf{q}_\sigma$ are the rows of M^{ils}. They depend only on the array geometry but not on the measurements d_σ, and may be termed *generalised reciprocal vectors*, see below. Linearity in the data applies also to any aspect of the model that can be expressed through a linear operation acting on the model function. Since differential operators like grad or curl are linear, least squares estimators of gradients and currents from linear magnetic field models based on multi-spacecraft vector measurements are linear in the data. The representation $\mathbf{a} = \sum_\sigma \mathbf{q}_\sigma d_\sigma$ facilitates error analyses and

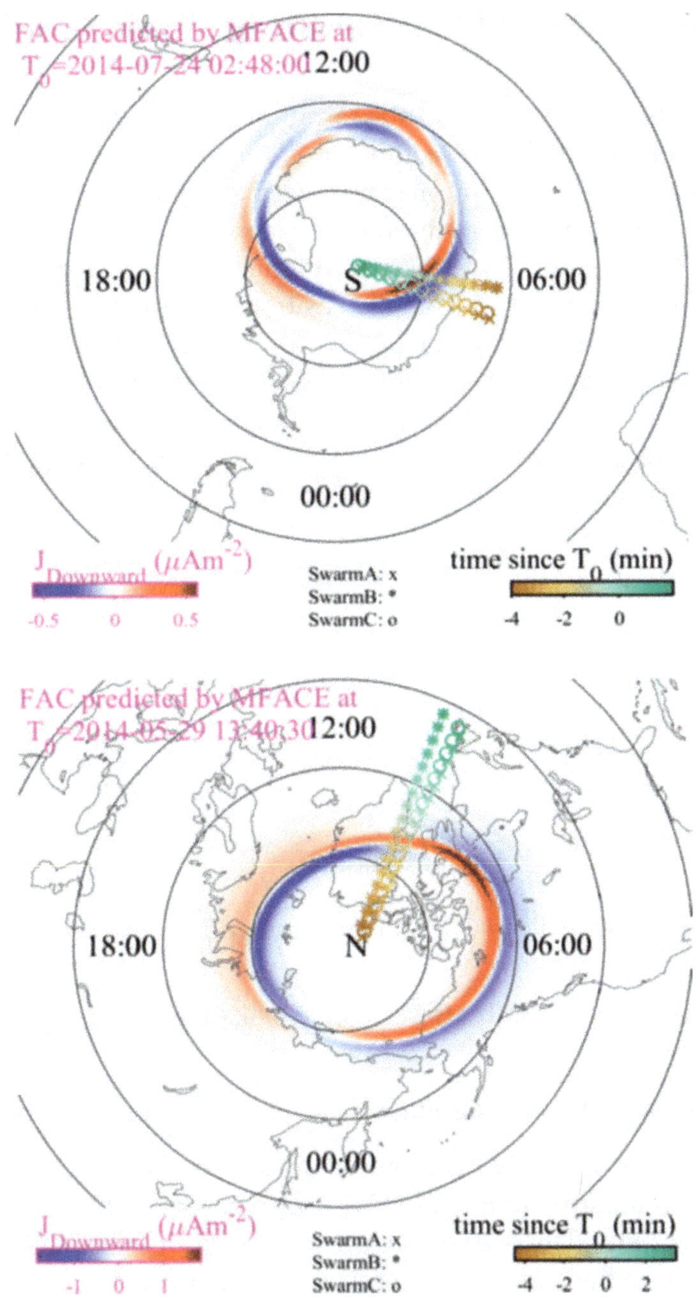

Fig. 4.1 Sample output of MFACE, an empirical model of auroral FAC based on 10 years of CHAMP magnetic field measurements. Model results are generated for 02:48:00 UT on 24 July 2014 and for 13:40:30 UT on 29 May 2014, corresponding to the centre times of two Swarm auroral crossings used for demonstrating FAC analysis techniques in Sects. 4.3.2 and 4.4.1

comparisons of the linear least squares technique with linear estimators derived form other principles such as finite differencing or boundary integrals, see Sect. 4.2.4.

4.2.2 *Local LS Estimators of Spatial Gradients*

The position vectors in an array of S spacecraft are denoted as $\mathbf{r}_\sigma, \sigma = 1, \ldots, S$, and the difference vectors are $\mathbf{r}_{\sigma\tau} = \mathbf{r}_\tau - \mathbf{r}_\sigma$. The average position or mesocentre $\mathbf{r}_* = (1/S) \sum_\sigma \mathbf{r}_\sigma$ may be chosen to coincide with the origin. In such a mesocentric coordinate system, the position tensor $\mathsf{R} = \sum_\sigma (\mathbf{r}_\sigma - \mathbf{r}_*)(\mathbf{r}_\sigma - \mathbf{r}_*)^\mathsf{T}$ simplifies to $\mathsf{R} = \sum_\sigma \mathbf{r}\,\mathbf{r}^\mathsf{T}$. Note that the volumetric tensor introduced by Harvey (1998) differs by a factor of $1/S$ from the position tensor defined here.

To estimate the gradient vector of a scalar field h, we consider the model $h(\mathbf{r}) = h_* + (\mathbf{r} - \mathbf{r}_*) \cdot \nabla h$ (or $h(\mathbf{r}) = h_* + \mathbf{r} \cdot \nabla h$ in a mesocentric frame). The function is linear in its four parameters, namely, the value h_* of h at the mesocentre and the three components of the gradient ∇h. Using S measurements h_σ at positions $\mathbf{r}_\sigma$, the least squares estimates are $h_* \simeq (1/S) \sum_\sigma h_\sigma$ and

$$\nabla h \;\simeq\; \mathbf{g} \;=\; \sum_\sigma \mathbf{q}_\sigma h_\sigma \;. \tag{4.2}$$

The local LS estimate of the gradient matrix of a vector field $\mathbf{B}$ is given by

$$\nabla \mathbf{B} \;\simeq\; \mathsf{G} \;=\; \sum_\sigma \mathbf{q}_\sigma \mathbf{B}_\sigma^\mathsf{T} \;. \tag{4.3}$$

The vectors $\mathbf{q}_\sigma$ are solutions of

$$\mathsf{R}\mathbf{q}_\sigma = \mathbf{r}_\sigma \tag{4.4}$$

Vogt et al. (2008b). Based on the rank of the (3×3) position tensor R, we distinguish two cases.

Invertible position tensor

If $S \geq 4$, and the position vectors are not all in one plane, the position tensor is non-singular (full rank 3), and we obtain $\mathbf{q}_\sigma = \mathsf{R}^{-1}\mathbf{r}_\sigma$ (Vogt et al. 2008b). The vectors $\mathbf{q}_\sigma$ can be understood as *generalised reciprocal vectors*, because in the special case $S = 4$ they coincide with the (tetrahedral) reciprocal vectors (Chanteur 1998; Chanteur and Harvey 1998; Vogt et al. 2008b) defined through

$$\mathbf{k}_\rho \;=\; \frac{\mathbf{r}_{\sigma\tau} \times \mathbf{r}_{\sigma\nu}}{\mathbf{r}_{\sigma\rho} \cdot (\mathbf{r}_{\sigma\tau} \times \mathbf{r}_{\sigma\nu})} \tag{4.5}$$

where $(\rho, \sigma, \tau, \nu)$ is a cyclic permutation of $(1, 2, 3, 4)$.

Singular position tensor, planar spacecraft array

If $S \geq 3$, and the spacecraft are in all one plane but not co-linear, the position tensor has only rank 2. Measurements allow to determine the component $\nabla_p h$ of the

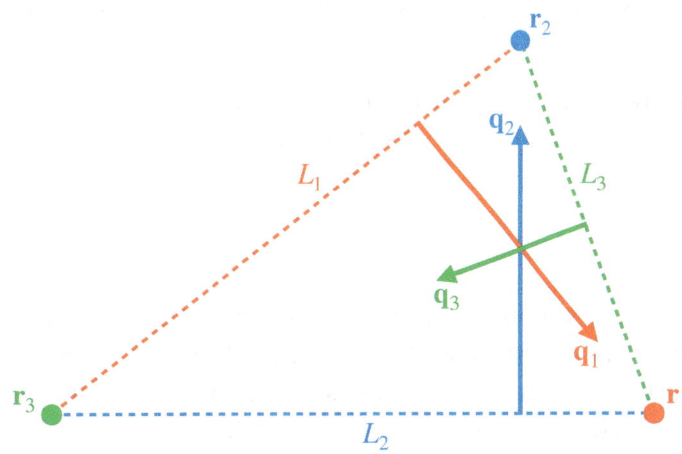

Fig. 4.2 Geometry of planar reciprocal vectors for a three-spacecraft configuration. Each vector $\mathbf{q}_\sigma$ is perpendicular to the line segment L_σ facing satellite no. σ at position vector $\mathbf{r}_\sigma$. The length $|\mathbf{q}_\sigma|$ is inversely proportional to the distance between L_σ and $\mathbf{r}_\sigma$

gradient in the plane spanned by the three spacecraft (in-plane or perpendicular gradient) but not the component $\nabla_n h$ normal to that plane (out-of-plane or normal component). Additional information in the form of geometrical or physical assumptions (conditions, constraints) is required to determine $\nabla_n h$. The vectors $\mathbf{q}_\sigma$ are the minimum-norm solutions of $\mathsf{R}\mathbf{q}_\sigma = \mathbf{r}_\sigma$ (Vogt et al. 2013), and may be termed *planar reciprocal vectors*. In the special case $S = 3$, they can be written in the form (Vogt et al. 2009)

$$\mathbf{q}_\sigma = \frac{\mathbf{n} \times \mathbf{r}_{\tau\nu}}{|\mathbf{n}|^2}, \quad \sigma = 1, 2, 3, \tag{4.6}$$

where (σ, τ, ν) is a cyclic permutation of $(1, 2, 3)$, $\mathbf{n} = \mathbf{r}_{12} \times \mathbf{r}_{13} = \mathbf{r}_{\sigma\tau} \times \mathbf{r}_{\sigma\nu}$, and the corresponding unit vector is $\hat{\mathbf{n}} = \mathbf{n}/|\mathbf{n}|$. The geometry of planar reciprocal vectors for a three-spacecraft configuration is sketched in Fig. 4.2. The relationships of planar reciprocal vectors to the eigenvalues and eigenvectors of the volumetric tensor $(1/S)\mathsf{R}$ are discussed in detail by Shen et al. (2012b).

The singular position tensor case is most relevant for the Swarm mission: here the gradient vector cannot be resolved fully from the measurements, and additional information has to be taken into account to reconstruct its out-of-plane component. Constraints may in principle be incorporated in the least squares framework using Lagrange multipliers. The approach chosen by Vogt et al. (2009) is based on geometrical considerations, and offers the possibility to choose between three types of constraints: (1) gradient parallel to a given direction $\hat{\mathbf{e}}$, (2) gradient perpendicular to a given direction $\hat{\mathbf{e}}$, (3) the physical structure is stationary in a reference frame moving with a known velocity relative to the spacecraft array. Of particular importance for studies of field-aligned currents is a fourth constraint that combines the force-free condition $\mathbf{B} \times (\nabla \times \mathbf{B}) = \mathbf{0}$ with $\nabla \cdot \mathbf{B} = 0$ to estimate the full magnetic gradient matrix from spacecraft measurements in one plane (Shen et al. 2012a; Vogt et al. 2013).

4.2.3 Local LS Estimators of Electric Currents

In the invertible position tensor case (spacecraft are not all in one plane), the local LS estimate of the curl $\nabla \times \mathbf{B}$ can be written as

$$\nabla \times \mathbf{B} \simeq \mathbf{c} = \sum_{\sigma} \mathbf{q}_{\sigma} \times \mathbf{B}_{\sigma} \,. \tag{4.7}$$

In the singular position tensor case (spacecraft are in one plane but not co-linear), one may incorporate an adequate constraint to reconstruct the full gradient matrix $\nabla\mathbf{B}$ first, and then take its skew-symmetric part $\frac{1}{2}(\nabla\mathbf{B} - \nabla\mathbf{B}^{\mathsf{T}})$ to obtain the curl (Vogt et al. 2009). If the curl is known to be parallel to a given direction as in the case of auroral FACs, one may also start from the density j_n of the normal current (i.e. the component perpendicular to the plane spanned by the spacecraft positions) that is fully determined by the measurements through

$$\mu_0 j_n = (\nabla \times \mathbf{B})_n \simeq c_n = \hat{\mathbf{n}} \cdot \sum_{\sigma} \mathbf{q}_{\sigma} \times \mathbf{B}_{\sigma} = \sum_{\sigma} (\hat{\mathbf{n}} \times \mathbf{q}_{\sigma}) \cdot \mathbf{B}_{\sigma} = \sum_{\sigma} (\hat{\mathbf{n}} \times \mathbf{q}_{\sigma}) \cdot \mathbf{B}_{p,\sigma} \tag{4.8}$$

Here $\hat{\mathbf{n}}$ is a unit vector normal to the spacecraft plane, and $\mathbf{B}_{p,\sigma}$ are the planar components of the measured vectors $\mathbf{B}_{\sigma}$. The field-aligned current density is then given by $j_{\parallel} \simeq j_n / \hat{\mathbf{n}} \cdot \hat{\mathbf{B}}_0$ where $\hat{\mathbf{B}}_0$ is the direction of the ambient magnetic field.

4.2.4 Related Local Estimators of Gradients and Currents

In preparation of the Cluster mission, gradient analysis methods were derived from several different principles such as discretised boundary integration (Dunlop et al. 1988), spatial interpolation (Chanteur 1998), and least squares estimation (Harvey 1998). For the Swarm mission, curl estimators based on finite differences (FD) and also on discretised boundary integrals (BI) were developed by Ritter and Lühr (2006), see also Shen et al. (2012a) and Ritter et al. (2013). Although the underlying principles differ, the resulting estimators may still turn out to be identical. Chanteur and Harvey (1998) demonstrated that in the regular (tetrahedral) four-spacecraft case, spatial interpolation yields the same analysis scheme as unconstrained least squares. Considering virtual four-point almost planar configurations relevant for the Swarm mission, (Vogt et al. 2013) compared different estimators for the normal component of the curl (corresponding to the radial current density) and found that the FD and BI estimators are algebraically identical.

Inter-comparisons of gradient and curl analysis schemes are facilitated by the observation that (to our knowledge) all the proposed multi-spacecraft methods yield estimators that are linear in the data. Using general results from linear algebra, (Vogt et al. 2008b) demonstrated that the problem of linear and consistent four-point gradient estimation has a unique solution. The same argumentation can be applied to

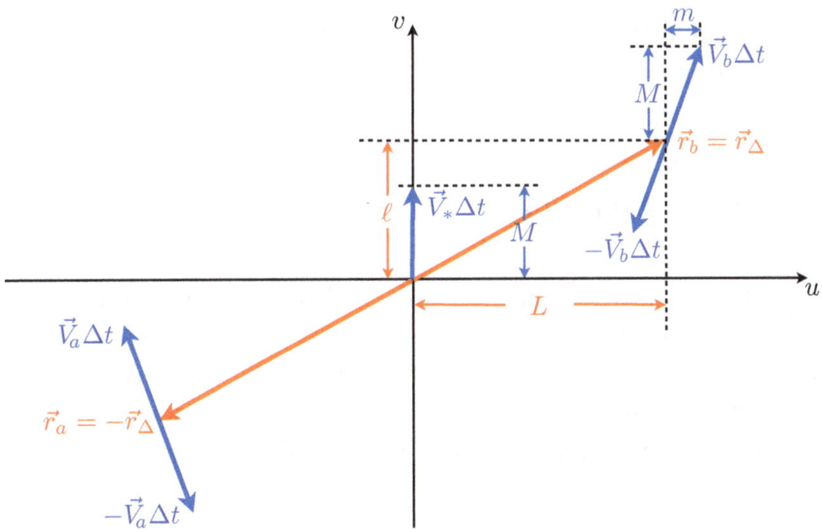

Fig. 4.3 Virtual four-point configuration produced by two satellites on close orbits such as SwA (corresponding to satellite a) and SwC (satellite b). The shape of the four-point array can be characterised by the dimensionless parameters $\mu = M/L$ (stretching parameter, ratio of along-track to across-track distance), $\lambda = \ell/L$ (skewness parameter, ratio of relative offset 2ℓ to across-track distance $2L$), and $\varepsilon = m/M$ (deviation of the two velocity directions from the perfectly parallel case). From Vogt et al. (2013), Fig. 4.1

three-spacecraft arrays: the number of free parameters (mesocentre value h_* and two components of the planar gradient $\nabla_p h$) is the same as the number of measurements (values h_σ at three-spacecraft positions), therefore there is a unique three-point linear and consistent estimator for the planar gradient that can be expressed explicitly, e.g., using the formalism of Vogt et al. (2009).

All linear estimators for the gradient of a scalar variable h can be represented in the form $\nabla h \simeq \sum_\sigma \mathbf{p}_\sigma h_\sigma$ with a specific set of vectors $\mathbf{p}_\sigma$, termed canonical base vectors by Vogt et al. (2013). For vector measurements, we obtain $\nabla \mathbf{B} = \sum_\sigma \mathbf{p}_\sigma \mathbf{B}_\sigma^\mathsf{T}$. The corresponding linear curl estimator can then be written in the form $\nabla \times \mathbf{B} = \sum_\sigma \mathbf{p}_\sigma \times \mathbf{B}$. The FD/BI and LS estimators for planar four-point configurations produced from the SwA–SwC pair yield canonical base vectors that differ only in terms proportional to a small configurational parameter $\varepsilon = m/M \sim 10^{-2}$ with m and M as in Fig. 4.3, see Vogt et al. (2013) for a detailed description. The FD/BI estimator of the normal curl component $(\nabla \times \mathbf{B})_n$ can be written in the compact form

$$\mu_0 j_n = (\nabla \times \mathbf{B})_n \simeq \frac{1}{2A}\left[\left(\mathbf{B}_c^+ - \mathbf{B}_a^-\right)\cdot\left(\mathbf{r}_a^+ - \mathbf{r}_c^-\right) - \left(\mathbf{B}_a^+ - \mathbf{B}_c^-\right)\cdot\left(\mathbf{r}_c^+ - \mathbf{r}_a^-\right)\right] \tag{4.9}$$

where the A is the modulus of the oriented area

$$\mathbf{A} = \frac{1}{2}\left(\mathbf{r}_a^- - \mathbf{r}_a^+\right)\times\left(\mathbf{r}_c^+ - \mathbf{r}_a^+\right) + \frac{1}{2}\left(\mathbf{r}_c^+ - \mathbf{r}_c^-\right)\times\left(\mathbf{r}_a^- - \mathbf{r}_c^-\right), \tag{4.10}$$

see also Appendix B in Vogt et al. (2013). Subscripts a and c indicate SwA and SwC, respectively, and superscripts $\pm$ denote the two time-shifted measurements.

4.2.5 Errors and Limitations

Spatial gradient estimates produced from multi-point measurements are affected by several errors and limitations: (a) measurement errors, (b) positional errors, (c) imperfections of the assumed linear model. In the planar spacecraft array case when additional information has to be considered to reconstruct the normal gradient, and (d) uncertainties in the imposed geometrical or physical conditions give rise to additional errors.

Measurement errors

Random uncertainties produced by limitations of the experimental setup (instrumental noise) are called measurement errors or physical errors. They can be quantified by means of error covariances $\langle \delta h_\sigma \, \delta h_\tau \rangle$ for scalar observables, and $\langle \delta \mathbf{B}_\sigma \, \delta \mathbf{B}_\tau^\mathsf{T} \rangle$ for vectors such as the magnetic field. Representations such as $\nabla h \simeq \mathbf{g} = \sum_\sigma \mathbf{q}_\sigma h_\sigma$ for linear gradient estimators offer a coherent framework for an assessment of the resulting uncertainties (Chanteur 1998; Vogt and Paschmann 1998; Vogt et al. 2008b, 2009, 2013). The special case of isotropic and uncorrelated measurement errors yields $\langle \delta h_\sigma \, \delta h_\tau \rangle = \delta_{\sigma\tau} (\delta h)^2$ where $\delta_{\sigma\tau}$ is the Kronecker delta symbol, and δh is a measure of instrumental sensitivity, resulting in the parameter covariant matrix

$$\langle \delta \mathbf{g} \, \delta \mathbf{g}^\mathsf{T} \rangle = (\delta h)^2 \mathsf{Q} \tag{4.11}$$

with the tensor $\mathsf{Q} = \sum_\sigma \mathbf{q}_\sigma \mathbf{q}_\sigma^\mathsf{T}$ (reciprocal tensor). For planar arrays of spacecraft positions, $\mathbf{g}$ is an estimator of the in-plane gradient component. The square magnitude error is given by the trace

$$\langle |\delta \mathbf{g}|^2 \rangle = (\delta h)^2 \, \mathrm{trace}(\mathsf{Q}) = (\delta h)^2 \sum_\sigma |\mathbf{q}_\sigma|^2 \, . \tag{4.12}$$

In both the invertible and the singular position tensor case, the term $\mathrm{trace}(\mathsf{Q}) = \sum_\sigma |\mathbf{q}_\sigma|^2$ is the square of an inverse length scale and can be understood as an error amplification factor that depends on the geometry (extension and shape) of the spacecraft array, see Vogt et al. (2008b, 2009, 2013). Normalisation using the mean square inter-spacecraft distance $(1/S) \sum_\sigma |\mathbf{r}_\sigma|^2$ yields a scaled version of the geometrical error amplification factor. The case of gradient matrix estimators $\nabla \mathbf{B} \simeq \mathsf{G} = \sum_\sigma \mathbf{q}_\sigma \mathbf{B}_\sigma^\mathsf{T}$, based on vector measurements is discussed in detail by Chanteur (1998) and Vogt et al. (2009).

Using the same approach and assumptions, the accuracy of linear curl estimators $\nabla \times \mathbf{B} \simeq \mathbf{c} = \sum_\sigma \mathbf{q}_\sigma \times \mathbf{B}_\sigma$ is studied in Appendix B. The parameter covariance matrix is given by

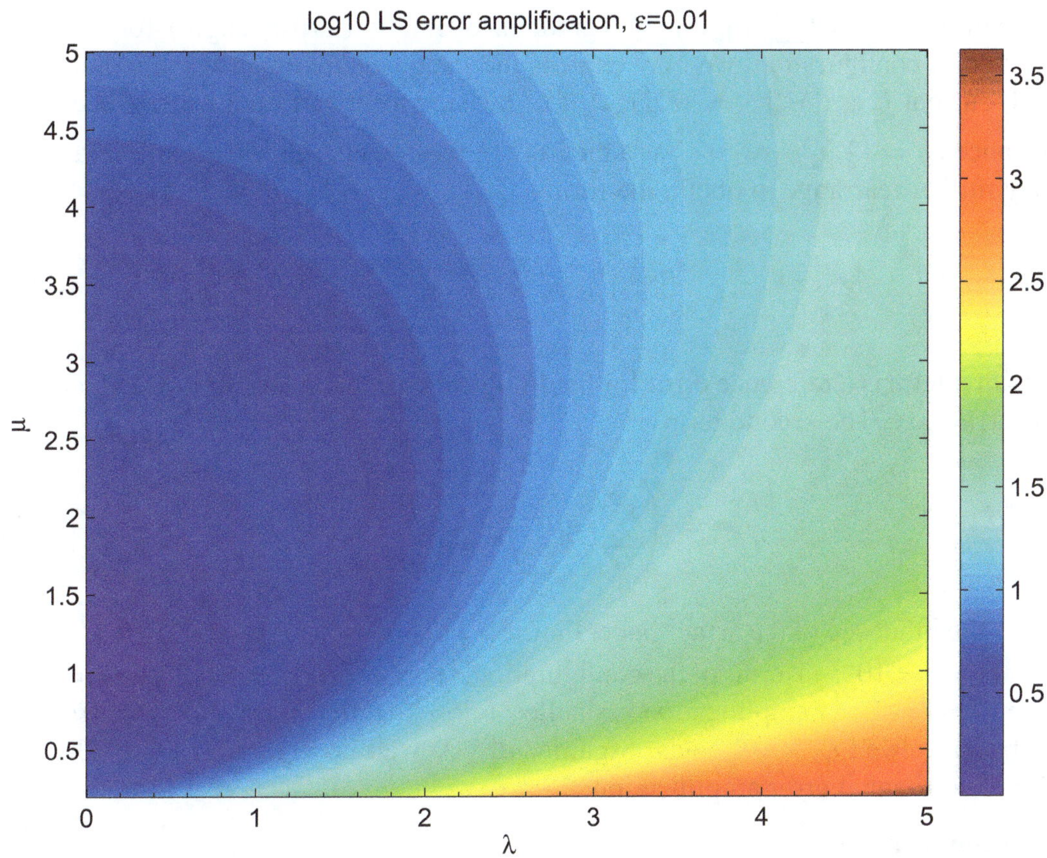

Fig. 4.4 Logarithm of the ratio L_r^2/L_q^2 as defined by Eq. (4.16), an effective error amplification factor controlled by the shape of the planar four-point configuration sketched in Fig. 4.3, as a function of the stretching parameter $\mu = M/L$ and the skewness parameter $\lambda = \ell/L$. From Vogt et al. (2013), Fig. 3

$$\left\langle \delta\mathbf{c}\,\delta\mathbf{c}^{\mathsf{T}} \right\rangle = (\delta B)^2 \left\{ \sum_\sigma |\mathbf{q}_\sigma|^2 \mathsf{E} - \sum_\sigma \mathbf{q}_\sigma \mathbf{q}_\sigma^{\mathsf{T}} \right\} = (\delta B)^2 \left\{ \mathrm{trace}(\mathsf{Q})\mathsf{E} - \mathsf{Q} \right\} \quad (4.13)$$

where E denotes the identity matrix. For a planar spacecraft array with normal unit vector $\hat{\mathbf{n}}$ we obtain

$$\left\langle |\delta c_n|^2 \right\rangle = (\delta B)^2 \sum_\sigma |\mathbf{q}_\sigma|^2 . \quad (4.14)$$

This case applies to Swarm field-aligned current estimates with the vectors $\mathbf{q}_\sigma$ being the canonical base vectors of the respective LS or FD/BI estimator (Vogt et al. 2013).

Defining $L_q = \left(\sum_\sigma |\mathbf{q}_\sigma|^2\right)^{-1/2}$ (gradient estimation error length) for a planar four-point configuration, we may express the mean square error of the radial (normal) current j_n as $\langle|\delta j_n|^2\rangle = (\delta B)^2/(\mu_0^2 L_q^2)$. Using the mean square inter-spacecraft distance $L_r = \left(\frac{1}{4}\sum_\sigma |\mathbf{r}_\sigma|^2\right)^{1/2}$ as a measure of the extent of the spacecraft array, one may further rearrange to obtain the form

$$\langle|\delta j_n|^2\rangle = \frac{(\delta B)^2}{\mu_0^2 L_r^2}\frac{L_r^2}{L_q^2}\,. \tag{4.15}$$

The first term is a reference error for field-aligned current density controlled only by the array size. The second term

$$\frac{L_r^2}{L_q^2} = \left(\sum_\sigma |\mathbf{q}_\sigma|^2\right)\cdot\frac{1}{4}\left(\sum_\sigma |\mathbf{r}_\sigma|^2\right) \tag{4.16}$$

gives the influence of the array shape. Figure 4.4 displays the logarithm of L_r^2/L_q^2 for $m/M = \varepsilon = 10^{-2}$ in terms of the configurational parameters $\mu = M/L$ and $\lambda = \ell/L$ (see Fig. 4.3). Error amplification is smallest (close to unity) for equal-sided ($\mu \approx 1$) and non-skewed ($\lambda \approx 0$) quads. Significantly skewed configurations ($\lambda \gtrsim 3$) give rise to substantial error amplification.

Positional errors

Random uncertainties in spacecraft positions are called positional errors or geometrical errors. Isotropic and uncorrelated positional errors can be incorporated in the parameter covariances obtained from considering only measurement errors by replacing $(\delta h)^2 \rightarrow (\delta h)^2 + |\nabla h|^2(\delta r)^2$ (Vogt and Paschmann 1998; Vogt et al. 2009). In the case of magnetic gradient and electric current estimates based on Swarm magnetic field data, positional errors have a much smaller impact than measurement errors because $|\nabla\mathbf{B}|\delta r \ll \delta B$ and can thus be neglected. This statement remains valid if positional inaccuracies in virtual four-point configurations imposed by time-shift errors δt (in the order of ms) are taken into account, then $\delta r \sim V\delta t$ where V is the spacecraft speed.

Model imperfections

Nonlinear variations of the observable over the spatial region covered by the spacecraft array cause deviations from the assumed linear model, and the spatial gradient is not perfectly uniform. In contrast to the statistical nature of measurement errors and positional errors which become less important for larger inter-spacecraft separation distances, gradient estimation errors due to deviations from linearity tend to increase with spacecraft separations (Robert et al. 1998). In the auroral zone the problem implies that current structures with variation scales (sheet widths) smaller than the array extension cannot be resolved and are effectively smeared out.

Uncertainties of imposed conditions

In the planar spacecraft array case (position tensor has rank 2) where only the in-plane component can be directly estimated from the measurements, the constraint equations used to reconstruct the normal gradient may not be perfectly satisfied, producing additional errors. The quality of the normal gradient can be assessed by means of error indicators (Vogt et al. 2009, 2013). The assumption that the full gradient is aligned with a given direction $\hat{\mathbf{e}}$ (parallel constraint) can lead to large uncertainties if $|\hat{\mathbf{e}} \times \hat{\mathbf{n}}|$ is small. The normal gradient estimate resulting from the perpendicular constraint (full gradient perpendicular to a given direction $\hat{\mathbf{e}}$) should be taken with care if the error indicator $|\hat{\mathbf{e}} \cdot \hat{\mathbf{n}}|$ is small. An error indicator for the force-free case that should not become too small is $|\hat{\mathbf{B}}_0 \cdot \hat{\mathbf{n}}|$. Since in the auroral zone the magnetic field forms a small angle with the radial vector, the error indicator for the virtual four-point configuration constructed from Swarm dual-spacecraft positions is close to unity and thus well-behaved. Larger uncertainties are expected at low latitudes.

4.3 Multi-spacecraft Array Techniques in Practice

The multi-point array techniques of Sect. 4.2 rest on the choice of canonical base vectors. For local LS estimators in the invertible position tensor case and a mesocentric coordinate frame, the canonical base vectors are generalised reciprocal vectors $\mathbf{q}_\sigma = \mathsf{R}^{-1}\mathbf{r}_\sigma$. Estimators of the magnetic gradient matrix $\nabla\mathbf{B}$ and the curl vector $\nabla \times \mathbf{B}$ are given by

$$\nabla\mathbf{B} \simeq \mathsf{G} = \sum_\sigma \mathbf{q}_\sigma \mathbf{B}_\sigma^\mathsf{T} , \tag{4.17}$$

$$\nabla \times \mathbf{B} \simeq \mathbf{c} = \sum_\sigma \mathbf{q}_\sigma \times \mathbf{B}_\sigma . \tag{4.18}$$

Practical aspects of four-spacecraft LS estimators were discussed, e.g., by Chanteur and Harvey (1998), Vogt et al. (2008b), and Vogt (2014).

This section is concerned with the implementation and applications of local LS estimators for the planar array case, i.e. three-spacecraft arrays and virtual four-point configuration constructed from the positions of SwA and SwC. Then the canonical base vectors $\mathbf{q}_\sigma$ are minimum-norm solutions of $\mathsf{R}\mathbf{q}_\sigma = \mathbf{r}_\sigma$, and Eqs. (4.17) and (4.18) produce the in-plane gradient and the normal curl component directly from the measurements. The three-spacecraft LS gradient estimator was tested by Vogt et al. (2009) and applied to Cluster pressure measurements in the magnetail. The dual-satellite LS FAC estimator was validated by Vogt et al. (2013) using Cluster observations of a force-free plasma structure in the solar wind that had previously been studied and characterised in detail by means of multi-spacecraft timing analysis (Vogt et al. 2011). Below in Sect. 4.3.2 we present selected applications of the three-spacecraft and the dual-satellite LS FAC estimator to Swarm magnetic field measurements, after discussing the implementation of local LS estimators for planar arrays in Sect. 4.3.1.

4.3.1 Implementation of Planar Multi-point Array Estimators

Local LS gradient and/or curl estimation using measurements of a planar spacecraft array involves the following steps.

Construction of canonical base vectors

In the planar array case with the position vectors of all spacecraft located in one plane, we first compute the eigenvalues and eigenvectors of the position tensor R, and then construct the pseudo-inverse Q from the two largest eigenvalues ρ_1, ρ_2 and the corresponding eigenvectors $\hat{\mathbf{e}}_1$ and $\hat{\mathbf{e}}_2$ as

$$\mathsf{Q} = \rho_1^{-1}\hat{\mathbf{e}}_1\hat{\mathbf{e}}_1^{\mathsf{T}} + \rho_2^{-1}\hat{\mathbf{e}}_2\hat{\mathbf{e}}_2^{\mathsf{T}} . \tag{4.19}$$

The eigenvalues are assumed to be in descending order, $\rho_1 \geq \rho_2 \geq 0$, and $\rho_3 = 0$ because the spacecraft array is planar. The canonical base vectors are $\mathbf{q}_\sigma = \mathsf{Q}\mathbf{r}_\sigma$ (in a mesocentric frame). The procedure works for both the virtual four-point configurations formed by positions of SwA and SwC as well as for three-spacecraft arrays. In the latter case the canonical base vectors are planar reciprocal vectors (Vogt et al. 2009) that can also be computed using Eq. (4.6). For a thorough discussion of volumetric tensor $(1/S)\,\mathsf{R}$ eigenvectors and eigenvalues in the three-spacecraft case, see Appendix D of Shen et al. (2012b).

Estimation of planar gradient and/or normal current components

Array magnetic field data $\mathbf{B}_\sigma$ allow to estimate directly the in-plane component of the gradient matrix as $\nabla_p \mathbf{B} = \sum_\sigma \mathbf{q}_\sigma \mathbf{B}^{\mathsf{T}}$, and also the normal component c_n of the curl:

$$c_n = \hat{\mathbf{n}} \cdot \sum_\sigma \mathbf{q}_\sigma \times \mathbf{B}_\sigma = \sum_\sigma \left(\hat{\mathbf{n}} \times \mathbf{q}_\sigma\right) \cdot \mathbf{B}_\sigma . \tag{4.20}$$

The normal (out-of-plane) current density is given by $j_n = c_n/\mu_0$.

Quality indicators of planar gradient and normal current estimates

The stability of the (constrained) matrix inversion that yields the canonical base vectors is controlled by the effective condition number $\mathrm{CN}(\mathsf{R}) = \rho_1/\rho_2$ of the position tensor R.

Error amplification due to the array shape is controlled by the ratio of square length scales L_r^2/L_q^2 defined by Eq. (4.16), and directly related to the condition number through

$$\frac{L_r^2}{L_q^2} = \frac{1}{4}\left(2 + \mathrm{CN}(\mathsf{R}) + \frac{1}{\mathrm{CN}(\mathsf{R})}\right), \tag{4.21}$$

$$\mathrm{CN}(\mathsf{R}) = \exp\left[\operatorname{arcosh}\left(2\frac{L_r^2}{L_q^2} - 1\right)\right], \tag{4.22}$$

see Appendix C for a proof. Hence, $\mathrm{CN}(\mathsf{R})$ and L_r^2/L_q^2 contain essentially the same information, and for moderately large values are related linearly: $\mathrm{CN}(\mathsf{R}) \simeq 4L_r^2/L_q^2 - 2$. When also the array size is taken into account, error amplification is measured by the square inverse length scale $L_q^{-2} = \sum_\sigma |\mathbf{q}_\sigma|^2$. The uncertainty of both planar gradient and normal curl estimation is simply given by $\delta B/L_q$. Hence in addition to $\mathrm{CN}(\mathsf{R})$ also L_q should be computed and checked to assess the quality of the estimated derivative.

Construction of the full gradient matrix and/or the full current vector

In order to obtain the full gradient, the component normal to the spacecraft plane has to be constructed in addition to the planar gradient estimate. This can be achieved by means of suitable constraint equations as discussed in Sect. 4.2.2, see also Vogt et al. (2009). The curl vector $\nabla \times \mathbf{B}$ can then be read directly from the components of the skew-symmetric part of $\nabla\mathbf{B}$. In the special case of the force-free condition, the current is parallel to the ambient magnetic field $\mathbf{B}_0$, and the field-aligned current density $j_\parallel$ can be computed directly from the normal current density j_n through $j_\parallel \simeq j_n/\hat{\mathbf{n}} \cdot \hat{\mathbf{B}}_0$. The construction of normal gradients and planar curl components should be critically assessed using error indicators as discussed in Sect. 4.2.5.

4.3.2 Application to Swarm Auroral Crossings

To demonstrate local LS estimation of FACs for planar spacecraft arrays, we select two auroral crossings of the Swarm satellites when SwB was close enough to the SwA–SwC pair for the application of three-spacecraft techniques.

Figure 4.5 shows magnetic field measurements of the three Swarm satellites in the Southern hemisphere on 24 July 2014, 02:44–02:50 UT, together with different FAC estimates and quality indicators. Clearly visible are negative parallel (downward) currents between ~02:47:30 and ~02:48:00 followed by positive parallel (upward) current until ~02:48:30. The dual-satellite LS FAC estimates are very close to the Level-2 FAC product J_L2_AC apart from several smaller-scale deviations. They are due to the fact that the Level-2 dual-satellite FAC product is based on filtered Swarm magnetic field observations whereas here the least squares estimator processed unfiltered data as input. The smaller-scale deviations are much less pronounced if the dual-satellite LS estimate of the FAC profile is also computed after application of a suitable filter. The output produced by the three-spacecraft LS estimator, also based on unfiltered magnetic field measurements, also shows smaller-scale variability but otherwise follows the other two profiles quite well apart from an apparent time shift,

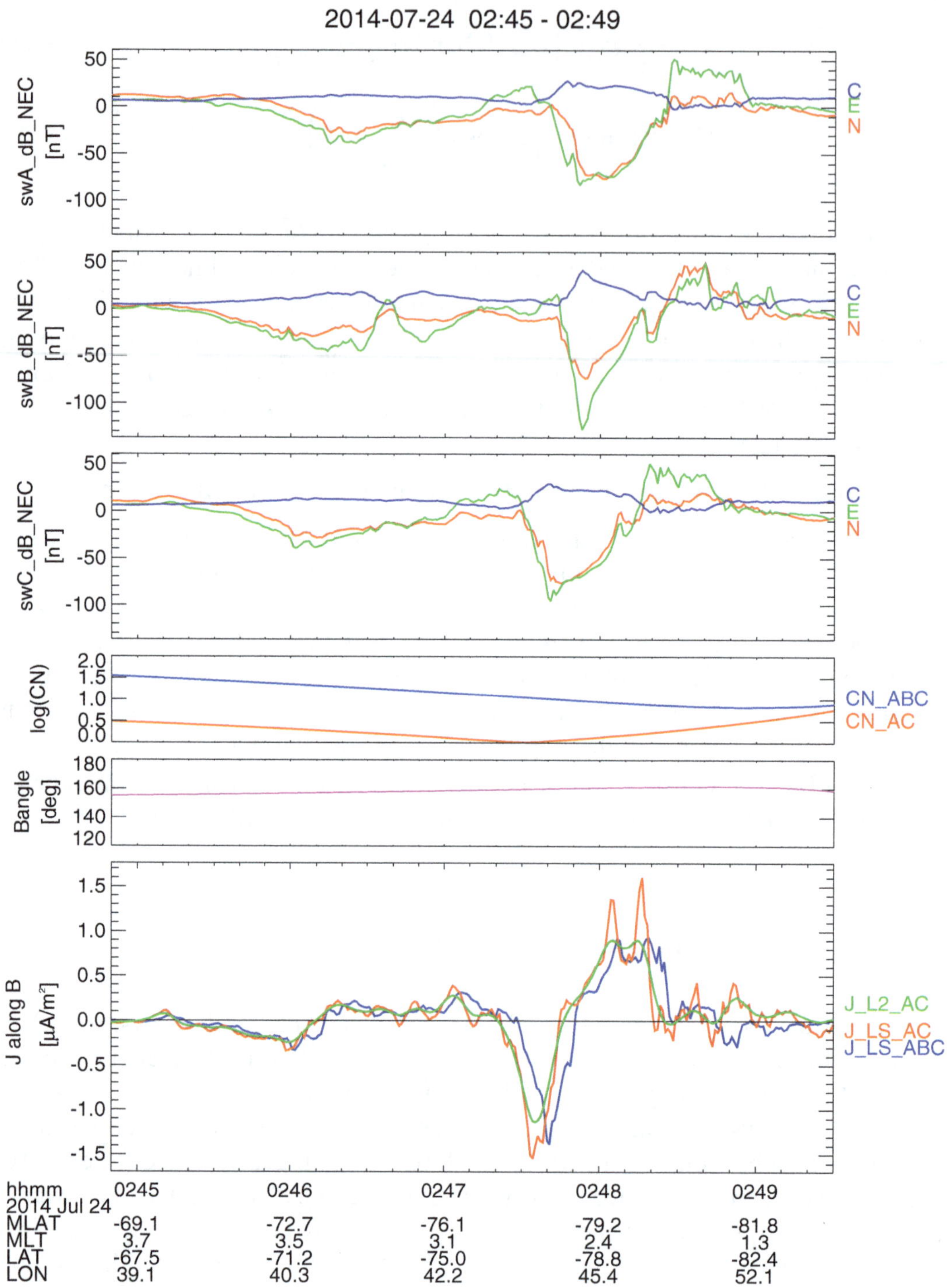

Fig. 4.5 Panels 1–3: magnetic field measurements of SwA, SwB, and SwC on 24 July 2014, 02:44–02:50 UT. Panel 4: logarithmic condition numbers for the three-spacecraft array (CN_ABC) and the virtual four-point configuration generated by SwA and SwC (CN_AC). Panel 5: angle between the ambient magnetic field direction $\hat{\mathbf{B}}_0$ and the normal vector of the three-spacecraft plane. Panel 6: comparison of the dual-satellite (virtual four-point) LS FAC estimator J_LS_AC and the three-spacecraft LS FAC estimator J_LS_ABC with the Level-2 dual-satellite FAC product J_L2_AC

caused by different mesocentres of the three-spacecraft array and the four-point configuration. Condition numbers for the current structure crossing are moderate (below 5 for the dual-satellite estimator, and not much larger than 10 for the three-spacecraft array). The angle between the ambient magnetic field and the normal direction of the three-spacecraft plane assumes tolerable values far from 90°.

The geometry of the current structure can be further studied using minimum variance analysis (MVA), discussed in more detail in Sect. 4.4 where the analysis procedure and key variables are explained. An important MVA parameter is the eigenvalue ratio that can be interpreted as a measure of planarity. For sufficiently large eigenvalue ratios we can think of the current structure as a sheet, with the eigenvector to the largest eigenvalue being tangential to the sheet. The auroral crossing considered here yields large eigenvalue ratios 29, 20 and 37 for SwA, SwB and SwC, respectively, and sheet orientations that are very consistent within a few degrees for all three satellites.

Data from the second Swarm auroral crossing on 29 May 2014, 13:36–13:44 UT, are displayed in Fig. 4.6. The largest current densities are observed between $\sim$13:39:30 and $\sim$13:41:30. Again apart from smaller-scale deviations due to differences in filtering of the input magnetic field measurements, the two dual-satellite FAC estimates (both based on data from SwA and SwC) are very similar. The FAC profile produced by the three-spacecraft LS estimator differs significantly at around 13:41, despite reasonable values of the quality indicators. Closer inspection of the SwB magnetic field profile reveals a substructure at 13:41 that is not present in the measurements of SwA and SwC, indicating non-uniform currents on the inter-spacecraft separation scale that are inconsistent with the linear model assumption. Eigenvalue ratios are 11, 16, 13 and thus somewhat smaller than for the first crossing, and also the sheet orientations obtained from single-spaceraft MVA show larger differences up to about 10 degrees.

4.4 Single-Spacecraft Multi-scale Analysis

Satellite measurements of the magnetic field allow the study of planar geospace structures such as current sheets or boundary layers through the eigenvalues and eigenvectors of the data covariance matrix. This type of principal axis decomposition is known as principal component analysis (PCA) or empirical orthogonal function (EOF) analysis in the statistical literature, and as minimum variance analysis (MVA) in space physics (Sonnerup and Cahill 1967). MVA can be derived using constrained least squares estimation (Sonnerup and Scheible 1998) and is usually applied to the entire geospace structure of interest. (Bunescu et al. 2015, 2017) introduced a multi-scale version by applying the MVA procedure using a range of sliding windows, thus producing local estimates of key MVA parameters such as the eigenvalue ratio and the angle characterising sheet orientation. The novel multi-scale version of MVA is described below in Sect. 4.4.2, followed by an application to Swarm magnetic field measurements in Sect. 4.4.3. The starting point of our discussion are the principles of MVA as summarised in Sect. 4.4.1.

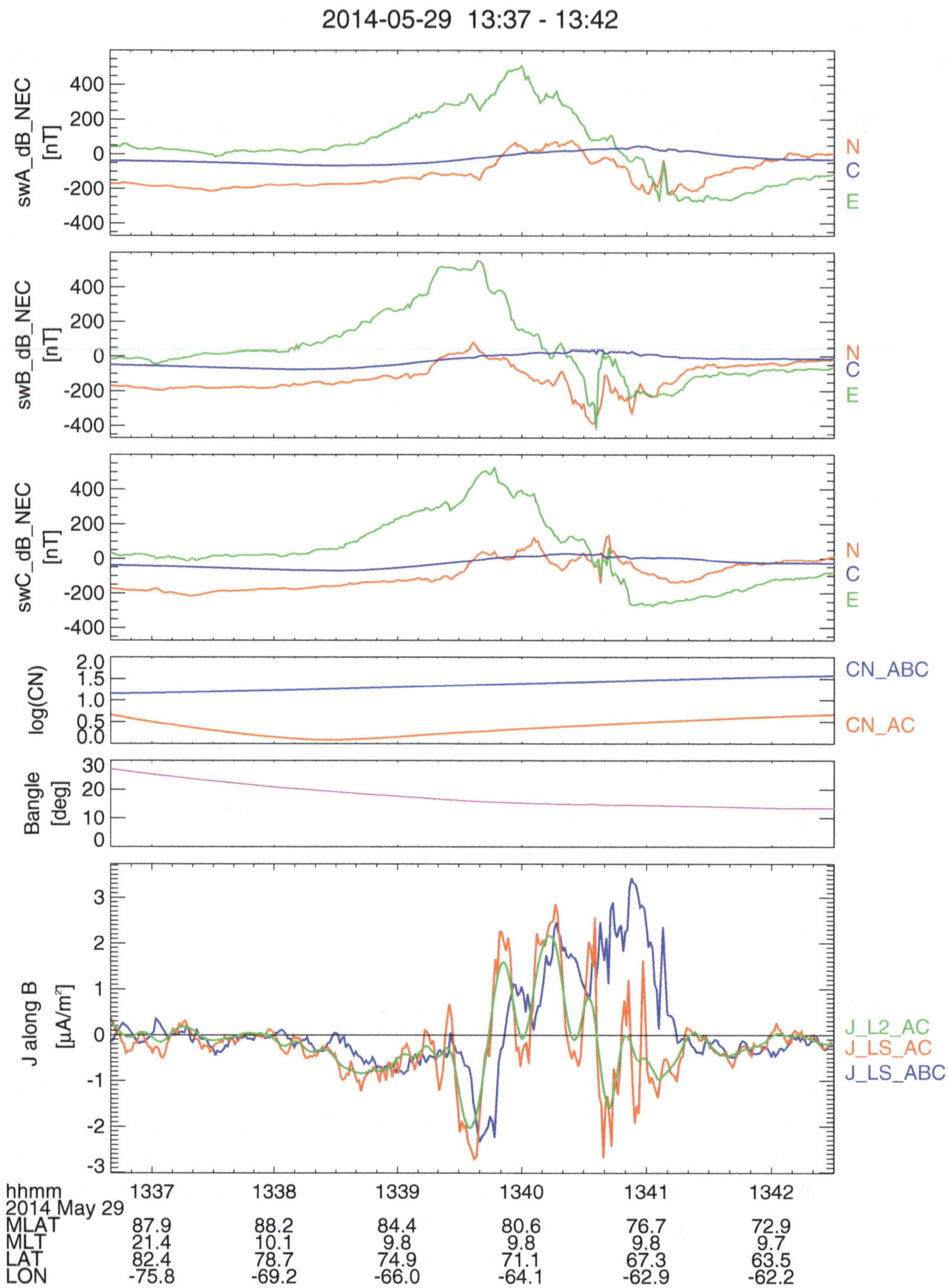

Fig. 4.6 Panels 1–3: magnetic field measurements of SwA, SwB, and SwC on 29 May 2014, 13:36–13:44 UT. Panel 4: logarithmic condition numbers for the three-spacecraft array (CN_ABC) and the virtual four-point configuration generated by SwA and SwC (CN_AC). Panel 5: angle between the ambient magnetic field direction $\hat{\mathbf{B}}_0$ and the normal vector of the three-spacecraft plane. Panel 6: comparison of the dual-satellite (virtual four-point) LS FAC estimator J_LS_AC and the three-spacecraft LS FAC estimator J_LS_ABC with the Level-2 dual-satellite FAC product J_L2_AC

4.4.1 MVA Applied to Auroral Current Sheets

Since the magnetic field $\mathbf{B}$ is solenoidal (divergence-free), planar magnetic structures varying only in one spatial direction $\hat{\mathbf{n}}$ satisfy

$$0 = \nabla \cdot \mathbf{B} = \hat{\mathbf{n}} \cdot \nabla B_n \,, \tag{4.23}$$

thus $B_n = \hat{\mathbf{n}} \cdot \mathbf{B}$ is constant. Assuming that any observed variability along $\hat{\mathbf{n}}$ is due to sufficiently small random errors, the eigenvector to the smallest eigenvalue of the data covariance matrix is a proxy of $\hat{\mathbf{n}}$. MVA applied to magnetic field data is sometimes termed MVAB. The method can be used also for other conserved plasma variables, see Sonnerup and Scheible (1998).

The quality of $\hat{\mathbf{n}}$ estimates is associated with eigenvalue ratios. In the case of auroral FAC sheets, magnetic perturbations are in the plane perpendicular to the ambient magnetic field. The problem reduces to two spatial dimensions with two relevant eigenvalues $\lambda_1 \geq \lambda_2$ and two eigenvectors $\hat{\mathbf{e}}_1, \hat{\mathbf{e}}_2$. The eigenvalue ratio λ_1/λ_2 can be understood as a measure of planarity and should be sufficiently large. The sheet orientation is given by tangential vectors $\hat{\mathbf{B}}_0$ (direction of the ambient magnetic field) and $\hat{\mathbf{e}}_1$, and the normal unit vector $\hat{\mathbf{n}} = \hat{\mathbf{e}}_2$. The orientation of auroral current sheets can be concisely characterised by the (inclination) angle formed by the sheet normal with magnetic north, or the spacecraft velocity vector (approximately geographic north for polar orbiting satellites such as CHAMP or Swarm).

4.4.2 Multi-scale Field-Aligned Current Analyzer

The multi-scale and continuous (local) variant of MVAB introduced by Bunescu et al. (2015, 2017), termed MS-MVA, can be summarised as follows:

- A range of window widths w with linear resolution dw is defined.
- At each time t of the magnetic field series, MVA is applied to an array of data segments of width w within a predefined range and centred at t, thus yielding a series of key MVA parameters $\lambda_1 = \lambda_{\max}$, $\lambda_2 = \lambda_{\min}$, $R_\lambda = \lambda_1/\lambda_2$ (eigenvalue ratio), and an inclination angle. All parameters are functions of time t and scale w.
- In addition to these MVA parameters, the derivative of the largest eigenvalue $\lambda_1 = \lambda_{\max}$ with respect to scale w is computed numerically to yield $\partial_w \lambda_{\max}$.
- The continuous and multi-scale MVA parameters are displayed as functions of time t and scale w in a suitable two-dimensional graphical representation, either as a contour plot and/or using an appropriate colour bar. Important scales are found to show up well in colour plots of $\partial_w \lambda_{\max}$.

MS-MVA was validated using synthetic data and measurements of the Cluster and FAST satellites.

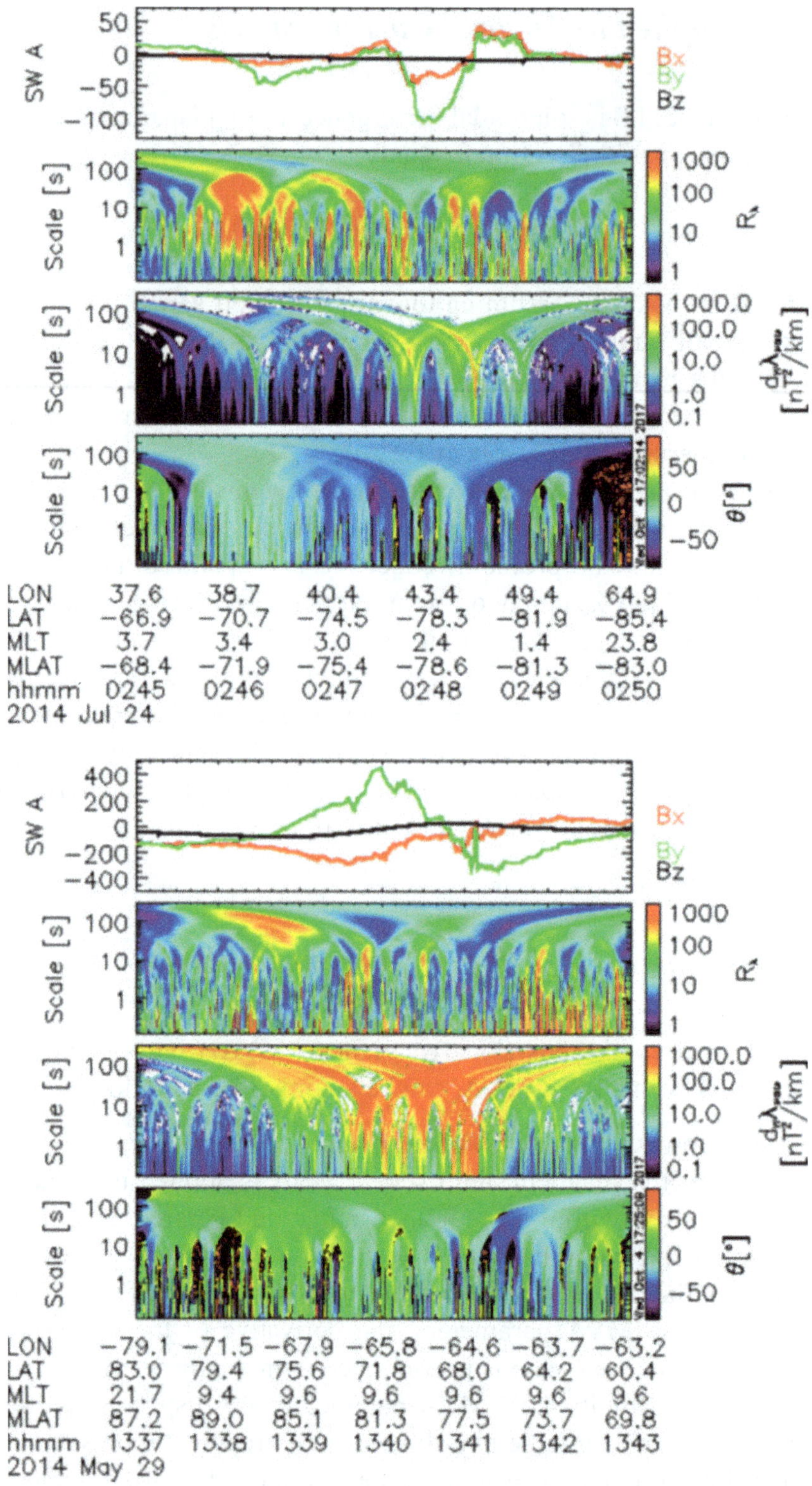

Fig. 4.7 Swarm magnetic measurements and MS-MVA parameters for two selected auroral crossings. Panels 1–4: 24 July 2014, 02:45-02:50 UT. Panels 5–8: 29 May 2014, 13:37-13:43 UT. Shown are the magnetic field measurements of SwA (panels 1 and 5), the eigenvalue ratio R_λ (panels 2 and 6), the R_λ (panels 2 and 6), the derivative $\partial_w \lambda_{\max}$ (panels 3 and 7) and the current sheet inclination (panels 4 and 8)

4.4.3 Application of MS-MVA to Swarm Auroral Crossings

Figure 4.7 shows the MS-MVA results for the two Swarm auroral crossings considered already in Sect. 4.3.2. In both cases, MS-MVA was applied to magnetic field data from SwA (panels 1 and 5), here given in the mean-field-aligned (MFA) coordinate frame (note that in Figs. 4.5 and 4.6 the magnetic field was displayed in NEC coordinates). The eigenvalue ratio R_λ is displayed in panels 2 and 6. The current sheet inclination is shown in panels 4 and 8. The multi-scale nature of the FAC sheets is visualised very clearly in the panels 3 and 7 showing $\partial_w \lambda_{\max}$.

4.5 Summary

The local least squares approach to the estimation of spatial derivatives from multi-spacecraft magnetic field measurements yields a generic framework for the analysis for auroral FACs and their errors. This report reviewed the underlying principles, estimation procedures, uncertainties, limitations, and practical aspects. The array geometry defines the position tensor with eigenvalues controlling the quality of gradient and curl estimates. Linear estimators can be uniquely qualified through their set of canonical base vectors, facilitating error analysis and comparison with alternative approaches. In planar spacecraft array configurations, reconstruction of the full gradient and curl vectors requires additional information that can be supplemented in the form of geometrical or physical constraints. The multi-scale nature of auroral currents can be investigated using a multi-scale version of the well-established minimum variance analysis. Analysis techniques were illustrated using selected auroral crossings of the Swarm satellites.

Acknowledgements Financial support by the Deutsche Forschungsgemeinschaft in the context of the DFG Special Programme SPP 1788 *DynamicEarth* through grant VO 855/4-1 is acknowledged. The authors thank the International Space Science Institute in Bern, Switzerland, for supporting the ISSI Working Group *Multi-Satellite Analysis Tools, Ionosphere*, from which this chapter resulted. The editors thank Chao Shen for his assistance in evaluating this chapter.

Appendix

Appendix A: General Linear Least Squares Modeling

Least squares modelling can be characterised as a statistical technique to find the parameters of a model (m) that gives the best approximation of a given data set (d) of S measurements contaminated by random errors (residuals r). The measurements d_σ form the components of a data vector $\mathbf{d}$ that can be understood as an object in S-dimensional data space $\mathscr{D}$. The corresponding model predictions yield another vector

m, and the best approximation in the least squares sense is given by minimising the total square deviation $\chi^2 \propto |\mathbf{d} - \mathbf{m}|^2 = |\mathbf{r}|^2$, where the square norm derives from a scalar product that may be designed to account for non-constant errors and possibly correlated observations through an error covariance matrix. In this sense, the best model vector minimises the (square) distance to the data vector. Furthermore, the best model satisfies the *orthogonality principle*: the residual vector $\mathbf{r} = \mathbf{d} - \mathbf{m}$ is orthogonal to the space $\mathscr{M}$ formed by all admissible model vectors. The effective dimension of the model space is the number N of model parameters.

Suppose all parameters $a_1, a_2, \ldots, a_N$ enter the model function m linearly, then $m = \sum_\nu a_\nu f_\nu$ with basis functions f_ν, and the a_ν are called amplitudes. The space $\mathscr{M}$ of all admissible models forms a linear subspace of the data space $\mathscr{D}$. Casting the model parameters into an amplitude vector **a**, and the predictions of individual basis functions into a $S \times N$ matrix M (design matrix), the model vector can be written in the form $\mathbf{m} = \mathsf{M}\mathbf{a}$. Parameter estimation is reduced to a linear inverse problem. In the overdetermined case ($S > N$), the solution is given by the so-called normal equations

$$\mathbf{a} = \left(\mathsf{M}^\mathsf{T}\mathsf{M}\right)^{-1}\mathsf{M}^\mathsf{T}\mathbf{d} = \mathsf{M}^{\text{ils}}\mathbf{d} \quad \text{with} \quad \mathsf{M}^{\text{ils}} = \left(\mathsf{M}^\mathsf{T}\mathsf{M}\right)^{-1}\mathsf{M}^\mathsf{T} . \tag{4.24}$$

The matrix M^{ils} is the pseudo-inverse of M in the least squares sense. The problem simplies further in the case of mutually orthogonal basis functions, then the amplitudes are given by $a_\nu = \left(\mathbf{f}_\nu/|\mathbf{f}_\nu|^2\right) \cdot \mathbf{d}$ where a vector $\mathbf{f}_\nu$ comprises the predictions of the basis function f_ν, typically obtained through evaluation at the S independent variables (usually spatial coordinates, possibly combined with auxiliary parameters) corresponding to the measurements.

Appendix B: Accuracy of Linear Curl Estimators

Consider a linear estimator of the form

$$\mathbf{c} = \nabla \times \mathbf{B} = \sum_\sigma \mathbf{q}_\sigma \times \mathbf{B}_\sigma \tag{4.25}$$

with vectors $\mathbf{q}_\sigma$ that are functions of the spacecraft positions but do not depend on the measurements $\mathbf{B}_\sigma$. The associated linear variation is given by

$$\delta\mathbf{c} = \sum_\sigma \left(\mathbf{q}_\sigma \times \delta\mathbf{B}_\sigma + \delta\mathbf{q}_\sigma \times \mathbf{B}_\sigma\right) . \tag{4.26}$$

In the case of Swarm, the typical positional inaccuracy δr and instrumental error δB are such that the second term can be dropped because in the auroral zone $|\nabla\mathbf{B}|\delta r \ll \delta B$, see Vogt et al. (2013).

Defining the tensor X_σ through $\mathsf{X}_\sigma \mathbf{u} = \mathbf{q}_\sigma \times \mathbf{u}$ yields $\delta \mathbf{c} = \sum_\sigma \mathsf{X}_\sigma \delta \mathbf{B}_\sigma$. The parameter covariance tensor is then given by

$$\left\langle \delta \mathbf{c}\, \delta \mathbf{c}^\mathsf{T} \right\rangle = \left\langle \left(\sum_\sigma \mathsf{X}_\sigma \delta \mathbf{B}_\sigma \right) \left(\sum_\nu \mathsf{X}_\nu \delta \mathbf{B}_\nu \right)^\mathsf{T} \right\rangle = \sum_{\sigma,\nu} \mathsf{X}_\sigma \left\langle \delta \mathbf{B}_\sigma \delta \mathbf{B}_\nu^\mathsf{T} \right\rangle \mathsf{X}_\nu^\mathsf{T} \quad (4.27)$$

Assuming isotropic and uncorrelated errors $\left\langle \delta \mathbf{B}_\sigma\, \delta \mathbf{B}_\nu^\mathsf{T} \right\rangle = (\delta B)^2 \delta_{\sigma\nu} \mathsf{E}$, we may write

$$\left\langle \delta \mathbf{c}\, \delta \mathbf{c}^\mathsf{T} \right\rangle = (\delta B)^2 \sum_\sigma \mathsf{X}_\sigma \mathsf{X}_\sigma^\mathsf{T} \quad (4.28)$$

Since $\mathsf{X}_\sigma \mathbf{u} = \mathbf{q}_\sigma \times \mathbf{u}$, $\mathsf{X}_\sigma^\mathsf{T} \mathbf{u} = -\mathbf{q}_\sigma \times \mathbf{u}$, and thus

$$\mathsf{X}_\sigma \mathsf{X}_\sigma^\mathsf{T} \mathbf{u} = -\mathbf{q}_\sigma \times (\mathbf{q}_\sigma \times \mathbf{u}) = \left\{ |\mathbf{q}_\sigma|^2 \mathsf{E} - \mathbf{q}_\sigma \mathbf{q}_\sigma^\mathsf{T} \right\} \mathbf{u} \quad (4.29)$$

we obtain

$$(\delta B)^{-2} \left\langle \delta \mathbf{c}\, \delta \mathbf{c}^\mathsf{T} \right\rangle = \sum_\sigma |\mathbf{q}_\sigma|^2 \mathsf{E} - \sum_\sigma \mathbf{q}_\sigma \mathbf{q}_\sigma^\mathsf{T} = \mathrm{trace}(\mathsf{Q})\mathsf{E} - \mathsf{Q} \quad (4.30)$$

where $\mathsf{Q} = \sum_\sigma \mathbf{q}_\sigma \mathbf{q}_\sigma^\mathsf{T}$.

The error covariance of the curl component $c_n = \hat{\mathbf{n}} \cdot \mathbf{c} = \hat{\mathbf{n}}^\mathsf{T} \mathbf{c}$ orthogonal to a given unit vector $\hat{\mathbf{n}}$ can be expressed in the form

$$\begin{aligned} (\delta B)^{-2} \left\langle |\delta c_n|^2 \right\rangle = (\delta B)^{-2} \left\langle \delta c_n\, \delta c_n \right\rangle &= (\delta B)^{-2} \hat{\mathbf{n}}^\mathsf{T} \left\langle \delta \mathbf{c}\, \delta \mathbf{c}^\mathsf{T} \right\rangle \hat{\mathbf{n}} \\ = \mathrm{trace}(\mathsf{Q}) \hat{\mathbf{n}}^\mathsf{T} \hat{\mathbf{n}} - \hat{\mathbf{n}}^\mathsf{T} \mathsf{Q} \hat{\mathbf{n}} &= \sum_\sigma |\mathbf{q}_\sigma|^2 - \sum_\sigma |\hat{\mathbf{n}} \cdot \mathbf{q}_\sigma|^2 \\ = \sum_\sigma |\hat{\mathbf{n}} \times \mathbf{q}_\sigma|^2 . \end{aligned} \quad (4.31)$$

In the planar spacecraft array case, and if $\hat{\mathbf{n}}$ is the normal unit vector to the spacecraft plane, $\hat{\mathbf{n}} \cdot \mathbf{q}_\sigma = 0$ and thus $\left\langle |\delta c_n|^2 \right\rangle = (\delta B)^2 \sum_\sigma |\mathbf{q}_\sigma|^2$.

Appendix C: Condition Number of a Planar Position Tensor

In order to show that the product of $L_q^{-2} = \sum_\sigma |\mathbf{q}_\sigma|^2$ (inverse square gradient estimation error length) and $L_r^2 = \frac{1}{4} \left(\sum_\sigma |\mathbf{q}_\sigma|^2 \right)^{1/2}$ (mean square inter-spacecraft distance) is in the planar spacecraft array case related to the effective condition number of the position tensor R through Eq. (4.22)

$$\mathrm{CN}(\mathsf{R}) = \exp\left[\operatorname{arcosh}\left(2\frac{L_r^2}{L_q^2} - 1\right)\right] , \tag{4.32}$$

we work in a mesocentric coordinate frame to express the position vectors $\mathbf{r}_\sigma$ and the position tensor $\mathsf{R} = \sum_\sigma \mathbf{r}_\sigma \mathbf{r}_\sigma^\mathsf{T}$, thus $L_r^2 = \frac{1}{4}\mathrm{trace}(\mathsf{R})$. The eigenvalues $\rho_1 \geq \rho_2 \geq \rho_3$ and corresponding eigenvectors $\hat{\mathbf{e}}_1, \hat{\mathbf{e}}_2, \hat{\mathbf{e}}_3$ of R yield the alternative representation $\mathsf{R} = \sum_k \rho_k \hat{\mathbf{e}}_k \hat{\mathbf{e}}_k^\mathsf{T}$. Since the spacecraft configuration is assumed to be planar, R has rank 2, thus the third eigenvalue ρ_3 is zero,

$$\mathsf{R} = \rho_1 \hat{\mathbf{e}}_1 \hat{\mathbf{e}}_1^\mathsf{T} + \rho_2 \hat{\mathbf{e}}_2 \hat{\mathbf{e}}_2^\mathsf{T} , \tag{4.33}$$

and $L_r^2 = \frac{1}{4}\mathrm{trace}(\mathsf{R}) = \frac{1}{4}(\rho_1 + \rho_2)$. The effective condition number $\mathrm{CN}(\mathsf{R})$ in the construction of the pseudo-inverse

$$\mathsf{Q} = \rho_1^{-1} \hat{\mathbf{e}}_1 \hat{\mathbf{e}}_1^\mathsf{T} + \rho_2^{-1} \hat{\mathbf{e}}_2 \hat{\mathbf{e}}_2^\mathsf{T} \tag{4.34}$$

is $\mathrm{CN}(\mathsf{R}) = \rho_1/\rho_2$. We further note that by defintion of $\mathbf{q}_\sigma = \mathsf{Q}\mathbf{r}_\sigma$,

$$\sum_\sigma \mathbf{q}_\sigma \mathbf{q}_\sigma^\mathsf{T} = \sum_\sigma \mathsf{Q}\mathbf{r}_\sigma \mathbf{r}_\sigma^\mathsf{T} \mathsf{Q}^\mathsf{T} = \mathsf{Q}\mathsf{R}\mathsf{Q}^\mathsf{T} = \mathsf{Q} \tag{4.35}$$

because the operator $\mathsf{Q} = \mathsf{Q}^\mathsf{T}$ acts only in the spacecraft plane (spanned by the two eigenvectors $\hat{\mathbf{e}}_1$ and $\hat{\mathbf{e}}_2$), and QR yields the identity operation on that plane. Therefore,

$$\rho_1^{-1} + \rho_2^{-1} = \mathrm{trace}(\mathsf{Q}) = \sum_\sigma |\mathbf{q}_\sigma|^2 = L_q^{-2} , \tag{4.36}$$

and

$$\frac{L_r^2}{L_q^2} = \frac{1}{4}(\rho_1 + \rho_2)\left(\rho_1^{-1} + \rho_2^{-1}\right) = \frac{1}{4}\left(2 + \frac{\rho_1}{\rho_2} + \frac{\rho_2}{\rho_1}\right) = \frac{1}{4}\left(2 + \mathrm{CN}(\mathsf{R}) + \frac{1}{\mathrm{CN}(\mathsf{R})}\right) . \tag{4.37}$$

With $x = \ln \mathrm{CN}(\mathsf{R}) = \ln(\rho_1/\rho_2)$ we obtain

$$\frac{L_r^2}{L_q^2} = \frac{1}{4}\left(2 + \mathrm{e}^x + \mathrm{e}^{-x}\right) = \frac{1}{2}(1 + \cosh x) , \tag{4.38}$$

and finally $\cosh\left[\ln \mathrm{CN}(\mathsf{R})\right] = 2L_r^2/L_q^2 - 1$.

References

Amm, O., H. Vanhamäki, K. Kauristie, C. Stolle, F. Christiansen, R. Haagmans, A. Masson, M.G.G.T. Taylor, R. Floberghagen, and C.P. Escoubet. 2015. A method to derive maps of ionospheric conductances, currents, and convection from the Swarm multisatellite mission. *Journal of Geophysical Research (Space Physics)* 120: 3263–3282.

Balikhin, M.A., S. Schwartz, S.N. Walker, H.S.C.K. Alleyne, M. Dunlop, and H. Lühr. 2001. Dual-spacecraft observations of standing waves in the magnetosheath. *Journal of Geophysical Research* 106: 25395–25408.

Bauer, T.M., M.W. Dunlop, B.U. Ö Sonnerup, N. Sckopke, A.N. Fazakerley, and A.V. Khrabrov. 2000. Dual spacecraft determinations of magnetopause motion. *Geophysical Research Letters* 27: 1835–1838.

Bunescu, C., O. Marghitu, D. Constantinescu, Y. Narita, J. Vogt, and A. Blăgău. 2015. Multiscale field-aligned current analyzer. *Journal of Geophysical Research* 120: 9563–9577.

Bunescu, C., O. Marghitu, J. Vogt, D. Constantinescu, and N. Partamies. 2017. Quasiperiodic field-aligned current dynamics associated with auroral undulations during a substorm recovery. *Journal of Geophysical Research* 122: 3087–3109.

Chanteur, G. 1998. Spatial interpolation for four spacecraft: Theory. *ISSI Scientific Reports Series* 1: 371–393.

Chanteur, G., and C.C. Harvey. 1998. Spatial interpolation for four spacecraft: Application to magnetic gradients. *ISSI Scientific Reports Series* 1: 349–369.

Chanteur, G. and F. Mottez. 1993. Geometrical tools for Cluster data analysis. In *Proceedings International conference "Spatio-Temporal Analysis for Resolving Plasma Turbulence (START)", Aussois, 31 Jan.–5 Feb. 1993* pp. 341–344. ESA WPP-047.

De Keyser, J., F. Darrouzet, M.W. Dunlop, and P.M.E. Décréau. 2007. Least-squares gradient calculation from multi-point observations of scalar and vector fields: Methodology and applications with Cluster in the plasmasphere. *Annals of Geophysics* 25: 971–987.

Dudok de Wit, T., V.V. Krasnoselskikh, S.D. Bale, M.W. Dunlop, H. Lühr, S.J. Schwartz, and L.J.C. Woolliscroft. 1995. Determination of dispersion relations in quasi-stationary plasma turbulence using dual satellite data. *Geophysical Research Letters* 22: 2653–2656.

Dunlop, M.W., A. Balogh, K.-H. Glassmeier, and P. Robert. 2002. Four-point Cluster application of magnetic field analysis tools: The Curlometer. *Journal of Geophysical Research* 107: 1384.

Dunlop, M.W., D.J. Southwood, K.-H. Glassmeier, and F.M. Neubauer. 1988. Analysis of multipoint magnetometer data. *Advances in Space Research* 8: 273–277.

Harvey, C.C. 1998. Spatial gradients and the volumetric tensor. *ISSI Scientific Reports Series* 1: 307–322.

He, M., J. Vogt, H. Lühr, and E. Sorbalo. 2014. Local time resolved dynamics of field-aligned currents and their response to solar wind variability. *Journal of Geophysical Research* 119: 5305–5315.

He, M., J. Vogt, H. Lühr, E. Sorbalo, A. Blagau, G. Le, and G. Lu. 2012. A high-resolution model of field-aligned currents through empirical orthogonal functions analysis (MFACE). *Geophysical Research Letters* 39: 18105.

Lotko, W., B.U.O. Sonnerup, and R.L. Lysak. 1987. Nonsteady boundary layer flow including ionospheric drag and parallel electric fields. *Journal of Geophysical Research* 92: 8635–8648.

Lühr, H., J. Park, J.W. Gjerloev, J. Rauberg, I. Michaelis, J.M.G. Merayo, and P. Brauer. 2015. Field-aligned currents' scale analysis performed with the Swarm constellation. *Geophysical Research Letters* 42: 1–8.

Lyons, L.R. 1980. Generation of large-scale regions of auroral currents, electric potentials, and precipitation by the divergence of the convection electric field. *Journal of Geophysical Research* 85: 17–24.

Lysak, R.L. 1990. Electrodynamic coupling of the magnetosphere and ionosphere. *Space Science Reviews* 52: 33–87.

Neubauer, F.M., and K.-H. Glassmeier. 1990. Use of an array of satellites as a wave telescope. *Journal of Geophysical Research* 95: 19115–19122.

Paschmann, G. and P.W. Daly. 1998. *Analysis methods for multi-spacecraft data, no. SR-001 in ISSI Scientific Reports*. ESA Publ. Div., Noordwijk, Netherlands.

Paschmann, G. and P.W. Daly. 2008. *Multi-spacecraft analysis methods revisited, no. SR-008 in ISSI Scientific Reports*. ESA Publ. Div., Noordwijk, Netherlands.

Paschmann, G., S. Haaland, and Treumann, R. 2002. Auroral plasma physics. *Space Science Reviews*, 103.

Pinçon, J.L., and F. Lefeuvre. 1991. Local characterization of homogeneous turbulence in a space plasma from simultaneous measurements of field components at several points in space. *Journal of Geophysical Research* 96: 1789–1802.

Reigber, C., H. Lühr, and P. Schwintzer. 2002. Champ mission status. *Advances in Space Research* 30: 129–134.

Ritter, P., and H. Lühr. 2006. Curl-B technique applied to Swarm constellation for determining field-aligned currents. *Earth, Planets, and Space* 58: 463–476.

Ritter, P., H. Lühr, and J. Rauberg. 2013. Determining field-aligned currents with the Swarm constellation mission. *Earth, Planets, and Space* 65: 1285–1294.

Robert, P., M.D. Dunlop, A. Roux, and G. Chanteur. 1998. Accuracy of Current Density Estimation. *ISSI Scientific Reports Series* 1: 395–418.

Russell, C.T., J.T. Gosling, R.D. Zwickl, and E.J. Smith. 1983. Multiple spacecraft observations of interplanetary shocks ISEE three-dimensional plasma measurements. *Journal of Geophysical Research* 88: 9941–9947.

Shen, C., Z.J. Rong, and M. Dunlop. 2012a. Determining the full magnetic field gradient from two spacecraft measurements under special constraints. *Journal of Geophysical Research* 117: 10217.

Shen, C., Z.J. Rong, M.W. Dunlop, Y.H. Ma, X. Li, G. Zeng, G.Q. Yan, W.X. Wan, Z.X. Liu, C.M. Carr, and H. Rème. 2012b. Spatial gradients from irregular, multiple-point spacecraft configurations. *Journal of Geophysical Research* 117: A11207.

Sonnerup, B.U.O., and L.J. Cahill Jr. 1967. Magnetopause structure and attitude from explorer 12 observations. *Journal of Geophysical Research* 72: 171–183.

Sonnerup, B.U.Ö., and M. Scheible. 1998. Minimum and maximum variance analysis. *ISSI Scientific Reports Series* 1: 185–220.

Vogt, J. 2002. Alfvén wave coupling in the auroral current circuit. *Surveys in Geophysics* 23: 335–377.

Vogt, J. 2014. Analysis of data from multi-satellite geospace missions. *Handbook of Geomathematics* (eds. W. Freeden, M.Z. Nashed, and T. Sonar), Springer, Berlin, Heidelberg, pp. 1–28.

Vogt, J., A. Albert, and O. Marghitu. 2009. Analysis of three-spacecraft data using planar reciprocal vectors: Methodological framework and spatial gradient estimation. *Annals of Geophysics* 27: 3249–3273.

Vogt, J., S. Haaland, and G. Paschmann. 2011. Accuracy of multi-point boundary crossing time analysis. *Annals of Geophysics* 29: 2239–2252.

Vogt, J., and G. Haerendel. 1998. Reflection and transmission of Alfvén waves at the auroral acceleration region. *Geophysical Research Letters* 25: 277–280.

Vogt, J., Y. Narita, and O.D. Constantinescu. 2008a. The wave surveyor technique for fast plasma wave detection in multi-spacecraft data. *Annals of Geophysics* 26: 1699–1710.

Vogt, J., and G. Paschmann. 1998. Accuracy of plasma moment derivatives. *ISSI Scientific Reports Series* 1: 419–447.

Vogt, J., G. Paschmann, and G. Chanteur. 2008b. Reciprocal vectors. *ISSI Scientific Reports Series* 8: 33–46.

Vogt, J., E. Sorbalo, M. He, and A. Blagau. 2013. Gradient estimation using configurations of two or three spacecraft. *Annals of Geophysics* 31: 1913–1927.

Chapter 5
Multi-spacecraft Current Estimates at Swarm

Malcolm Wray Dunlop, J.-Y. Yang, Y.-Y. Yang, Hermann Lühr and J.-B. Cao

Abstract In this chapter the application of the curlometer technique to various regions of the inner magnetosphere and upper ionosphere and for special circumstances of sampling is described. The basic technique is first outlined, together with the caveats of use, covering: the four-spacecraft technique, its quality factor and limitations; the lessons learnt from Cluster data, together with issues of implementation, scale size and stationarity, and description of the key regions covered by related methodology. Secondly, the application to the Earth's ring current region is outlined, covering: the application of Cluster crossings to survey the ring current; the use of the MRA (magnetic rotation analysis) method for field curvature analysis; the use of THEMIS (Time History of Events and Macroscale Interactions during Sub-storms mission) three-spacecraft configurations to sample the ring current, and future use of MMS (Magnetospheric MultiScale mission) and Swarm data, i.e. the case of small separations. Thirdly, the application of the technique to the low altitude regions covered by Swarm is outlined, covering: the extension of the method to stationary signals; the use of special configurations and adjacent times to achieve 2, 3, 4, 5 point analysis; the use of the extended 'curlometer' with Swarm close configurations to compute 3-D current density, and a brief indication of the computation of current sheet orientation implied by 2-spacecraft correlations. Fourthly, the direct coordination of Cluster and Swarm to check the scaling and coherence of field-aligned currents (FACs) is outlined.

M. W. Dunlop (✉) · J.-Y. Yang · J.-B. Cao
School of Space and Environment, Beihang University, Beijing 100191, P.R. China
e-mail: malcolm.dunlop@stfc.ac.uk; m.w.dunlop@rl.ac.uk

M. W. Dunlop
RAL Space, Rutherford Appleton Laboratory, STFC-UKRI, Chilton, UK

Y.-Y. Yang
The Institute of Crustal Dynamics, CEA, Beijing 100085, P.R. China

H. Lühr
GFZ, Telegrafenberg, 14473 Potsdam, Germany

M. W. Dunlop and H. Lühr (eds.), *Ionospheric Multi-Spacecraft Analysis Tools*, ISSI Scientific Report Series 17,
https://doi.org/10.1007/978-3-030-26732-2_5

5.1 Introduction

In this chapter, we discuss the performance and lessons learnt from the application of the multi-spacecraft curlometer technique to the regions of the upper ionosphere and inner magnetosphere and discuss the adaptability of the method to situations where full spatial coverage cannot be achieved. Firstly, we review the method and secondly we consider its use both in situ in the ring current and in regions covered by field-aligned currents. The standard method used to interpret the Swarm dual-spacecraft data is discussed in detail in Chap. 6.

5.2 Basic Application of the Curlometer

Although fundamentally, currents in plasmas are set up by the differential motion of charged particles, in practice this approach requires that the whole particle 3D particle velocity distribution is measured for all ion species and electrons, where the currents are derived from computation of the bulk moments from the distribution functions. These measurements often rely on spacecraft spin (and thus have limited time resolution), or arrays of detectors, to obtain the full distribution. Recent results from the Magnetospheric Multiscale (MMS) mission (Burch et al. 2016), covering primarily the outer magnetosphere, however, benefit from distributions measured at high-time resolution, in addition to multi-point sampling at small (several km) spatial distances. Nevertheless, previous space missions have suffered from low cadence or incomplete particle measurements (examples of previous use of particle moments can be found in Henderson et al. 2008; Petrukovich 2015). In fact, the generation of currents in highly conducting space plasmas produce magnetic fields that modify the existing magnetic field (e.g. the Earth's internal magnetic field). As a result of quasi-neutrality, for highly conducting plasmas the displacement current, $\mu_0\varepsilon_0\partial \boldsymbol{E}/\partial t$, where $\boldsymbol{E}$ is the electric field, can be neglected in Maxwell's equations Russell et al. 2016) so that the magnetic field generated by currents can be given by Ampère's law:

$$\mu_0 \boldsymbol{J} = \mathbf{curl}(B)$$

where $\boldsymbol{J}$ is the current density and $\boldsymbol{B}$ the magnetic field. This relation enables us to derive the strength and orientation of currents directly from the magnetic field and its gradients, but requires measurements at multiple spatial positions, simultaneously. Magnetic fields in space can typically be measured with higher accuracy and higher time resolution than particle moments. Multiple spacecraft flying in formation, and each measuring the magnetic field, allow a linear estimate of Ampère's law (above) to be made, i.e. the electric current density from $\mathbf{curl}(\boldsymbol{B})$. The method, termed the curlometer technique (Dunlop et al. 1988), was introduced to utilize four-point measurements in space (the minimum to return a vector estimate of current density); anticipating the four-spacecraft Cluster mission (Escoubet et al. 2001). Although

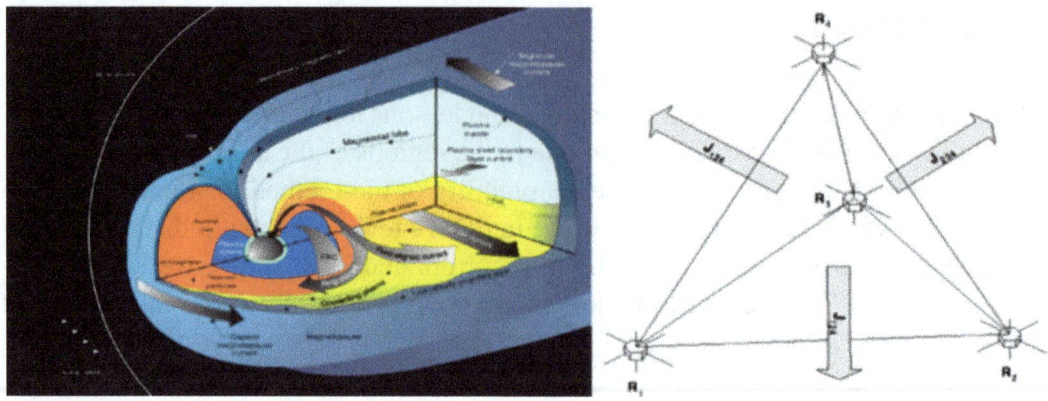

Fig. 5.1 Depiction of the large-scale currents in the Earth's magnetosphere (adapted from Kivelson and Russell (1995), left) and the curlometer concept (after Dunlop et al. 1988, right)

robust and reliable in many regions of the magnetosphere (such as the magnetopause, the magnetotail and the ring current), the accuracy of the method is limited by uncertainties in the spacecraft separation distances, timing and the form (scale size) of the current structure; as well as accuracy of the magnetic field measurements (see Fig. 5.1).

5.2.1 *Four-Spacecraft Technique: Quality Factor and Limitations*

The curlometer method estimates the average current density within the tetrahedral configuration formed by the individual spacecraft positions and measured magnetic fields, based on the assumption that: **curl(*B*)** = μ_0***J***. The basic method provides all components of the electric current density by either estimating linear approximations to the spatial gradients needed for **curl(*B*)**, or by using the linear (discrete), integral form of Ampère's law ($\mu_0 \int \boldsymbol{J}.\mathrm{ds} = \int B.\mathrm{dl}$). The latter provides an estimate of the average current density, J_{1ij}, normal to the face 1ij of the tetrahedron (see the right hand side of Fig. 5.1) from:

$$\mu_0 < \boldsymbol{J} > .(\Delta \boldsymbol{R}_i \wedge \Delta \boldsymbol{R}_j) = \Delta \boldsymbol{B}_i.\Delta \boldsymbol{R}_j - \Delta \boldsymbol{B}_j.\Delta \boldsymbol{R}_i, \tag{5.1}$$

e.g. $\mu_0 < \boldsymbol{J} >_{123}.(\Delta \boldsymbol{r}_{12} \hat{} \Delta \boldsymbol{R}_{13}) = \Delta \boldsymbol{B}_{12}.\Delta \boldsymbol{R}_{13} - \Delta \boldsymbol{B}_{13}.\Delta \boldsymbol{R}_{12}$ (division of the magnitude $\mu_0 \Delta \boldsymbol{r}_{12} \hat{} \Delta \boldsymbol{R}_{13}$ gives J explicitly).

Here, $\Delta \boldsymbol{B}_{1i}$, $\Delta \boldsymbol{R}_{1i}$ (which are the differences in the measured magnetic field and positions from spacecraft 1 to each of the others) have been shortened in notation to $\Delta \boldsymbol{B}_i$, $\Delta \boldsymbol{R}_i$, which represent the vector difference in the measured magnetic field vector and spatial position vector. Any reference spacecraft can be used and of course each of the four faces of the tetrahedron gives an estimate of the current density normal to that face and in principle can be used independently. Three faces of the tetrahedron combine to provide three non-coplanar components of the current (which can then be used to construct the average vector current density in any system of coordinates), while the fourth is redundant (see below).

The average value of div$\boldsymbol{B}$ over the volume of the tetrahedron can also be estimated from $< div(B) > \left|\Delta \boldsymbol{R}_i . \Delta \boldsymbol{R}_j \wedge \Delta \boldsymbol{R}_k\right| = \left|\sum_{cyclic} \Delta \boldsymbol{B}_i . \Delta \boldsymbol{R}_j \wedge \boldsymbol{R}_k\right|$,

e.g. $<div(\boldsymbol{B})>_{1234}(\Delta\boldsymbol{R}_{12} \cdot \Delta\boldsymbol{R}_{13} \wedge \Delta\boldsymbol{R}_{14}) = \Delta\boldsymbol{B}_{12} \cdot \Delta\boldsymbol{R}_{13} \wedge \Delta\boldsymbol{R}_{14} + \Delta\boldsymbol{B}_{13} \cdot \Delta\boldsymbol{R}_{14} \wedge \Delta\boldsymbol{R}_{12} + \Delta\boldsymbol{B}_{14} \cdot \Delta\boldsymbol{R}_{12} \wedge \Delta\boldsymbol{R}_{13}$,

Here, any non-zero value indicates the approximate size of the neglected non-linear gradients (Dunlop et al. 2002, 2016, 2018; Robert et al. 1998 and see also Haaland et al. 2004), indirectly (since div$\boldsymbol{B} = 0$ exactly). The relative orientation and spatial scale of the current structures sampled, compared with the shape and size of the spacecraft configuration ultimately control the error in the linear estimates through the neglected non-linear gradients in $\boldsymbol{B}$. This error is unknown in absolute terms, but usually can be indirectly monitored through the quality parameter, Q = $|\text{div}B|/|\textbf{curl}(B)|$, where small values of Q are desirable, since div$\boldsymbol{B}$ should ideally be zero. For highly non-regular tetrahedral shapes, however, the use of this indicator is less certain and typically Q is used only to indicate when the curlometer estimate is likely to be bad. In addition, the overall uncertainty in the linear approximation of both div$\boldsymbol{B}$ and **curl***B* also results from measurement errors in the magnetic field, spatial position and inter-spacecraft timing.

The redundant face of the tetrahedron can also indicate uncertainty in the calculation. The choice of three faces used in the calculation can be cycled using different reference spacecraft in Eq. 5.1 so that four different results can be obtained, verifying the sensitivity of the estimate for each Cartesian component of $\boldsymbol{J}$ (this then provides a way to assess uncertainty independently to the estimate of div$\boldsymbol{B}$, and hence Q). Directly estimating the linear spatial gradients for each current density component; for example, from the dyadic of B, is identical to this original form of the curlometer (see Chap. 4), but the error handling (and hence relevant quality parameters) is slightly different and in the implementation of the equations above, the estimate is often self-stabilizing for common current structures such as sheets and tubes (Dunlop et al. 2002; Robert et al. 1998), because of the closed form of the integral equations.

Clearly, it will be the case that certain forms of structure may align with one or more components of the current density, often resulting in more accurate determination of that component. In fact, non-regular tetrahedral spacecraft configurations preferentially access some directions in space more accurately than others, and hence sample the components of the current preferentially (depending on alignment, or misalignment, to the dominant current direction and its form). When sampling large-scale currents at the magnetopause or magnetotail, for example, the effect of irregular configurations can be a significant drawback; but it can also be a benefit to the measurement of currents which are highly directed to the form of the background magnetospheric regions. Examples of these are the ring current, where the dominant current is azimuthal (Zhang et al. 2011) and field aligned currents (FACs), where one face of the tetrahedron formed by the spacecraft can be used to estimate the component of the current which is closest to the FAC direction (Dunlop et al. 2015b). In general, however, the effect of spatial structure cannot be separated from that arising from temporal behaviour on time scales smaller than the natural convection time across the spacecraft array (see Sect. 5.2.2 for more discussion).

5.2.2 Cluster Lessons: Implementation, Scale Size and Stationarity

The implementation of the method from Eq. 5.1 is done with reference to the right hand side of Fig. 5.1, where it can be seen that the size of the current components perpendicular to each face of the tetrahedron can be found from the terms on the right hand side of Eq. 5.1 and the normals to each face (and hence the orientation of these current components) are obtained from $(\Delta \boldsymbol{R}_i \wedge \Delta \boldsymbol{R}_j)/|(\Delta \boldsymbol{R}_i \wedge \Delta \boldsymbol{R}_j)|$ on the left hand side of Eq. 5.1. Cartesian components can then be found from projections of three of these components onto X, Y, Z coordinates in the usual way. In order to make this computation from four-spacecraft data, the time series need to be interpolated onto a common timeline.

For the method to perform well, in general, the spatial configuration needs to be small compared to the characteristic scale size of the current structure to minimize the effect of the non-linear gradients (i.e. non-measured gradients in the current density) and therefore accurately compute actual current density. This condition, however, is limited by timing errors between spacecraft and the effect of the measurement errors in $\boldsymbol{B}$ and $\boldsymbol{R}$, which become more significant at smaller spatial scales. Thus, smaller tetrahedral scales require higher absolute accuracy in $\boldsymbol{B}$ and $\boldsymbol{R}$, and rapid temporal behaviour requires higher cadence and accuracy of the measurement times. For example, for the Cluster mission, the spacecraft separations were typically larger than a few 100 km and linearization errors dominated errors in the estimates, since measurement uncertainty (~0.1 nT in $\boldsymbol{B}$; a few km for $\boldsymbol{R}$ and millisecond timing) was low by comparison (for currents greater than a few nAm^{-2}). In the Cluster regime, therefore, Q is a reasonable quality indicator. By comparison, the typical spacecraft separations accessed by the MMS mission (Burch et al. 2016) are 10's of km so that the curlometer is likely to be more often in the linear regime where errors due to gradients in the current density are small. On these smaller spatial scales, however, the measurement errors could become significant unless the currents measured are large.

There is a further consideration on the calculation of the magnetic field differences in the context of the linear approximation. Although the method can be applied to the measured $\boldsymbol{B}$ point by point in time, the background field may itself contain strong non-linear gradients even though the background current density is zero (this is the case for the internal geomagnetic field, arising from the IGRF, dominated by the Earth's dipole field). In that case, the neglect of non-linear gradients in the linear estimates can imply non-physical (i.e. not real) currents (the effect is significant in the inner magnetosphere and therefore affects the ring current calculation, and the calculation of FACs at low orbit for example). Thus, implied currents from the estimate will arise when the curlometer is applied to a geomagnetic model field in which the current is zero (such as the IGRF). This effect, first noted in Dunlop et al. (2002) was analysed in the context of the ring current more recently by Grimald et al. (2012). The curlometer can be applied to residual measured fields which result after first subtracting a zero-current model field from the measured magnetic field, however, and this minimises such effects. Indeed it is necessary in the high field regions of the magnetosphere and ionosphere (Yang et al. 2016; Dunlop et al. 2015). The use of a defined model field containing no current density is usually sufficient and of course it is usually desirable that any de-trending of the data in this way should not remove any real currents which may be present. Nevertheless, this does provide a way to separately analyse different current systems (such as large and small-scale effects) where some part of the current system is known or can be accurately modelled.

5.2.3 Key Regions Covered by Related Methodology

The curlometer method has proved to be robust and has been successfully applied in many different regions of the Earth's magnetosphere, such as the magnetopause (e.g. Dunlop et al. 2002, 2005; Haaland et al. 2004; Panov et al. 2006); the magnetotail current sheet (e.g. Runov et al. 2006; Nakamura et al. 2008; Narita et al. 2013); the ring current and inner magnetosphere (e.g. Vallat et al. 2005; Shen et al. 2014; Yang et al. 2016); field-aligned currents (FAC, e.g. Forsyth et al. 2008; Shi et al. 2010), and other transient signatures (e.g. Roux et al. 2015; Xiao et al. 2004; Shen et al. 2008), as well as to structures in the solar wind (e.g. Eastwood et al. 2002).

Methods which more generally estimate the gradient of the magnetic field have also been extensively applied to multi-spacecraft magnetic field measurements (Chanteur 1998; Harvey 1998; Vogt et al. 2008; Shen and Dunlop 2008) and methods based on calculation of the magnetic rotation rates to estimate field line curvature, as well as the current density directly, have also been developed (Shen et al. 2007, Dunlop and Eastwood 2008) and can be readily applied to irregular configurations of 3–5 spacecraft (Shen et al. 2012a, b). In addition, equivalent methods have been developed based on the use of planar reciprocal vectors (Vogt et al. 2008, and Chap. 4) and the method of least squares (e.g. DeKeyser et al. 2007; Hamrin et al. 2008). Although these methods have been predominantly applied to the outer magnetospheric regions (dayside magnetopause, magnetotail and lobes), where the influence of the Earth's internal field is weak and temporal fluctuations are often dominant, recently there have been a number of studies using multi-spacecraft estimates of current density in the inner magnetospheric regions and ring current (Vallat et al. 2005; Zhang et al. 2011; Shen et al. 2014), and in regions supporting field aligned currents (Marchaudon et al. 2009; Shi et al. 2010, 2011).

These methods (including the basic 'curlometer' above) are applicable to situations where less than four-spacecraft are closely grouped (Dunlop et al. 2016, 2018). With less than four-spacecraft, only a partial estimate of a single component of $\boldsymbol{J}$

(the component normal to the plane of the spacecraft) can be made unless other data sets are used in conjunction with the magnetic field, such as the plasma moments, or where assumptions in the behaviour of the currents can be made (e.g. stationarity of the field, known FACs or force free structures) in certain regions such as at low Earth orbit (e.g. Vogt et al. 2009, 2013; Shen et al. 2012a, b; Ritter and Lühr 2013, Dunlop et al. 2015a, b). Nevertheless, until the launch of the three Swarm spacecraft in 2014, altitudes at low-Earth orbit (LEO) had not benefited from multi-point measurements. In fact, the method can be generalised under certain assumptions of stationarity (see the applications below discussed in Sect. 5.4 and Chaps. 4 and 6).

The above outline represents the key issues revealed from the standard application of the method in many regions of near-Earth space since the launch of Cluster (see for instance Dunlop and Eastwood, (2008), for a review, and Dunlop et al. (2016, 2018), for a summary of the practical experience gained). Ready to use implementations of the curlometer method can be obtained from the Cluster Science archive (http://www.cosmos.esa.int/web/csa/software), and see the technical note by Middleton and Masson (2016).

5.3 Use of Cluster and THEMIS for in Situ Ring Current Surveys

The terrestrial ring current (RC) extends significantly in both latitude (−30 to 30 deg) and radially from about 2–7 R_E as a result of its variability in strength and location.

5.3.1 Application of Cluster Crossings to Survey the Ring Current

The curlometer has been applied to the RC using the in situ measurements of Cluster (Dandouras et al. 2018; Dunlop et al. 2018), which generally sampled the RC as the spacecraft passed through perigee during all phases of the mission. During the early to middle phase (2001–2009), the polar Cluster orbit passed normally through the ring current (at radial distances of ~4–4.5 R_E), as was first reported by Vallat et al. (2005) and later extended to full azimuthal coverage in local time (see the left hand side of Fig. 5.2) by Zhang et al. (2011). These studies found that the stability of the current density for each pass was such that the orientation of the Cluster configuration typically allows the azimuthal (ring plane) component, J_ϕ, to be estimated accurately, since the configuration often aligns perpendicularly to the ring plane. Cluster in fact also often samples FACs adjacent to the RC, where the alignment of J_N (the current component normal to the plane of the configuration) to $J_\|$ is significant, so that the occurrence of R2 FACs, which connect through the RC, can be inferred, in principle.

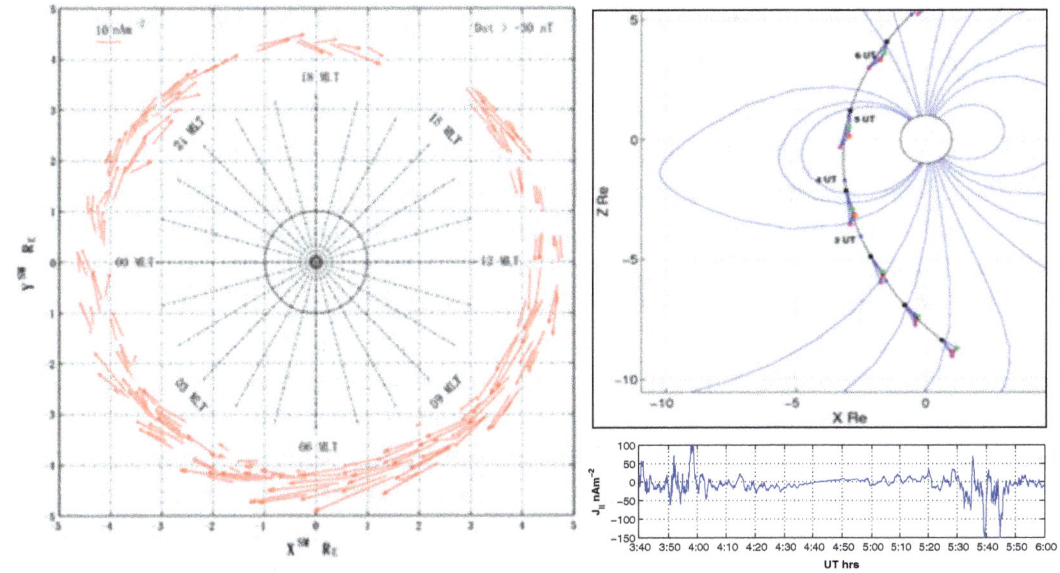

Fig. 5.2 (left) Full azimuth scan (magnetic local time, MLT) of ring current (RC) passes, plotted in solar magnetic (SM) coordinates (dipole aligned), and between -30° to 30° latitude (Zhang et al. 2011). The length and direction of the vectors represent ten minute averages of the current density obtained from the curlometer. The measurements represent non-storm (Dst > −30 nT) values of the RC. The RC strength is seen to increase with MLT on the dawn-side (03–12 MLT) and is a little suppressed on the dusk side. (right) The top plot shows a Cluster orbit relative to the magnetic field lines for a night-side perigee orientation where the configuration is oriented suitably to access the azimuthal ring current component and also is aligned normal to the field lines on exit from the RC. The lower plot shows the corresponding FA current density

The right hand side of Fig. 5.2 shows an orbit of Cluster, where the orientation of the Cluster configuration is suitable for sampling stable RC measurements as well as for the R2 FAC orientation as the spacecraft exit north and south of the ring plane. The lower panel of Fig. 5.2 shows that the field aligned component is zero within the core RC region (4:30–5:00 UT), but then starts showing FAC signatures at the RC edges and beyond. For large-scale current systems, such as the ring current, where the direction of the main current can be assumed, or for systems of known FACs, where the dominant current is along the known magnetic field direction, one face of the tetrahedron (i.e. three of the spacecraft) can often be selected to align closely to the main current. The crossing positions shown on the left hand side of Fig. 5.2 in fact statistically map to the equatorward boundary of the Auroral zone determined for Kp < 2 (non-storm activity) by Xiong et al. (2014), the R2 FAC intensity peaks.

5.3.2 *Use of the Magnetic Rotation Analysis (MRA) Method for Field Curvature Analysis*

In the in situ ring current, it is suitable to subtract the IGRF from the measured data in order to form magnetic field residuals for computation of the current density. This has been discussed by Shen et al. (2014), who also computed the magnetic field line curvature (using the MRA method) for a number of storm-time events. That study also showed a strengthening of the RC with storm activity, behaviour which was also reflected in the field-line curvature estimates. The magnetic field-line (MFL) curvature results are shown in Fig. 5.3 for the storm time events of Shen et al. (2014) as well as for the combined database of events with Zhang et al. (2011). The plots show that, based on estimates of the MFL curvature for events with sym-H activity levels down to around −50 nT, the dawn and night side (green and black) strengths are higher (smaller curvature) than the dayside and dusk events. At higher activity the situation is less clear, but the night side (black) population shows higher strengths (lower Rc) generally (this could be a result of sub-storm phase effects in this region). Nevertheless, this broadly is consistent with the curlometer statistics of Zhang et al. (2011).

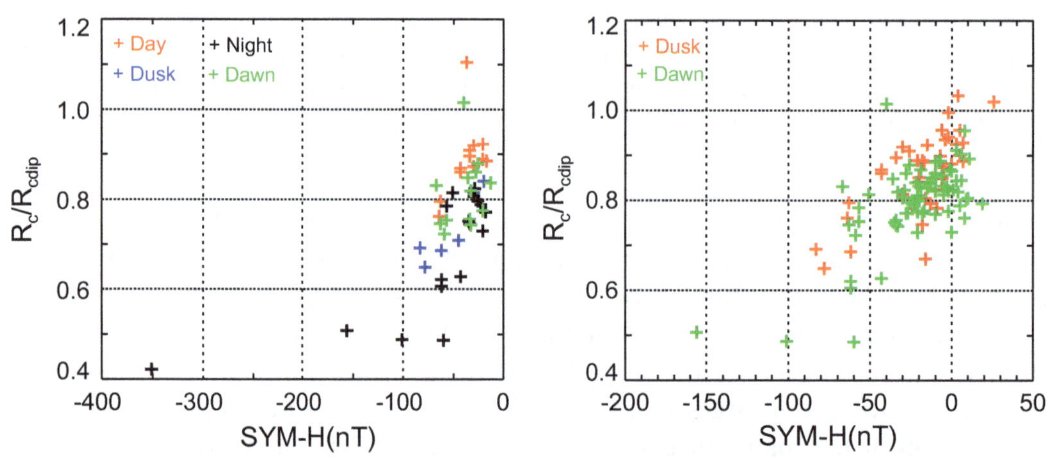

Fig. 5.3 Plots of MRA analysis of the magnetic field line curvature (Rc) estimates against sym-H activity. (left) The results for Cluster storm time RC crossings (after Shen et al. 2014), where the data are classified by local time sector (red, day; blue, dusk; green, dawn; black, night). (right) The combined statistics for both storm and non-storm conditions binned only with respect to two wide LT ranges: red (dawn + night) and green (dusk + day)

5.3.3 *Use of THEMIS Three-Spacecraft Configurations to Sample the Ring Current*

A further study has been carried out by Yang et al. (2016), using an application of the method of Shen et al. (2012) to the group of three magnetospheric THEMIS spacecraft, which often came into a close, 3-spacecraft configuration at times when they were providing coverage through the Earth's ring current. The geometry of the configurations is shown on the left hand side of Fig. 5.4 to illustrate that the current component normal to the configuration can be readily obtained, but generally takes some angle relative to the J_ϕ component (in the X,Y plane of the magnetic dipole equator). This geometry using only three-spacecraft is therefore useful in circumstances like this where the expected main current component is known.

The THEMIS azimuthal coverage is summarized on the right hand side of Fig. 5.4, where radial coverage is achieved from about 4 up to 12 R_E (i.e. a much larger range than for Cluster). The results were classified according to sub-storm phase and show both a clear radial profile for the recovery phase of each sub-storm period, and resolve the Eastward reverse current on the inner edge of the RC. The Eastward/westward ring current boundary is found to be at L = 5 on average, but is sensitive to storm activity. This coverage is unfortunately limited on the dawn-side so that LT trends are not well resolved in that sector. There is also a bias to strong storm events near midnight LT.

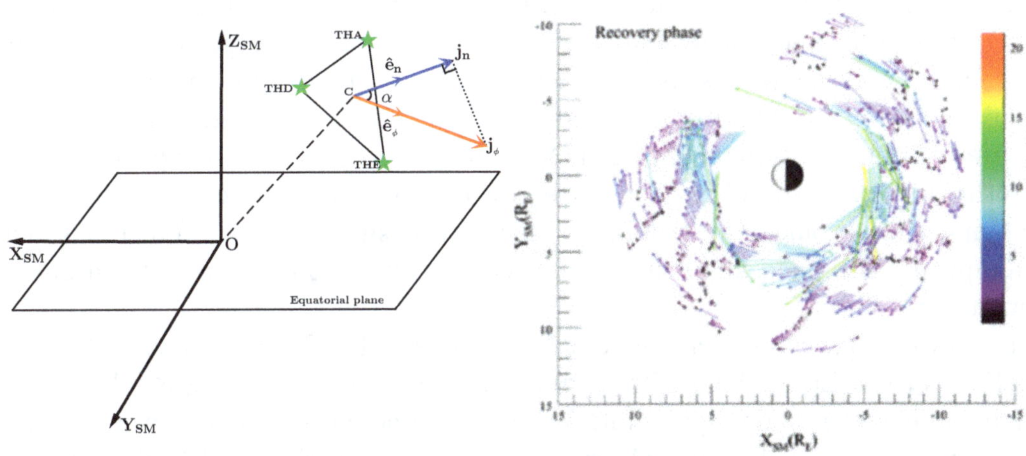

Fig. 5.4 (left) Application of the method of Shen et al. (2012) for the case of three THEMIS spacecraft (from Yang et al. 2016). (right) THEMIS coverage of the RC for current densities projected into the ring plane and for all storm activities during the recovery phase (i.e. outside the main phase of the storm)

5.3.4 Future Use of MMS and Swarm: Small Separations

In fact, the MMS spacecraft, launched recently in 2015, also cover the ring current and comparisons may be made to these THEMIS measurements for the much smaller MMS separation scales, as well as the complementary Cluster coverage through the RC. All of these missions: MMS, THEMIS and Cluster are currently operating and can potentially extend these ring current surveys for the period covered by Swarm. The use of Swarm measurements can be carried out in two ways: either for event studies where conjunctions with both the RC and R2 FACs can be found with Swarm, or by statistically comparing in situ RC strength with corresponding RC terms in the model field derived from Swarm data (perhaps to show LT sensitivity).

5.4 Multi-spacecraft Analysis for Swarm: FACs

The multi-spacecraft techniques are applied as purely spatial estimates, point by point in time, but some magnetospheric phenomena are better suited to allow the temporal and spatial variations to be disentangled than others, depending on their properties and sampling resolution. In low altitude regions, however, studies have, until the launch of the three Swarm (Friis-Christensen et al. 2008) spacecraft (labelled A, B and C), relied on estimates arising from single spacecraft data, for which separation of temporal and spatial variations is particularly difficult (Lühr et al. 2015) unless some knowledge about the key properties of both large and small-scale field aligned currents (FACs) is assumed. As we describe below, these conditions can be relaxed in different ways depending on the spacecraft coverage achieved by the array of spacecraft. In principle, the Swarm mission provided the opportunity to perform direct estimates of current density through the multi-spacecraft techniques, removing at least some of the ambiguity arising from single-spacecraft methods, but then typically is limited to the larger scale currents with some degree of stationary properties.

The Swarm spacecraft were launched on 22 November 2013 into circular, polar, low-Earth orbits (LEO). After 17th April 2014, a final configuration was achieved in which the Swarm A and C spacecraft fly close together at ~100–150 km separation and at a mean altitude of about 481 km (giving an orbital period of ~94 min). Swarm B, is flying at a slightly higher orbit of ~531 km, with a slightly different orbital period of ~95 min (see Dunlop et al. 2015), where its orbital plane drifts relative to A and C. At the start of science operations, Swarm B was initially aligned with the orbital planes of A and C and the three-spacecraft flew close to each other at regular periods every few days. The polar orbits take the Swarm spacecraft through the auroral regions and across the polar cap at high latitudes and sample all local times in about 132 days (spacecraft A/C), similar to the coverage of the CHAMP spacecraft (Reigber et al. 2002).

In this high field region magnetic residuals are computed by first subtracting a high-resolution internal field model before application of the curlometer. If it is assumed that the magnetic field signals associated with the currents do not significantly evolve in time over the durations needed (~10–20 s) for the spacecraft to move to adjacent positions (such assumptions of stationarity of the field can typically be made at low Earth orbits and often actual time dependence can be mitigated by suitable low pass filtering of the data), then these adjacent positions can be combined with the current A, B, C positions to produce added measurement points, as depicted on the left-hand side of Fig. 5.5 (see full description in Sect. 5.4.1). Thus, the multi-point curlometer can in principle be applied from selections of any combination of positions (extending the work of Ritter et al. 2006, 2013), also described in Chap. 6). Different configurations of the combined positions can be characteristically either 2-D or 3-D and are associated with different effective mean times for the measurements so that the different choices of combinations of time-shifted spacecraft positions can test the temporal stability as well as the validity of the field-aligned component. Furthermore, in a similar manner to the choice of tetrahedral face in the standard application described in Sect. 5.3, the choices can indicate the stability and accuracy of the estimate (see also the discussion in Chap. 8). These considerations have been explored in two recent papers (Dunlop et al. 2015, 2015b), who estimated the current density at Swarm altitudes using 2–4 spacecraft positions, and showed that coordinated field aligned current (FAC) signatures can exist at both Cluster and Swarm and we provide an account of the methodology below in Sects. 5.4.1–5.4.3.

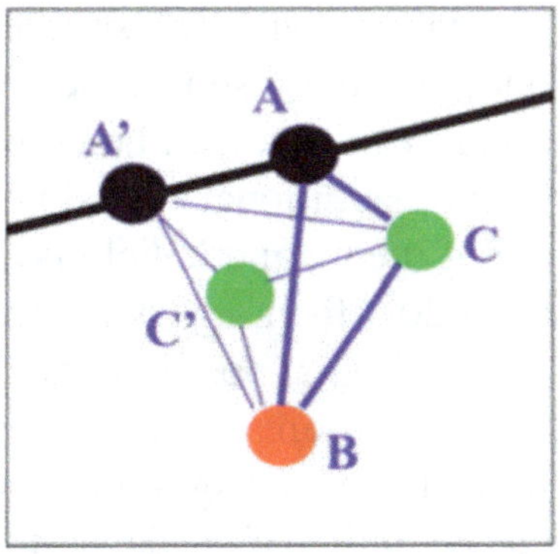

Fig. 5.5 Orbit track of the three Swarm spacecraft (A, B, C) when these are grouped closely together. The positions A', C' are of these spacecraft a few seconds earlier. When these are combined, giving up to five positions, then the full curlometer can be applied and all components of the current density, ***J***, may be recovered (i.e. by using selected configurations of four spatial positions, e.g. A, A', B, C and using some mean time reference suitable for each configuration) (from Dunlop et al. 2015a, b)

5.4.1 *Method: Application of the Curlometer to Stationary Signals*

Chapter 6 outlines in detail the standard method of calculation of FACs, which uses the dual-spacecraft Swarm dataset and is applied to produce the standard Level 2 dual-spacecraft FAC data products using a specific implementation, and criteria on the field model used to construct residual magnetic field vectors. The dual-spacecraft method (Ritter et al. 2013), applies the curlometer in a modified form using only the two-spacecraft flying side by side, Swarm A and C, and therefore applies to most of the mission operations. The adaptation uses time shifted positions of A and C and seeks to optimise the alignment of the vertical current component linearly estimated from these positions to the FAC direction (i.e. to the expected background field direction).

Even when flying in close formation, the three Swarm spacecraft generally allow only a partial estimate of the current density (i.e. one component, normal to the plane of the configuration), unless these time shifted positions are used in combination with the current ABC positions. Figure 5.5 shows that the ABC positions, for some current time-step, are planar, but adding measurements at the positions (A' and C') for adjacent times (either earlier or later and typically shifted by 5–20 s to regularise the configuration) produces 3-D configurations. This provides a measurement set of up to 5 spatial points, from which the magnetic gradients (and curl B in particular) may be estimated. Other configurations result if the position of B is also time-shifted. It is a fundament assumption that temporal variations are slower than the sampling rate of the sampled magnetic structures at LEO, so that the magnetic field signals can be considered to be stationary for these short time periods and therefore that the adjacent positions of the spacecraft in time can be considered to sample the same current structure, thereby providing additional spatial positions.

Thus, if the spatial structure of the magnetic field is assumed to be stationary on short time scales of a few up to 20 s then adjacent (time-shifted) positions of the spacecraft can add to the number of spatial positions used to estimate the differences in the magnetic field between each position. These assumptions can be tested through the stability of the estimates for different configuration choices, but as mentioned the effect of temporal variations can be limited if the data is filtered to lower cadence. The curlometer method anyway resolves current only on the scale of the configuration and therefore the matched time scales are the convective scales across the spacecraft configuration. Furthermore, as mentioned earlier, at these low Earth altitudes, the internal geomagnetic field, which contains non-linear gradients, dominates, so that linear estimators of the currents should be applied to the residual fields, obtained following subtraction of the static (and current free) internal field (in the examples below we subtract the Chaos model field Olsen et al. 2014) in order to remove errors introduced by the neglect of the non-linear terms.

The method is applied to the possible configurations in Fig. 5.1, obtained by selecting different sets of positions, to produce 2, 3 and 4-spacecraft estimates of current density from the basic 3 spacecraft spatial configuration of Swarm. By selecting planar groups of 3 or 4-spacecraft (as well as using the spatial array ABC), a 'curlometer' estimate can be made such that three-spacecraft positions essentially form one face of the tetrahedron shown in the right hand panel of Fig. 5.1 (the use of four-spacecraft positions recovers the full curlometer estimate). There are, in fact, 4 sets of four-spacecraft, involving spacecraft B, that can be formed from the 5 positions shown (the array AA'CC' is nearly planar so is not a usable fifth grouping; see below). In the case of four-spacecraft the quality parameter from the linear estimate of div(**B**)/**curl**(**B**) can also be estimated (Dunlop et al. 1988) as described in Sect. 5.3.

Thus, if three Swarm spacecraft are close together, as shown here, different tetrahedral configurations may be selected so that the five positions provide some redundancy in the calculation; allowing the quality of the estimate to be tested. In this case however, account must be taken of the fact that each 4-point configuration has a slightly different barycentre (and therefore a different mean time associated with the estimate). A number of additional points should be noted:

1. Time shifting AC results in a nearly planar configuration of four positions (AA'CC'), where only one component of the current density is found so that the field-aligned current in particular is only obtained from a projection onto the field-aligned direction of the component normal to the spacecraft plane containing the spacecraft positions. This is true for the standard 'dual satellite' Level 2 (L2) data product FAC_TMS_2F in the Swarm dataset, calculated from two-spacecraft (A and C) (Ritter et al. 2013).
2. The current density estimate resulting from each group of spacecraft, relates to a particular barycentre (centre of volume of the configuration, see Harvey et al. 1998), so that the combination of different groups of spacecraft refer to slightly different mean times. This allows the degree of stationarity of the measurement to be probed, in principle, through different choices of spacecraft (which have different mean times for the corresponding barycentre). For example, three-spacecraft combinations (such as ACC', AA'C etc.) can be selected and refer to slightly different mean times.
3. For the special case of the configuration in Fig. 5.5, both four and three-spacecraft estimates can be cross-compared. The array of Swarm A, B and C produces a purely spatial estimate, but is typically in a slightly tilted plane to the A, C orbit tracks and additionally corresponds to the leading time of measurement.
4. We have chosen to show results below from the configurations formed through time shifting the positions of A and C (as indicated in Fig. 5.5), since this produces the best alignment of the configurations to the L2, 2-spacecraft parameter, and of course it is expected that in this high latitude region the dominant currents will be field–aligned and therefore approximately perpendicular to the basic plane formed by AA'CC'. The alignment is less critical for the four-spacecraft estimates, but it is still useful to use this choice for the L2 comparisons (see later discussion).
5. A more generalised method of constructing (or measuring) the configuration can be devised, as studied in terms of homogeneity scales in the least squares approach discussed by De Keyser et al. (2007) (and see references therein), and

this could be used in future applications and compared to other gradient methods (see conclusions). Here, we benefit from the special context of the Swarm orbit geometry and natural alignment expected of the main FACs.

It is therefore instructive to compare the estimates found from different combinations and below the comparison of the various estimates for a number of events are shown.

5.4.2 *Use of Special Configurations: 2-, 3-, 4-, 5-Point Analysis*

To test the application of the curlometer to the Swarm data, events can be selected from the first phase of science operations (April–August 2014), when the alignment of the orbits allowed the three-spacecraft to repeatedly come close together. The data used is 1 Hz Swarm level 1b data (https://earth.esa.int/web/guest/swarm/data-access), taken by the vector field magnetometers (VFM, Friis-Christensen et al. 2008), but where these data have had the Earth's static, internal field removed using the CHAOS-4plus model (e.g. Olsen et al. 2014), although this was updated with the CHAOS-6 model for the re-testing summarised in Chap. 8, very little difference in the estimates resulted from the updating of the CHAOS model. The curlometer method is then applied to the residual field data.

As shown on the spacecraft orbits in Figs. 5.6 and 5.7, the optimum time-shift to best match all methods and optimally regularise the spatial configuration of the spacecraft is found to be 20 s. A convention is adopted where the time-shifted positions, A', C' are labelled A' = An (for forward shift in time) or Ap (for backward shift in time), and similarly for C'. The field-aligned components of the multi-spacecraft, curlometer estimates are compared in the top set of panels in Figs. 5.6 and 5.7 to the 2-spacecraft, time-shifted method of Ritter et al. (2013) (shown as L2, from the Swarm Level 2 product FAC_TMS_2F, derived using VFM measurements). This highlights both compatibility with L2 and the differences observed between estimates. The set shows different spacecraft groups, selected from the enlarged constellations on the right hand side of Figs. 5.6 and 5.7. This $J_{\|}$ component is taken from the projection of J_{N} in each three point case (so depends on the actual orientation of the plane of the configuration) and is the field aligned component of the vector current in the case of the 3-D 4-point arrays. The first FAC signature in Fig. 5.6 has been analysed recently by Dunlop et al. (2015).

The top panel of the upper plots in Fig. 5.6 shows that AApCC$_{\mathrm{P}}$ (which is effectively a 2-D configuration) matches well to the signature obtained from the standard L2 2-spacecraft product. The line for L2 is broken for the interval 04:11–04:12:30 UT, indicating where the orbits crossover (first spacecraft B and then AC, as indicated in the left-hand orbit plots). The quality of the estimates is expected to be downgraded as the form of the spacecraft array changes dramatically. This in fact is reflected in the deviations between the estimates (a measure of quality) as this

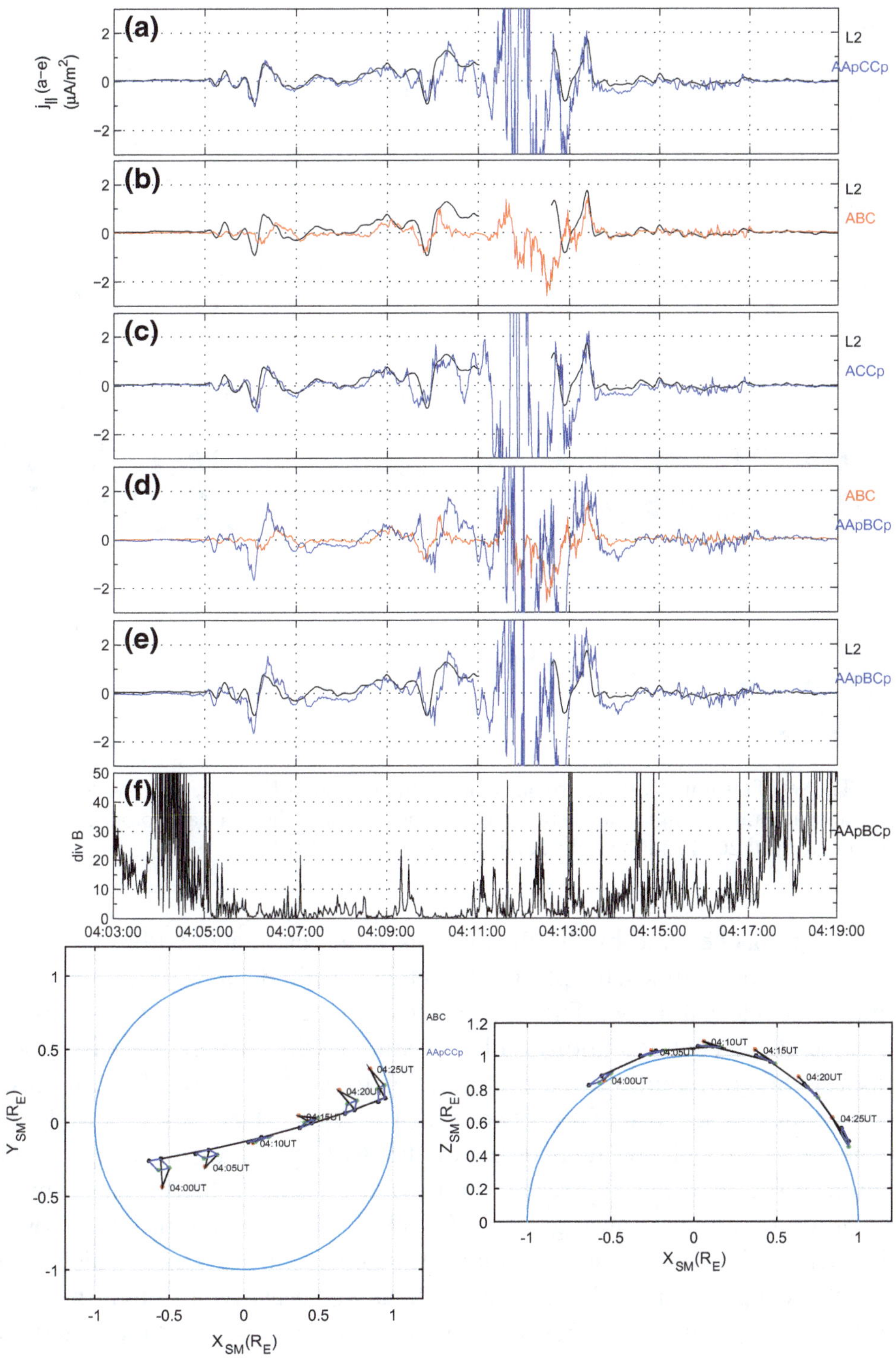

Fig. 5.6 The upper plot shows the set of curlometer estimates from selected configurations, for the 22 April 2014. From top to bottom, the panels show (the smoothed level 2 (L2) data product is shown as the black line): the 2-D 4-point array AApCCp; the three-spacecraft ABC spatial array (i.e. no time-shift, which has a different J_N direction); the three point array, ACCp; comparison of the 3-D 4-point array (AApBCp) with the ABC array and L2, respectively. The lowest panel shows the ratio div(B)/curl(B) (labelled divB for short) for AApBCp. The lower plots are orbit views (right, XY_{SM}, left, XZ_{SM}) with enlarged (x5) configurations as the Swarm spacecraft flew over the auroral zone and polar cap, where AC are time-shifted (from Dunlop et al. 2015)

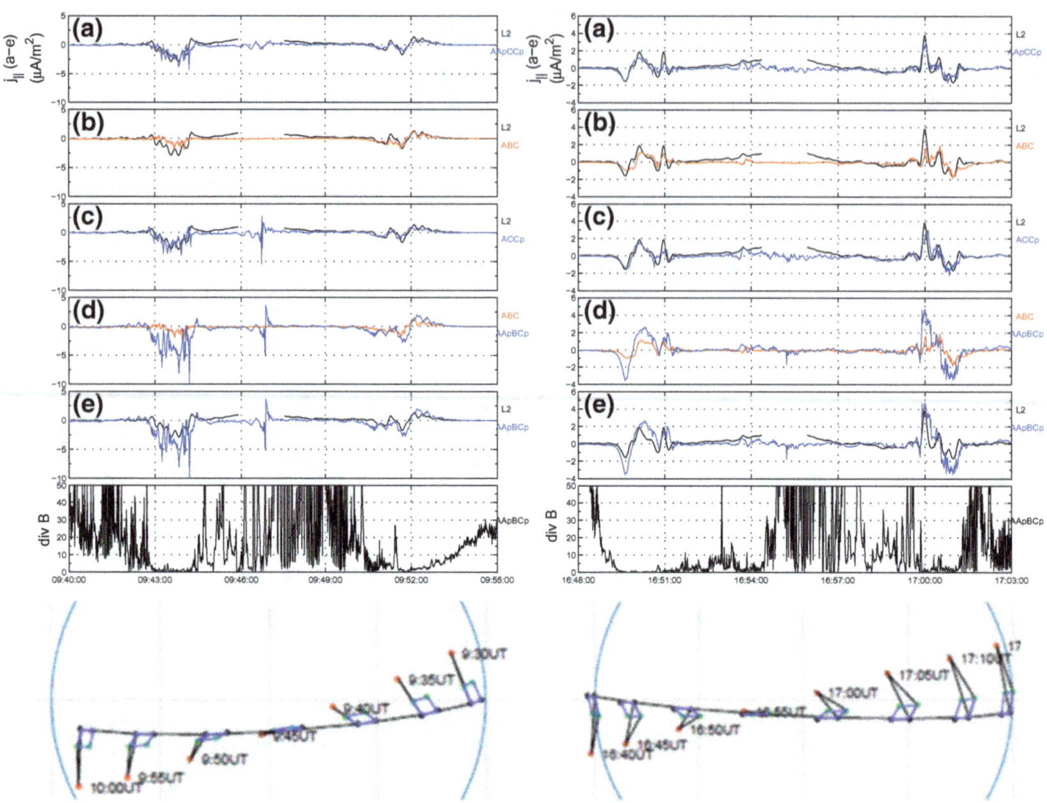

Fig. 5.7 The set of curlometer comparisons in the same format as Fig. 5.6, with the XY_{SM} projections of the spacecraft configurations shown in the lower plots. The left hand plots correspond to the pass on 28 April 2014 and the right-hand to that of 4 May 2014

interval is approached and by the fluctuations in the estimates within the interval. Nevertheless, at other times, the profiles match in amplitude and approximate timing throughout the interval shown. This is as expected since the spacecraft positions (and hence plane of each configuration) is closely similar (the L2 parameter is smoothed and has a slightly different, shifted configuration to AA'CC'). The effect of different time-shift choices has most effect between 04:10-04:11 UT and 04:12:30-04:14 UT.

In the second panel, the relative tilt of the ABC plane (as seen in the XZ_{SM} projection in Fig. 5.6) initially results in a lower amplitude FAC signal, since the projection of J_N is more misaligned to $J_{||}$, for this configuration. Note, however, that the ABC estimate is a purely spatial estimate (linear gradients) which slightly changes the sampling of the FAC. These effects again evolve through the interval as the ABC configuration changes such that both the planar orientation (and hence the projection $J_{||}$) and relative time of the corresponding barycentres changes along the orbit. The third panel shows that the use of only three positions, ACCp, also compares well with the L2 FAC signatures although some additional features are apparent. This estimate is sensitive to the choice of spacecraft positions since the effective barycentre time changes with each choice. Moreover, for ACCp as shown, the relative timing of the barycentre to that of the L2 estimate, changes along the orbit. In principle, this allows the quality of the L2 FAC estimate to be verified, i.e.

different choices for the three positions results in slightly different barycentre times relative to those of the L2 product, so that small temporal and spatial effects can in principle be revealed (these show up as small timing shifts in the traces because different configurations sample the FAC signature at slightly different times). There are 8 possible 3-spacecraft configurations which can be chosen relative to AApCCp or AAnCCn and these are summarised in Chap. 8.

The two lower panels in Fig. 5.2 show that the full curlometer estimate, arising from a 3-D 4-point array (here chosen as AApBCp) identifies the field-aligned signatures seen in the profile of the L2 product. In addition, the relative shifts in the profiles seen in the ABC estimate are confirmed to arise from that choice of configuration. The actual profile obtained for this 4-point estimate is sensitive to the choice of spacecraft positions and the resulting configuration, particularly for the field-aligned component which may align better with one or more of the faces of the configuration. The set of choices can therefore provide a further quality measure on the features observed, through the change in effective barycentre and different tetrahedral shape formed by each 4-position set. In principle, this can also be used to explore any effects of non-stationarity. For this event however, the profiles are broadly consistent. The $J_{\|}$ component for this configuration is actually obtained from the three vector components of current density (see Sect. 5.2) and so can be used even when the FAC direction deviates far from the plane of the spacecraft configurations. In fact, in both the lower panels, this estimate produces the highest amplitude FAC signal, suggesting that the L2 estimate (and that from the other three-spacecraft arrays) omits some of the actual FAC through the misalignment of J_N to the field aligned direction (see Sect. 5.4). The four-spacecraft estimate, despite incorporating two time-shifted positions, is in fact rather stable throughout the intervals tested and to those adjacent to the orbit crossovers in particular. It therefore provides additional coverage of signatures when the L2 product has low quality.

The bottom panel shows the standard estimate of div(**B**)/curl(**B**), which is obtainable for each choice of 4 positions (5 sets for each time-shift direction: forward and back). This parameter is only one measure of the quality of the linear gradients (see De Keyser et al. 2008), but it can be seen that this remains low throughout the intervals containing FAC signatures (where there is significant current). As such, this parameter does not absolutely indicate accuracy, but, as a rule of thumb, a value of 30% has previously been used as a threshold indicator of reliable current estimates. The value is large and shows high fluctuations in the regions at either end of the interval where the current is zero and also shows a significant increase during the period of orbit crossover (04:11–04:12:30 UT). This is therefore taken as a good indication of the overall quality of the estimates, although it only represents a measure of the non-linear gradients, not the absolute error or the effect of time dependence. The further choice of 4-spacecraft from the five (time-shifted) positions allows us to probe the effects of both the different spatial coverage (shape of the 4-spacecraft tetrahedron) achieved and time dependence through the different barycentre positions.

Two other events are shown in Fig. 5.7 for intervals extending across the polar passes in each case. On the left-hand side, corresponding to a pass on the 28 April 2014, the profiles are remarkably matched yet the increased amplitude of FACs caught by the four-spacecraft estimate is pronounced, particularly for the signature between 09:42–09:45 UT. Note that div(**B**) is again very small during the key intervals of FACs, but shows low quality through the times adjacent to the orbit crossover, while the current density is very small. Similarly in the right panels, corresponding to the pass of the 4 May 2014, nearly all of the fine structure of the FACs is reproduced in all estimates providing confidence that these features are well represented. Again the amplitudes are caught best by the four-spacecraft estimates and in this case the ABC configuration is also well matched to the other profiles. The values of div(**B**) are similar to the previous event with low values during the FACs. In each event it is clear that despite the ABC estimate resulting from a pure spatial gradient, it suffers from the misalignment of the configuration plane to the true FAC direction so that the amplitude of this component is lower. The 3-D 4-point configurations appear to capture the true FAC amplitudes.

5.4.3 *Use of the Extended 'Curlometer' with Swarm Close Configurations: 3-D Current Density*

As described above, in the case of the basic, two-spacecraft method (and the other 2-D configurations) only the J_{N} current density component is obtained, so that any currents which are perpendicular to the magnetic field direction can only be seen through misalignment of the constellation plane to the background field. There may of course be errors introduced through lack of knowledge of the model field coordinates, but here the data defined coordinates are assumed. Figure 5.8 shows the vector current density estimates obtained from the 3-D 4-point estimates for key intervals during the 22 April and 4 May passes and for two different configuration choices. All components are in principle determined, so that $J_{\parallel}$ and $J_{\perp}$ can be directly computed. The main FAC signatures in each event are indicated and the characteristic forms for the other components can be seen in each case.

As indicated in Sect. 5.4.2, the FAC signatures broadly agree in form between each method, although the amplitudes of the 4-spacecraft curlometer are larger than the L2 parameter in some cases. A specific set of positions are shown in Fig. 5.8, based on time shifting the Swarm spacecraft A and C in order to remain as closely related to the L2 parameter as possible (as in Sect. 5.4.2). The character of the signatures does change slightly with each choice of four spatial positions (selected from the 5 positions depicted on the orbit plots of Figs. 5.6 and 5.7), but is most consistent for the optimum time shift giving the most regular configurations (in this case the value of div(**B**)/curl(**B**) is also predominantly insensitive to the choice of configuration). Indeed, the form of the signatures overall remains recognisable and in fact the change in each component seen for the different estimates between the top and bottom panels is less than 0.2 $\mu\mathrm{Am}^{-2}$ in the centre of the main current signatures (which range in magnitude from about 2–4 $\mu\mathrm{Am}^{-2}$). This represents a maximum error in the estimates of around 10% due to changing the configuration.

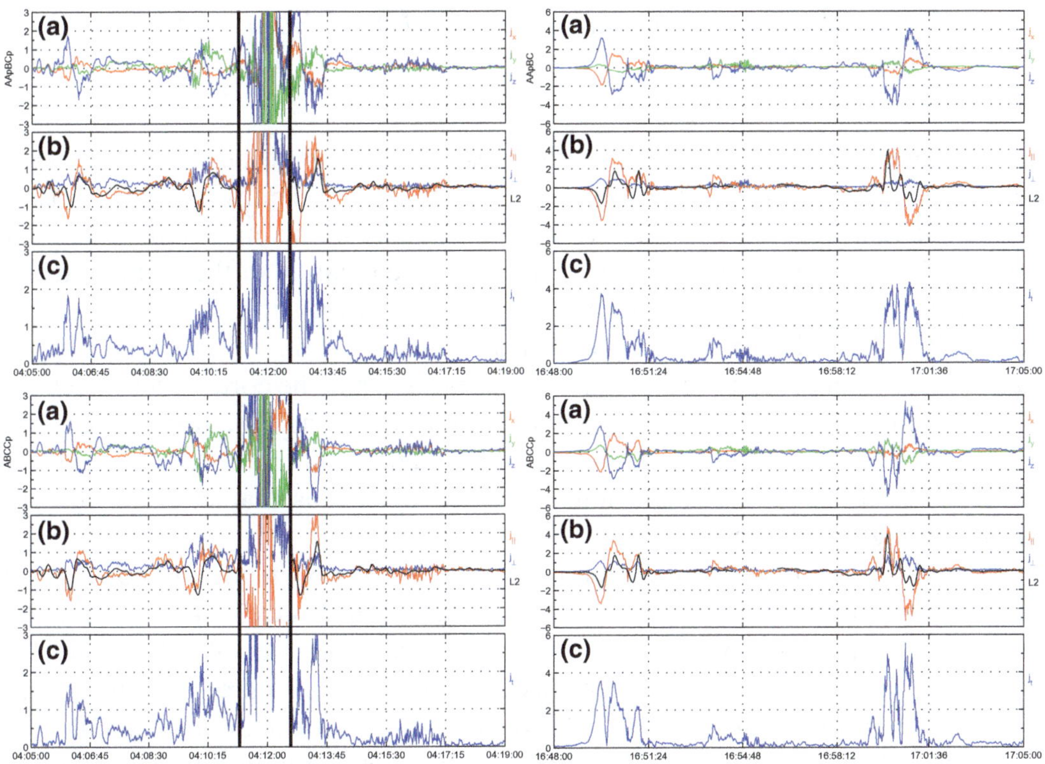

Fig. 5.8 Vector current density estimates obtained from the 4-point estimates for the key intervals during the 22 April (left) and 4 May (right) passes. For each event we plot the estimates from two configuration choices (top and bottom sets). For each plot, panel **a** (top) traces show the three SM components of the vector current; **b** (middle) traces show $J_{\|}$ and $J_{\perp}$, together with the L2 trace, and panel **c** (bottom) trace shows |J|. The estimates for the 22 April (left-hand panels) near the orbit crossover of Swarm A/C and B (between the vertical black lines) are unreliable due to distortion of the spacecraft configuration (from Dunlop et al. 2015)

For the interval on the 22 April we have also marked the region where the estimates will be unreliable due to the orbit crossover.

In each case, therefore, the perpendicular currents at either side of the FAC sheets are small but reach ~0.7 μAm^{-2} and thus are significant compared to the error. Furthermore, if the effect of changing the time-shift applied to the spacecraft positions is checked it is found that the perpendicular components remain significant although smaller time shifts, of 5 or 10 s, despite the severe distortion in the spatial configuration. For at least three of the signatures the perpendicular components show features which are consistent with the presence of perpendicular currents surrounding a FAC sheet (e.g. Gjerloev and Hoffman 2002; Liang and Liu 2007), i.e. reversals in the J_x and J_y components (SM coordinates) within the main FAC (see also Ritter et al. 2004; Wang et al. 2006). These signatures are not analysed but simply point out that the 4-spacecraft configuration can reveal perpendicular as well as parallel components. This was the first time these have been shown from direct measurements of the full vector current density at LEO, rather than as projections from single components (see Shore et al. 2013).

5.4.4 *Current Sheet Orientation Implied by 2-Spacecraft Correlations*

A complementary technique based on the cross-correlation of the single spacecraft estimates of FACs at each Swarm spacecraft has been recently introduced (Lühr et al. 2015; Yang et al. 2018), which results in estimates of the maximum correlation between sliding data intervals at two-spacecraft. The position of the maximum correlations found can be defined in terms of the required time shift (for example, from spacecraft A–C) and corresponding difference in the local time (this is given by the relative location of the spacecraft on the pair of orbits), and the correlation values can be computed as a function of local time (MLT) of the orbits. The results are found to depend on the level of filtering applied to the single spacecraft FAC data, and can identify different large-scale behaviour within broad regions characterised either by region 1 or region 2 currents.

An application of this information uses the positions of maximum correlations on the orbits to define the orientation of equivalent planer current sheets. These are estimated from the mid points of the maximum correlation intervals. If the sampled FACs are primarily large scale, then this can indicate the degree of order in the alignments (e.g. to the auroral oval). Conversely, if there are significant small-scale structures or fluctuations present then this will be reflected in unstable orientations and a lack of ordering. Figure 5.9 (adapted from Yang et al. 2018) shows the patterns in current sheet orientations found for intervals broadly associated with both region 1 and region 2 currents.

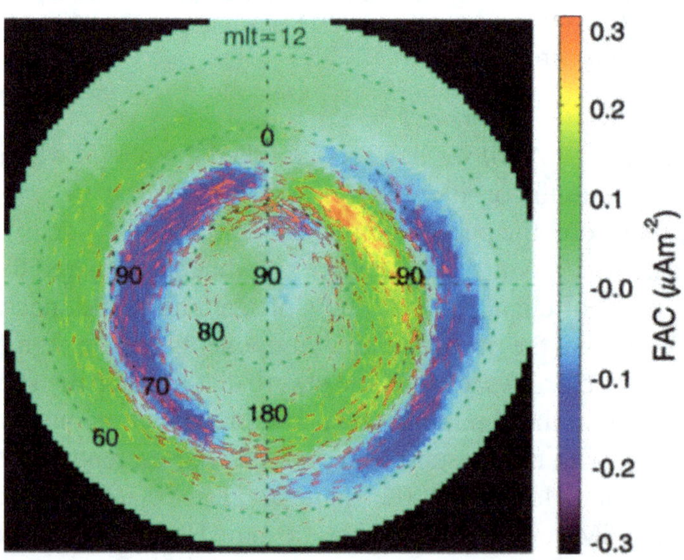

Fig. 5.9 Plot adapted from Yang et al. (2018), for a Northern hemisphere polar map, showing the average FACs for Swarm A and C data from 17th Apr. 2014 to 30th Apr. 2016, overlain with implied current sheet orientations for both higher latitude regions and lower latitude regions. These are plotted using lines of normalised length which connect the average Swarm A and C positions of those orbit segments producing the maximum correlations (drawn for 20 s filtered data)

5.5 Swarm-Cluster Coordination: FAC Scaling and Coherence

Although signatures of FACs, have been extensively reported (e.g. McPherron et al. 1973; Shiokawa et al. 1998; Cao et al. 2010), the form of the FACs, flowing along near-Earth field lines, has a highly time dependent signature, depending on their scale size, so that single spacecraft measurements cannot easily probe their nature. At more distant magnetospheric locations, multi-spacecraft analysis has been possible (e.g. Marchaudon et al. 2009; Slavin et al. 2008; Cheng et al. 2016), and indeed the distributed multi-spacecraft capability of AMPERE (Anderson et al. 2000), although limited in accuracy, is providing global features of FACs. The combined use of the multi-spacecraft Swarm and Cluster missions, however, allow the detailed resolution of individual FAC structure with close formations of spacecraft at both low and medium Earth orbits.

At the start of the Swarm operations in 2014, the four Cluster spacecraft were flying in tilted, eccentric orbits, with perigee heights ranging between 3 and 4 R_E. For several hours around perigee, Cluster passes through the Earth's ring current and often passes through the region of high latitude large-scale field aligned currents (FACs), particularly above the ring plane (magnetic equator). A study of possible coordination of Cluster operations with Swarm passes was carried out resulting in an adjustment of the Cluster configuration in order to optimise the Cluster tetrahedron as it passed through the Earth's ring current. This had the result that the configuration was also compact at perigee as large-scale FAC region was sampled. This special phase of Cluster operations therefore coincided with the first phase of Swarm operations when the three Swarm spacecraft repeatedly were grouped together as discussed in Sect. 5.3. The Swarm orbital planes drift relatively to Cluster at about 131 (spacecraft A/C) and 108 (spacecraft B) deg/year so that the alignment with the Cluster orbit slowly changes throughout the mission. This resulted in a number of multi-spacecraft conjunctions of Cluster with Swarm, during 2014, for which it was possible to match large-scale FAC signatures across different altitudes and identify the scaling properties.

One such conjunction is discussed below using the methodology described in Sect. 5.3. In addition, a recently developed method (Xiong et al. 2014; Xiong and Lühr 2014) is also used to identify crossings of the expected poleward and equatorward auroral boundaries (taken as the maximum gradient in R1 and R2 FAC power), where their statistical model of the form of these boundaries (derived from small and medium scale FAC using 10 years of CHAMP magnetic field data) is employed to help order both data sets. For the Cluster signatures, spin averaged data from the fluxgate magnetometers (FGM) (Balogh et al. 2001) is used.

5.5.1 Conjunction Characteristics

Figure 5.10 shows the combined locations of the spacecraft during the period of interest, which corresponds to the event described in Fig. 5.6. Cluster was flying from dawn to dusk through midnight local time during the few hours around 4:00 UT and passed from low to high invariant magnetic latitudes (MLAT) at around midnight local time, and at 2.5 R_E altitude, before falling again to low latitudes and passing through the magnetic equator (left panel). During part of this interval (~03:55–04:25 UT), the Swarm spacecraft flew through the auroral zone and across the polar cap in a close configuration (~100–150 km separation) of all three-spacecraft (see inset in the right hand panel). The Cluster configurations were such that three-spacecraft (1, 3 and 4) were close together (~1000 km separation) and in a plane nearly perpendicular to the magnetospheric field, while the fourth spacecraft, C2-red, lay further away (at ~5000 km separation from the others), allowing good resolution of the FACs at Cluster corresponding to the normal component to the 3-spacecraft plane. The Swarm configurations show the crossover of the orbits just after 04:10 UT (in a slightly different form to that shown in Fig. 5.6). The inset in the left hand panel also shows that Swarm and Cluster came into close magnetic alignment just before

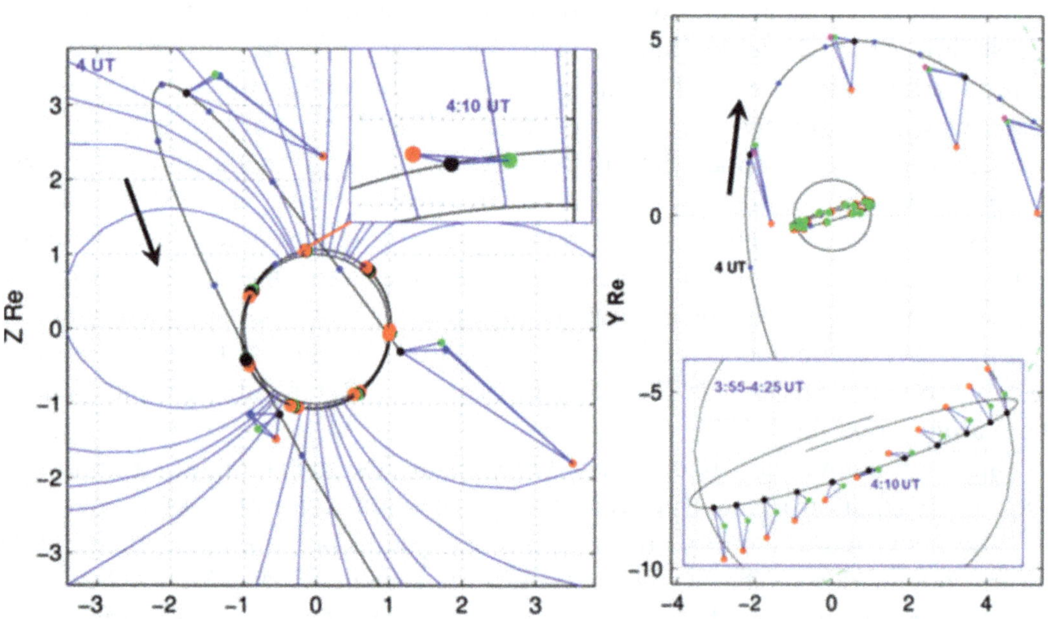

Fig. 5.10 Plot taken from Dunlop et al. (2015b), showing the orbits of Cluster and Swarm relative to the Earth in GSM coordinates on the 22 April 2014, projected into Z, X_{GSM} (left panel) and Y, X_{GSM} (right panel). The Earth is shown as a circle with model geomagnetic field lines shown for guidance in the Z, X projection. The Cluster configurations are shown enlarged by a factor of 3 relative to the orbit track of C1. The insets show zoomed views of the Swarm spacecraft configurations, projected into each plane, enlarged by a factor of 5 relative to the orbit of Swarm A. The Cluster colours are: C1-black, C2-red, C3-green and C4-blue, while the Swarm colours are: A-black, B-red and C-green

04:10 UT where Swarm flew under the magnetic footprint of Cluster, and shows the slightly higher altitude of the Swarm B spacecraft.

Figure 5.11 shows the ground mapped orbits of both Swarm and Cluster (we use ground magnetic footprints, from Tsyganenko, T89 model, (Tsyganenko 1989) to obtain a stable, relative position in both cases). These tracks confirm that the footprint of Swarm crosses the Cluster footprint between 4:05–04:07 UT. Also plotted are equatorward and poleward auroral boundaries, fitted statistically as ellipses to CHAMP data for different conditions by Xiong et al. (2014) and Xiong and Lühr (2014). These ellipses are plotted here for the times of the midpoints of the Swarm positions indicated by coloured dots on the Swarm orbit (Cluster positions are indicated by triangles). The method of Xiong et al. (2014), has been directly applied to the Swarm data for this interval, where the times indicated along the right hand Swarm track correspond to the actual estimates of the maximum gradient in FAC intensity. The mapped locations for Cluster 1, 3 and 4 cover the same scale as the Swarm array, so that the FACs are approximately covered on the same relative scale at each location.

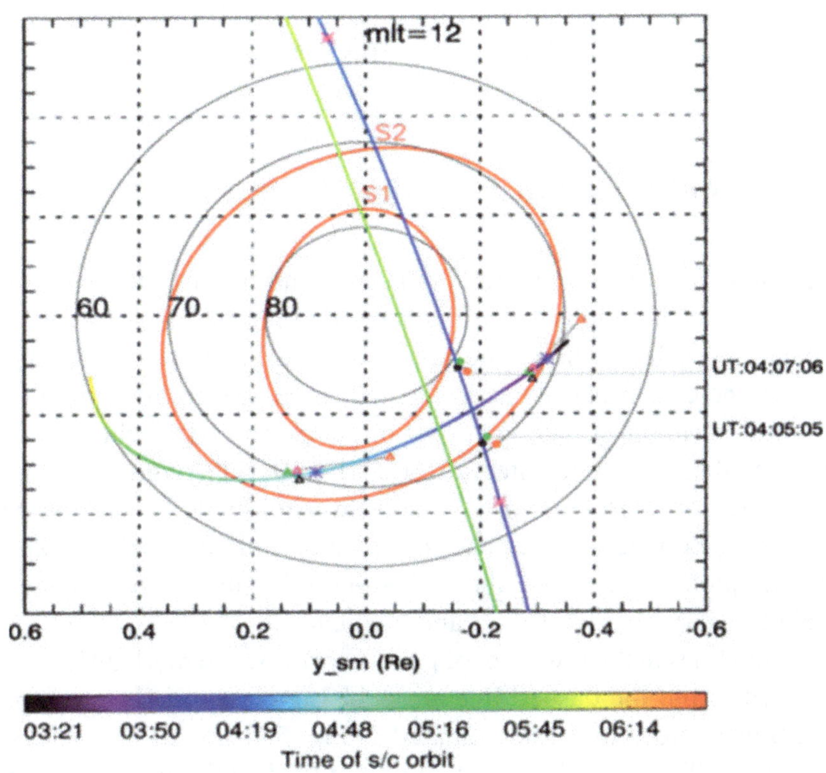

Fig. 5.11 Plot taken from Dunlop et al. (2015b), showing mapped footprints of Cluster and Swarm (where two passes are shown) in SM coordinates on the 22 April 2014, together with model, equatorward and poleward auroral boundaries, marked as S1 and S2 (referring to the expected locations of maximum R1 and R2 FA currents), which are taken from the fits in Eq. 5.8 of Xiong and Lühr (2014) after coordinate transformation. Also marked are times along the Swarm orbit for actual positions of the maximum gradients in FAC density (after Xiong et al. 2014)

The Cluster orbit is drawn from 03:40–05:00 UT, showing that Cluster moves from lower (~70°) to higher (~75°) MLAT, and back to lower MLAT, as it moves across MLT. Both the Cluster and Swarm tracks are colour coded with time to show that the orbits cross at about the same UT (Cluster crosses the MLT of Swarm a few minutes earlier than the Swarm flyover). The curlometer estimates for Cluster (not shown) confirm the spacecraft enter a region of negative parallel current, $J_{||}$, as it approaches and crosses the S2 boundary just after 03:45 UT (at MLAT = 71°) and as it approaches the boundary again, crossing just after 04:55UT and at MLAT = 67° (consistent with the connectivity of R2 large-scale currents for this dawn-side local time). At the higher MLAT = 75° positions, Cluster approaches the S1 boundary where it shows zero or positive $J_{||}$ (between ~04:00 and 04:15 UT).

Meanwhile for the first pass of Swarm (right track), the curlometer estimates for Swarm show the spacecraft also enter a region of FACs as the S2 boundary is crossed initially at 04:04 UT (MLAT = 71°). Signatures continue over the pass and are broadly consistent with the expected S1 and S2 boundaries as drawn, although other current systems are present (see Fig. 5.12, which shows the analysis of the period 03:40–04:20 UT, containing the first large-scale FAC at Cluster and the northern pass of Swarm). The terms large-scale and small-scale FACs below correspond to scale sizes at Swarm altitudes of >150 km and <150 km respectively.

5.5.2 *Analysis of Common FAC Signatures*

Figure 5.12 shows detailed estimates of the currents seen by Swarm using 1, 2, 3 and 4-spacecraft calculations taken from the configurations shown in Fig. 5.10. The third panel in each set shows the L2 product in black, the full 4 time-shifted spacecraft curlometer estimate in red, and the three-spacecraft estimate from the ABC configuration (in blue). The lower set of panels show the whole northern pass of Swarm (over the core time interval 04:03–04:19 UT), while the inset (upper panels) shows the first short burst of FACs on Swarm (during which Swarm crosses the Cluster orbit), between the ascending crossings of the S1/S2 boundaries (as indicated on the plot). Figure 5.11 shows that these boundary crossings correspond to the model boundary position for S2 and are near the position of the S1 boundary. The other vertical lines in the lower set of panels are drawn to mark different positions along the Swarm orbit: the dashed line at 04:11 UT corresponds to the position of the A/C and B orbit cross-over (as indicated in Fig. 5.10), while the A,C spacecraft cross at 04:12:30 UT, and the last vertical line at 04:14 UT corresponds to the first drop in size of the FACs (it also corresponds to the descending position of the S2 oval as drawn on Fig. 5.11).

These FAC estimates are best seen in the top set of panels, where the filtered single spacecraft signatures have similar, time-shifted profiles corresponding to the relative positions of spacecraft A, B and C as they cross the region: Swarm A and C are almost side-by-side, while Swarm B lags behind (~25 s). Swarm B, at a slightly higher altitude, sees a lower amplitude signature. In the lower panel, the 3-spacecraft

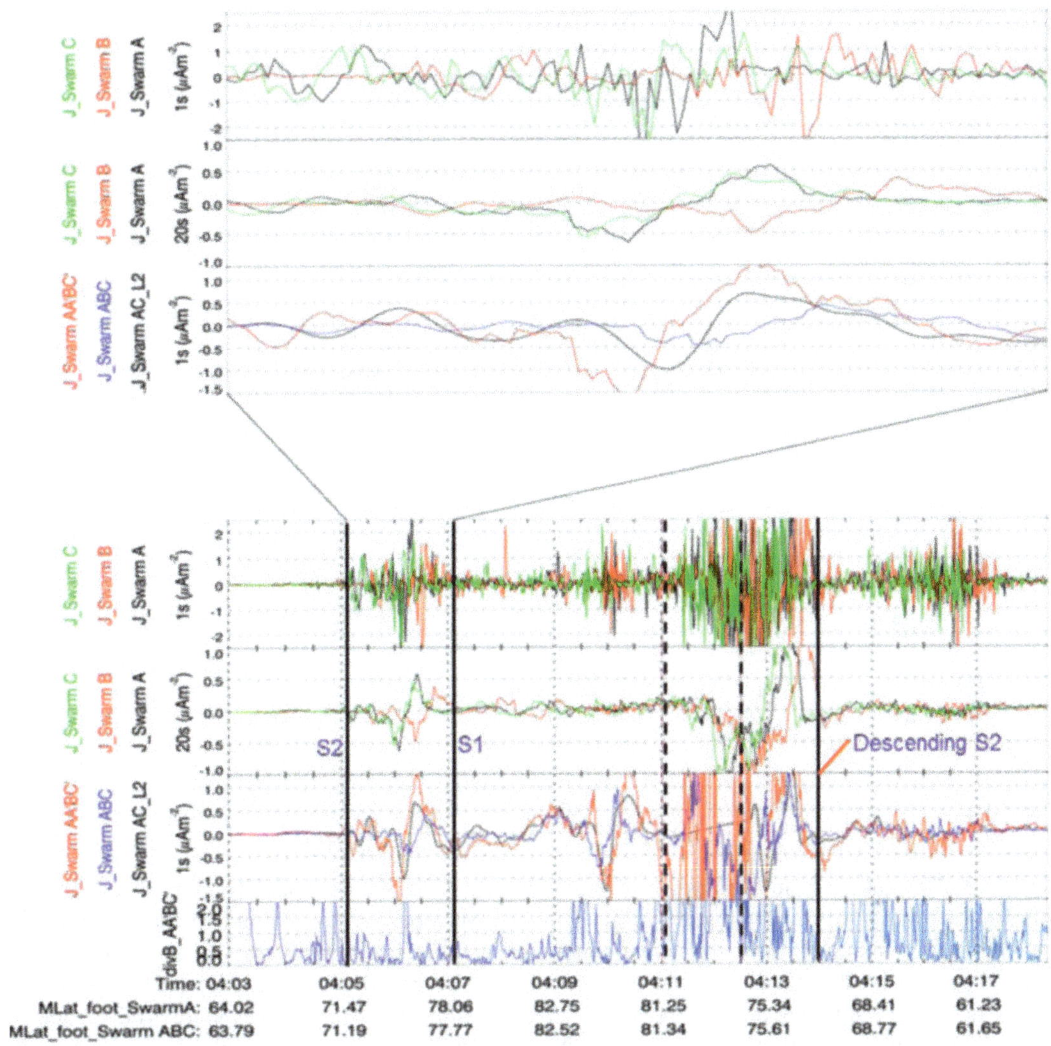

Fig. 5.12 Plot taken from Dunlop et al. (2015b), showing FAC estimates for Swarm for the whole northern pass (lower set) and for the intervals containing the ascending S2, S1 crossings (top set). In each set, the three panels represent (from the top): the level 2, single spacecraft estimates from dB/dt; the filtered single spacecraft estimates (20 s window), and the 2, 3 and 4-spacecraft curlometer estimates (as defined in Sect. 5.4). The ratios of div(B)/curl(B) obtained with the four-spacecraft estimate are also shown in the lowest panel, showing low values (high quality) for the key intervals containing the FACs

estimate (blue trace) gives the current normal to the ABC plane, which is tilted with respect to the FACs, so that its field aligned projection has low amplitude compared to that of the 4-point estimate (where the parallel component is taken from the full current density vector). The signatures of each estimate in the third panel are similar, but the profiles have relative time shifts since the barycentres of each are slightly different.

In Fig. 5.13 the Swarm and Cluster signatures are plotted in terms of their MLAT position to better compare these FAC features across altitude (~2–3 R_E). The Swarm plot corresponds to the time interval ~04:05–04:07:30 UT, containing the FACs seen

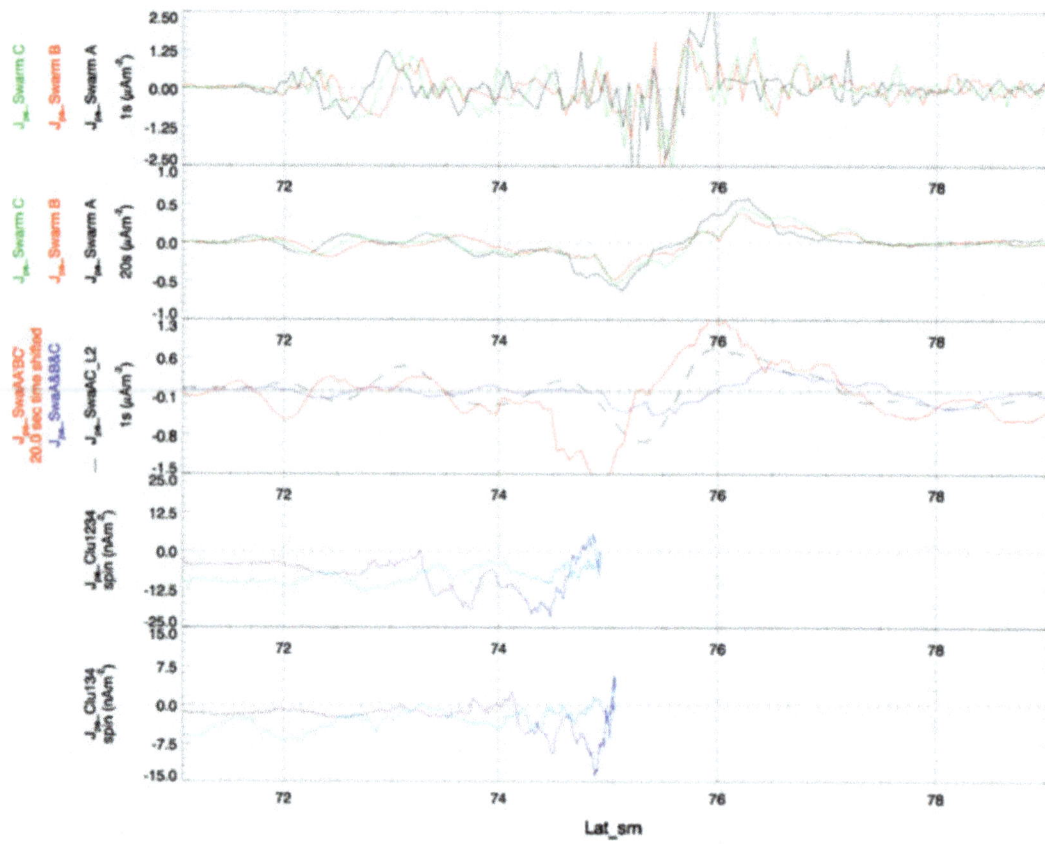

Fig. 5.13 Plot taken from Dunlop et al. (2015b), showing the mapped MLAT values (in SM coordinates) of Swarm and Cluster FACs. The top three panels correspond to the same set of FAC estimates as shown in the top set of panels in Fig. 4.3. The bottom two panels show Cluster FACs estimated by the full curlometer and from the face formed by C1, C3, C4 (which is almost perpendicular to the magnetic field), respectively. The light blue trace corresponds to the track of Cluster after the conjunction as it moves to lower latitudes, but far in local time from the Swarm track

between the S1 and S2 boundaries, while Cluster is plotted for the whole interval as it moves from MLAT = 71° and back again (03:30–04:40 UT). It needs to be borne in mind that there is a MLT difference between Cluster and Swarm which develops through the interval. Nevertheless, it can be seen that there are distinct features. For Swarm, the single spacecraft estimates contain two clear time independent signatures. The first is a large-scale (compared to the Swarm separations, i.e. approximately 150 km) feature between MLAT = 74° and 76° since it is sampled at the same position (MLAT) by each spacecraft as it flies through the FAC (the amplitude for the higher position of Swarm B is slightly reduced). The second (MLAT = 72° and 73°) is a small-scale structure since it is sampled at different MLAT positions by each spacecraft, implying limited (longitudinal) extent on the scale of the spacecraft separation.

It can further be seen in the third panel that the 3-spacecraft Swarm estimate derived from the ABC configuration (blue trace) broadly agrees with the Swarm

B signature in the second panel (i.e. that the estimate is driven by the swarm B location). Note that since the barycentres of the multi-spacecraft estimates differ slightly, there is no exact correspondence in MLAT. The Cluster signatures double back as the Cluster array moves from 71° to 75° (dark blue trace), and then back to 71° (light blue trace). The dark blue trace corresponds most closely in time with Swarm between 74° and 75° and shows a very similar profile in MLAT to the 4-spacecraft estimate from Swarm (red trace, third panel). The two Cluster estimates (from four and three-spacecraft calculation), serve to show the change in the effective location of the measurement when the inclusion of spacecraft 2 significantly shifts the effective mean position of Cluster.

Thus, the profiles show some similarity to Swarm and at similar MLAT positions, allowing for the slight shift in MLT between the spacecraft arrays. It is therefore probable that Cluster is sampling the same large-scale FAC as Swarm. The amplitudes of the FAC are also consistent: based on the expansion of the field lines, the field aligned current sheet cross-section should scale by about a factor of ~60 (ratio of field strengths) between the Swarm and Cluster altitudes. In fact, the amplitudes of current densities at Swarm and Cluster are ~1.3 μAm^{-2} and 20 nAm^{-2}, respectively, giving a ratio of ~65. The small-scale structure seen by Swarm at 73° would not be expected to map to Cluster positions in any coherent manner, and indeed there is no clear signature seen by Cluster at this position.

5.5.3 Other Events

A few other events during this phase of the Swarm mission have been analysed and show similar correspondences to those found above, even for intervals where there is a close conjunction with Cluster but no close grouping of the three Swarm spacecraft. In those cases particularly, but also generally, the Swarm signatures can be compared to the AMPERE model extracted for each event (see Chap. 7 for details of the AMPERE data taken from the Iridium spacecraft array) to clarify the characteristics of the large scale signatures. Unpublished analysis has suggested that filtered FAC signatures seen by Cluster can be mapped to the AMPERE altitudes, and can show similar large-scale features to the sampled AMPERE model currents, scaled to Cluster along the mapped orbit track.

5.6 Conclusions

Multi-spacecraft determination of currents, first from Cluster and later from the Swarm, Themis and MMS missions, has provided a wealth of new information about current structures in both the outer and inner magnetosphere. Despite the fact that for small-scale, time dependent current structures, the relative spacecraft separation and configuration, as well as magnetic field contributions from non-linear sources, constrain the applicability of the method, the curlometer has proven to be a reliable and robust tool to determine both partial and 3D currents over a wide parameter range. Furthermore, estimates made from linear gradients tend to return average quantities which can be compared to other methods to probe neglected temporal or spatial effects. In addition, the application of temporal filtering (to reduce data cadence) and the removal of contributions to the magnetic field which contain non-linear spatial gradients (e.g. through subtraction of known model fields), can mitigate errors arising from these physical effects. As well as a review of the method, two key applications have been shown: to the Earth's ring current and to FACs, which benefit from such spatial and temporal filtering. These applications also show how the method can be tailored to sampling by fewer than four-spacecraft, where not all components of the current density can be computed without further assumptions.

The Earth's ring current has been extensively sampled by both Cluster and THEMIS and this chapter has shown a few sample studies in this regard. It is not intended as a complete review, but serves too show examples of the application. Cluster has allowed a full azimuthal scan of ring current crossings in a limited radial range, while allowing the ring current to be sampled above and below the ring plane, where the spacecraft can sample the adjacent FACs. Both the curlometer and the MFL curvature techniques have been applied and show evidence of dawn-dusk asymmetry in the ring current density. Cluster (and also THEMIS) has obtained coverage of the ring current in the period since the Swarm launch and it is likely that Cluster crossings in this era can be coordinated with Swarm to probe the MLT behaviour. Swarm can measure the ring current contributions to the geomagnetic field via the geomagnetic field modelling. The three magnetospheric THEMIS spacecraft measure the ring current for a wider range of radial distance, but do not sample well the regions above and below the ring plane. Both Cluster (because of often distorted spacecraft configurations) and THEMIS (since only three-spacecraft are flying together) rely on the ring current component of the current density being well determined. THEMIS also reveals effects of storm phase influence in the near-tail region which overlaps with the night-side ring current.

This chapter has also described a development of the curlometer technique to the low altitude region covered by Swarm, where the adoption of the principle of time-shifted measurements (in order to increase the number of spatial measurements of the magnetic field) can produce stable 2-, 3-, 4-point estimates of the electrical current density (a more general application of the method described in Chap. 6) which can be cross-compared. A set of estimates from the various possible configurations of positions which are sensitive to the barycentre time and position can be found. The range of estimates which follow have been applied along with other related methods to a standard event set to show the variability in the quantities found (see Chap. 8). The methodology used in this chapter can define a standard variance in the range of estimates as one measure of overall quality, but also comparison of the barycentre shifts and sampling can probe the degree of temporal and spatial variations of the event sampled, since the variance across the different configurations is dependent on this. Effectively, cross-comparison verifies the temporal effects as well as the inference of field aligned currents from the measured current components. Increased stability of the signal measured from these alternative configuration choices gives confidence in the quality of the estimates and thus allows an increase in the coverage possible with valid estimates. Indeed the 3- and 4-spacecraft methods can show that the estimates can often be stable in intervals where the 2-spacecraft method breaks down. In the case of four-spacecraft configurations the full vector current can be resolved and provide a quality factor for the estimates directly. The spatial estimate of current requires the currents to be stable on time scales equivalent to the effective convection time of the spacecraft array across the structure. These time scales can be compared to the time shifts used and the time shift which produces the most regular configuration results in the optimal estimates of current density. Moreover, for the regions containing FACs, where the main current component is nearly perpendicular to the Swarm AC orbits (by operational design), it is natural to time shift the Swarm A and C positions. This produces configurations which relate most closely to the L2 FAC data product. Nevertheless, the estimate of div(**B**)/curl(**B**) was found to be relatively independent of the specific choice of 4-spacecraft configuration.

This chapter has also described how these methods have been applied to close magnetic conjunction between the Swarm and Cluster spacecraft arrays, where three Swarm spacecraft are flying in close formation and the Cluster configuration is elongated, but is well aligned to the background magnetic field direction. This data set therefore allows detailed, multi-spacecraft analysis of the magnetic field measurements at both the Swarm and Cluster locations. The combined data therefore identifies the spatial form of the large scale field-aligned current density at both low (~500 km altitude) Earth orbit and at higher altitudes of ~2.5 Re.

Acknowledgements We thank the ESA Swarm project for provision of the L1 data (https://earth.esa.int/web/guest/swarm/data-access) used here. This work is supported by the NSFC grants 41174141, 41431071, 40904042, 973 program 2011C B811404 and by NERC grants NE/P016863/1, NE/H004076/1. MWD is partly supported by STFC in-house research grant ST/M001083/1. YYY is sponsored by China State Scholarship Fund (201404910456) to visit at RAL. The authors thank the International Space Science Institute in Bern, Switzerland, for supporting the ISSI Working Group: "Multi-Satellite Analysis Tools, Ionosphere", from which this chapter resulted. The Editors thank Qinghe Zhang for his assistance in evaluating this chapter. Thanks to Colin Forsyth for early discussions.

References

Anderson, B.J., K. Takahashi, and B.A. Toth. 2000. Sensing Global Birkeland currents with Iridium engineering magnetometer data. *Geophysical Reseach Letters* 27: 4045–4048.

Balogh, A., C.M. Carr, M.H. Acuña, M.W. Dunlop, T.J. Beek, P. Brown, K.-H. Fornacon, E. Georgescu, K.-H. Glassmeier, J. Harris, G. Musmann, T. Oddy, and K. Schwingenschuh. 2001. The cluster magnetic field investigation: Overview of in-flight performance and initial results. *Ann. Geo Cluster special issue* 19: 1207–1218.

Burch, J.L., T.E. Moore, R.B. Torbert, and B.L. Giles. 2016. Magnetospheric multiscale overview and science objectives. *Space Science Reviews* 5: 1–17. https://doi.org/10.1007/s11214-015-0164-9.

Chanteur, G. 1998. Spatial interpolation for four spacecraft: Theory, in analysis methods for multispacecraft data', *ISSI Science Report*, 395–418, SR-001, Kluwer Academic Pub.

Cheng, Z.W., J.C. Zhang, J.K. Shi, L.M. Kistler, M. Dunlop, I. Dandouras, and A. Fazakerley. 2016. The particle carriers of field-aligned currents in the Earth's magnetotail during a substorm. *Journal of Geophysical Research: Space Physics* 121: 3058–3068. https://doi.org/10.1002/2015JA022071.

Cao, J.-B., C. Yan, M. Dunlop, H. Reme, I. Dandouras, T. Zhang, D. Yang, A. Moiseyev, S.I. Solovyev, Z.Q. Wang, A. Leonoviche, N. Zolotukhina, and V. Mishin. 2010. Geomagnetic signatures of current wedge produced by fast flows in a plasma sheet. *Journal Geophysical Research* 115: A08205. https://doi.org/10.1029/2009JA014891.

Dandouras, I., S. Rochel-Grimald, C. Vallat, M. Dunlop. 2018. Terrestrial ring current: a review of cluster results based on the curlometer. In *Electric currents in geospace and beyond* (eds. A. Keiling, O. Marghitu, and M. Wheatland), John Wiley & Sons, Inc, Hoboken, N.J., *AGU books*. https://doi.org/10.1002/9781119324522.ch8.

De Keyser, J., F. Darrouzet, M.W. Dunlop, and P.M.E. Decreau. 2007. Least-squares gradient calculation from multi-point observations of scalar and vector fields: Methodology and applications with Cluster in the plasmasphere. *Ann Geo's* 25: 971–987. https://doi.org/10.5194/angeo-25-971-2007.

Dunlop, M.W., D.J. Southwood, K.-H. Glassmeier, and F.M. Neubauer. 1988. Analysis of multipoint magnetometer data. *Advances in Space Research* 8: 273–277. https://doi.org/10.1016/0273-1177(88)90141-X.

Dunlop, M.W., et al. 2002. Four-point Cluster application of magnetic field analysis tools: The Curlometer. *Journal Geophysical Research* 107 (A11): 1384. https://doi.org/10.1029/2001JA005088.

Dunlop, M.W., and A. Balogh. 2005. Magnetopause current as seen by Cluster. *Annales Geophysicae* 23 (3): 901–907. https://doi.org/10.5194/angeo-23-901-2005.

Dunlop, M.W., and J.P. Eastwood. 2008. The Curlometer and other gradient based methods. In *Multi-spacecraft analysis methods revisited'*, eds. G. Paschmann and P.W. Daly, *ISSI Science Report*, **SR-008,** pp. 17–21, Kluwer Academic Pub.

Dunlop, M.W., Y.-Y. Yang, J-Y. Yang, H. Lühr, C. Shen, N. Olsen, Q.-H. Zhang, Y.V. Bogdanova, J.-B. Cao, P. Ritter, K. Kauristie, A. Masson and R. Haagmans. 2015. Multi-spacecraft current estimates at Swarm. *Journal of Geophysical Research* 120. https://doi.org/10.1002/2015ja021707.

Dunlop, M.W., J.-Y. Yang, Y.-Y. Yang, C. Xiong, H. Lühr, Y.V. Bogdanova, C. Shen, N. Olsen, Q.-H. Zhang, J.-B. Cao, H.-S. Fu, W.-L. Liu, C.M. Carr, P. Ritter, A. Masson, and R. Haagmans. 2015b. Simultaneous field-aligned currents at Swarm and Cluster satellites. *Geophysical Reseach Letters* 42: 3683–3691. https://doi.org/10.1002/2015GL063738.

Dunlop, M.W., S. Haaland, C.P. Escoubet, and X.-C. Dong. 2016. Commentary on accessing 3-D currents in space: Experiences from Cluster. *Journal Geophysical Research* 121: 7881–7886. https://doi.org/10.1002/2016JA022668.

Dunlop, M.W., S. Haaland, X-C. Dong, H. Middleton, P. Escoubet, Y-Y. Yang, Q-H Zhang, J-K. Shi and C.T. Russell. 2008. Multi-point analysis of current structures and applications: Curlometer technique. In *Electric currents in geospace and beyond* (ed. A. Keiling, O. Marghitu, and M. Wheatland), John Wiley & Sons, Inc, Hoboken, N.J., *AGU books*, 2018, https://doi.org/10.1002/9781119324522.ch4.

Escoubet, C.P., M. Fehringer, and M. Goldstein. 2001. Introduction: the Cluster mission. *Annales Geophysicae* 19: 1197–1200. https://doi.org/10.5194/angeo-19-1197-2001.

Eastwood, J.P., A. Balogh, M.W. Dunlop, and C.W. Smith. 2002. Cluster observations of the heliospheric current sheet and an associated magnetic flux rope and comparisons with ACE. *Journal Geophysical Research* 107 (A11): 1365. https://doi.org/10.1029/2001JA009158.

Forsyth, C. et al. 2008. Observed tail current systems associated with bursty bulk flows and auroral streamers during a period of multiple substorms. *Annales Geophysicae* 26. https://doi.org/10.5194/angeo-26-167-2008.

Friis-Christensen, E., H. Lühr, D. Knudsen, and R. Haagmans. 2008. Swarm—An earth observation mission investigating geospace. *Advances in Space Research* 41: 210–216. https://doi.org/10.1016/j.asr.2006.10.008.

Gjerloev, J.W., and R.A. Hoffman. 2002. Currents in auroral substorms. *Journal of Geophysical Research* 107 (A8). https://doi.org/10.1029/2001ja000194.

Grimald, S., I. Dandouras, P. Robert, and E. Lucek. 2012. Study of the applicability of the curlometer technique with the four Cluster spacecraft in regions close to Earth. *Annales Geophysicae* 30: 597–611. https://doi.org/10.5194/angeo-30-597-2012.

Haaland, S., B.Ö. Sonnerup, M. Dunlop, E. Georgescu, G. Paschmann, B. Klecker, and A. Vaivads. 2004. Orientation and motion of a discontinuity from Cluster curlometer capability: Minimum variance of current density. *Geophysical Research Letters* 31: L10804–L10804. https://doi.org/10.1029/2004GL020001.

Hamrin, M. et al. 2008. Ann. Geo.

Harvey, C.C. 1998. Spatial gradients and the volumetric tensor. In *Analysis methods for multispacecraft data', ISSI Science Report*, 395–418, SR-001, Kluwer Academic Pub.

Henderson, P.D. et al. 2008. The relationship between j and Pe in the magnetotail plasma sheet: Cluster observations. *Journal of Geophysical Research* 113: A07S31. https://doi.org/10.1029/2007ja012697, https://doi.org/10.1029/2007ja012697.

Kivelson, M., and C.T. Russell 1995. *Introduction to space physics*. Cambridge University Press.

Liang, J., and W.W. Liu. 2007. A MHD mechanism for the generation of the meridional current system during substorm expansion phase. *Journal Geophysical Research* 112: A09208. https://doi.org/10.1029/2007JA012303.

Lühr, H., J. Park, J.W. Gjerloev, J. Rauberg, I. Michaelis, J.M.G. Merayo, and P. Brauer. 2015. Field-aligned currents' scale analysis performed with the Swarm constellation. *Geophysical Research Letters* 42. https://doi.org/10.1002/2014gl062453.

Marchaudon, A., J.-C. Cerisier, M.W. Dunlop, F. Pitout, J.-M. Bosqued, and A.N. Fazakerley. 2009. Shape, size, velocity and field-aligned currents of dayside plasma injections: a multi-altitude study. *Annales Geophysicae* 27: 1251–1266. https://doi.org/10.5194/angeo-27-1251-2009.

McPherron, R.L., C.T. Russell, and M.P. Aubry. 1973. Satellite studies of magnetospheric substorms on August 15, 1968: 9. Phenomenological model for substorms. *Journal of Geophysical Research* 78 (16): 3131–3149. https://doi.org/10.1029/ja078i016p03131.

Middleton, H.R. and A. Masson. 2016. The Curlometer Technique: a beginner's guide, ESA Technical Note ESDC-CSA_TN_0001, http://www.cosmos.esa.int/web/csa/multi-spacecraft.

Nakamura, R., W. Baumjohann, M. Fujimoto, Y. Asano, A. Runov, C. Owen, A. Fazakerley, B. Klecker, H. Rème, and E. Lucek. 2008. Cluster observations of an ion-scale current sheet in the magnetotail under the presence of a guide field. *Journal Geophysical Research* 1978–2012: 113. https://doi.org/10.1029/2007JA012760.

Narita, Y., R. Nakamura, and W. Baumjohann. 2013. Cluster as current sheet surveyor in the magnetotail. *Annales Geophysicae* 31: 1605–1610. https://doi.org/10.5194/angeo-31-1605-2013.

Olsen, N., H. Lühr, C.C. Finlay, T.J. Sabaka, I. Michaelis, J. Rauberg, and L. Tøffner-Clausen. 2014. The CHAOS-4 Geomagnetic Field Model. *Geophysical Journal International* 197: 815–827. https://doi.org/10.1093/gji/ggu033.

Panov, E., J. Büchner, M. Fränz, A. Korth, Y. Khotyaintsev, B. Nikutowski, S. Savin, K.-H. Fornaçon, I. Dandouras, and H. Reme. 2006. Cluster spacecraft observation of a thin current sheet at the Earth's magnetopause. *Advances in Space Research* 37 (7): 1363–1372. https://doi.org/10.1029/2006GL026556.

Petrukovich, A., et al. 2015. Current sheets in the earth magnetotail: Plasma and magnetic field structure with cluster project observations. *Space Science Reviews* 188: 311–337. https://doi.org/10.1007/s11214-014-0126-7.

Reigber, C., H. Lühr, and P. Schwintzer. 2002. CHAMP mission status. *Advances in Space Research* 30 (2): 129–134.

Ritter, P., H. Lühr, A. Viljanen, O. Amm, A. Pulkinnen, and I. Sillanpää. 2004. Ionospheric currents estimated simultaneously from CHAMP satellite and IMAGE ground-based magnetic field measurements: A statistical study at auroral latitudes. *Annales Geophysicae* 22: 417–430.

Ritter, P., and H. Lühr. 2006. Curl-B technique applied to Swarm constellation for determining field-aligned currents. *Earth Planets Space* 58: 463–476.

Ritter, P., H. Lühr, and J. Rauberg. 2013. Determining field-aligned currents with the Swarm constellation mission. *Earth Planets Space* 65: 1285–1294. https://doi.org/10.5047/eps.2013.09.006.

Robert, P., M.W. Dunlop, A. Roux, and G. Chanteur. 1998. Accuracy of current density determination. In *Analysis methods for multispacecraft data', ISSI Science Report*, 395–418, SR-001, Kluwer Academic Pub.

Roux, A., P. Robert, D. Fontaine, O. LeContel, P. Canu, and P. Louarn. 2015. What is the nature of magnetosheath FTEs? *JGR* 120 (6): 4576–4595. https://doi.org/10.1002/2015JA020983.

Runov, et al. 2006. Local structure of the magnetotail current sheet: 2001 Cluster observations. *Annales Geophysicae* 24: 247–262. https://doi.org/10.5194/angeo-24-247-2006.

Russell, C.T., J.G. Luhmann, and R.J. Strangeway. 2016. *Space physics: An introduction*. 978-1107098824, Cambridge University Press.

Shen, C., X. Li, M. Dunlop, Q.Q. Shi, Z.X. Liu, E. Lucek, and Z.Q. Chen. 2007. Magnetic field rotation analysis and the applications. *Journal Geophysical Research* 112: A06211. https://doi.org/10.1029/2005JA011584.

Shen, C., and M.W. Dunlop, in Multi-spacecraft analysis methods revisited', eds. G. Paschmann and P.W. Daly, ISSI Science Report, **SR-008**, 27–30, Kluwer Academic Pub.

Shen, C. et al. 2008. Flattened current sheet and its evolution in substorms. *Journal of Geophysical Research* 113, A07S21. https://doi.org/10.1029/2007ja012812.

Shen, C., J. Rong, M. Dunlop, Y.H. Ma, G. Zeng, and Z.X. Liu. 2012a. Spatial gradients from irregular, multiple-point spacecraft configurations. *Journal Geophysical Research* 117: A11207. https://doi.org/10.1029/2012JA018075.

Shen, C., Z.J. Rong, and M. Dunlop. 2012b. Determining the full magnetic field gradient from two spacecraft measurements under special constraints. *Journal Geophysical Research* 117: A10217. https://doi.org/10.1029/2012JA018063.

Shen, C., et al. 2014. Direct calculation of the ring current distribution and magnetic structure seen by Cluster during geomagnetic storms. *Journal Geophysical Research* 119: 2458–2465. https://doi.org/10.1002/2013JA019460.

Shi, J.K. et al. 2010. South-north asymmetry of field-aligned currents in the magnetotail observed by Cluster. *Journal of Geophysical Research* 115. https://doi.org/10.1029/2009ja014446.

Shi, J.-K., J. Guo, M. Dunlop, T. Zhang, Z.X. Liu, E. Lucek, A. Fazakereley, H. Reme, and I. Dandouras. 2011. Inter-hemispheric asymmetry of the dependence of cusp location on dipole tilt: Cluster observations. *Journal Geophysical Research* 30 (21–26): 2012. https://doi.org/10.5194/angeo-30-21-2012.

Shiokawa, K., et al. 1998. High-speed ion flow, substorm current wedge and multiple Pi2 pulsations. *Journal Geophysical Research* 103 (A3): 4491–4507. https://doi.org/10.1029/97JA01680.

Shore, R.M., K.A. Whaler, S. Macmillan, C. Beggan, N. Olsen, T. Spain, and A. Aruliah. 2013. Ionospheric midlatitude electric current density inferred from multiple magnetic satellites. *Journal Geophysical Research* 118: 5813–5829. https://doi.org/10.1002/jgra.50491.

Slavin, J.A., G. Le, R.J. Strangeway, Y. Wang, S.A. Boardsen, M.B. Moldwin, and H.E. Spence. 2008. Space technology 5 multipoint measurements of near-Earth magnetic fields: Initial results. *Geophysical Reseach Letters* 35: L02107. https://doi.org/10.1029/2007GL031728.

Tsyganenko, N.A. 1989. A magnetospheric magnetic field model with a warped tail current sheet. *Planetary and Space Science* 37: 5–20.

Vallat, C., et al. 2005. First current density measurements in the ring current region using simultaneous multi-spacecraft CLUSTER-FGM data. *Annales Geophysicae* 23: 1849–1865. https://doi.org/10.5194/angeo-23-1849-2005.

Vogt, J., G. Paschmann, and G. Chanteur. 2008. Reciprocal vectors. In *Multi-Spacecraft analysis methods revisited,* eds. G. Paschmann and P.W. Daly, ISSI Science Report, **SR-008**, 27–30, Kluwer Academic Pub.

Vogt, J., A. Albert, and O. Marghitu. 2009. Analysis of three-spacecraft data using planar reciprocal vectors: Methodological framework and spatial gradient estimation. *Annales Geophysicae* 27: 3249–3273. https://doi.org/10.5194/angeo-27-3249-2009.

Vogt, J., E. Sorbalo, M. He, and A. Blagau. 2013. Gradient estimation using configurations of two or three spacecraft. *Annales Geophysicae* 31: 1913–1927. http://www.ann-geophys.net/31/1913/2013/.

Wang, H., S.Y. Ma, H. Lühr, Z.X. Liu, Z.Y. Pu, C.P. Escoubet, H.U. Frey, H. Réme, and P. Ritter. 2006. Global manifestations of a substorm onset observed by a multi-satellite and ground station network. *Annales Geophysicae* 24: 3491–3496. https://doi.org/10.5194/angeo-24-3491-2006.

Xiao et al. 2004. Inferring of flux rope orientation with the minimum variance analysis technique. *Journal of Geophysical Research* 109. https://doi.org/10.1029/2004ja010594.

Xiong, C., H. Lühr, H. Wang, and M.G. Johnsen. 2014. Determining the boundaries of the auroral oval from CHAMP field-aligned currents signatures—Part 1. *Annales Geophysicae* 32 (609–622): 2014. https://doi.org/10.5194/angeo-32-609-2014.

Xiong, C., and H. Lühr. 2014. An empirical model of the auroral oval derived from CHAMP field-aligned current signatures—Part 2. *Annales Geophysicae* 32 (623–631): 2014. https://doi.org/10.5194/angeo-32-623-2014.

Yang, J.-Y., M. Dunlop, C. Xiong, H. Lühr, Y-Y. Yang, J.-B. Cao, J Wild, L-Y Li, Y-D Ma, H-S Fu, W-L. Liu, and P. Ritter. 2018. Statistical correlations of field-aligned currents measured by Swarm. submitted to *Journal of Geophysical Research.*

Yang, Y.-Y., C. Shen, M. Dunlop, Z.-J. Rong, X. Li, V. Angelopolous, Z.-Q. Chen, G.-Q. Yan and Y. Ji. 2016. Storm time current distribution in the inner magnetospheric equator: THEMIS observations. *Journal of Geophysical Research* 120. https://doi.org/10.1002/2015ja022145.

Zhang, Q.-H., M.W. Dunlop, M. Lockwood, R. Holme, Y. Kamide, W. Baumjohann, R.-Y. Liu, H.-G. Yang, E.E. Woodfield, H.-Q. Hu, B.-C. Zhang, and S.-L. Liu. 2011. The distribution of the ring current: Cluster observations. *Ann. Geo.* 29: 1655–1662. https://doi.org/10.5194/angeo-29-1655-2011.

Chapter 6
Applying the Dual-Spacecraft Approach to the Swarm Constellation for Deriving Radial Current Density

Hermann Lühr, Patricia Ritter, Guram Kervalishvili and Jan Rauberg

Abstract One of the Swarm prime mission goals is the estimation of ionospheric currents. Of particular interest in this context are field-aligned currents (FACs). In order to improve our ability of determining FACs, two of the Swarm spacecraft are orbiting side-by-side separated only by 1.4° in longitude. This close-formation flight enables the application of Ampère's integral law to magnetic field measurements for estimating radial currents. From experience gained in space we can state that most reliable results are obtained in the auroral region. Here the spacing of the measurement quad and the size of current structures match best. In the vicinity of the poles, close to the orbital crossovers, spacecraft separations become too small for reliable gradient measurements. At low latitudes the separation becomes largest (~150 km). Here certain FAC features, e.g. associated with plasma instabilities and disturbances exist, which cannot be analysed reliably with the dual-SC approach. However, mid-latitude large-scale currents like the inter-hemispheric FACs can be recorded reliably by the Swarm mission. Besides presenting some measurement examples special emphasis is put on the discussion of underlying assumptions and on the limitations of the approach.

6.1 Introduction

Measurements of current distributions in space are normally based on the interpretation of magnetic fields they generate. In general, the magnetic field caused by an electric current can be predicted everywhere. But reversely, from single-satellite magnetic field measurements the current distribution cannot be deduced uniquely. For obtaining still useful results, for example the current geometry can serve as *a priori* information. Such assumptions are commonly based on physical considerations. In case magnetic field measurements from more points are available, well distributed in space, fewer assumptions are required. This argument has been an

H. Lühr (✉) · P. Ritter · G. Kervalishvili · J. Rauberg
GFZ, German Research Centre for Geosciences, Potsdam, Germany
e-mail: hluehr@gfz-potsdam.de

M. W. Dunlop and H. Lühr (eds.), *Ionospheric Multi-Spacecraft Analysis Tools*, ISSI Scientific Report Series 17,
https://doi.org/10.1007/978-3-030-26732-2_6

important motivation for deploying multiple-spacecraft constellations (e.g. 4 Cluster, 3 Space Technology 5 (ST-5), 3 Swarm, 4 Magnetospheric Multi-Scale (MMS)) flying in close formation.

For a local determination of the full current density vector simultaneous magnetic field measurements from at least four well-spaced points are needed. Realising this approach was one of the prime goals of the Cluster mission. The technique for estimating currents from the four Cluster spacecraft was developed and described before the mission (e.g. Dunlop et al. 1988), and later a number of studies made use of the so-called curlometer technique (e.g. Dunlop et al. 2002; Vallat et al. 2005). The same approach is now applied to MMS data (e.g. Zhao et al. 2016). Experience has shown that this 4-point method also has its limits (e.g. Dunlop et al. 2016). In particular it depends on a suitable spacing between the spacecraft. Most favourable is a regular tetrahedron. But the spacecraft formation evolves over an orbit, thus can only be optimal within a certain region. An important assumption of the method is that the magnetic field changes only linearly between the measurement points. A significant contribution to field curvature can come from the Earth's main field. Subtracting a suitable main field model from the magnetic field readings before interpretation will reduce that problem. Furthermore, it is requires that the inter-spacecraft separations have to be significantly smaller than the dimensions of the dominant current system. An attempt in that direction is realised by the MMS mission with spacecraft distances of only some tens of kilometres (e.g. Eastwood et al. 2016). However, in such cases demands are high on accurate positioning of the spacecraft and on the precision of the measurements. Estimates of typical FAC spatial scales have been derived from the three ST-5 spacecraft taking measurements at varying separations (e.g. Slavin et al. 2008).

When coordinated magnetic field measurements from less than four satellites are available assumptions on the current characteristics have to be made (e.g. Vogt et al. 2013). In the topside ionosphere, where the Swarm satellites orbit, the highly anisotropic conductivity distribution is an important constrain. The field-aligned conductivity is several orders of magnitudes higher than the transverse one. This implies that any current detected at that altitude is presumably field-aligned. Another assumption concerns the stationarity of current structures, including both their motion and temporal evolution. This requirement has to be related again to the orbital dynamics of the Swarm spacecraft. Any motion of a current sheet is typically much slower than the satellite velocity of 7.6 km/s. Similar care has to be taken when combining measurements in a solution from slightly different times. Larger-scale (>150 km) field-aligned current (FAC) structures, however, have been found to be stable typically for longer than one minute (e.g. Gjerloev et al. 2011; Lühr et al. 2015a).

Taking these characteristics into account, reliable FAC density estimates can be obtained in the ionosphere, especially at high latitudes, by a pair of spacecraft flying side-by-side like the Swarm A and C constellation. In the section to follow, we first provide a general introduction into the estimation of electric current density from magnetic field measurements. A detailed description of FAC determination from the lower pair of Swarm satellites is given in Sect. 6.3. For a selected number of cases we present examples of FAC observations in Sect. 6.4. The subsequent section provides a detailed discussion of uncertainties inherent to the actually implemented dual-SC method for FAC estimates. Finally we summarise the main points and draw conclusions.

6.2 Current Estimates from Satellites

The typical approach for deriving current estimates from magnetic field measurements is to make use of Ampère's law, the curl-B relation

$$curl\, \mathbf{B} = \mu_0 \mathbf{j} \tag{6.1}$$

where $\mathbf{j}$ is the current density, μ_0 is the permeability of free space and $\mathbf{B}$ the magnetic field generated by the current. For the vertical current component, j_Z, Eq. (6.1) reduces to

$$jz = \frac{1}{\mu_0}\left(\frac{\partial By}{\partial x} - \frac{\partial Bx}{\partial y}\right) \tag{6.2}$$

where Bx and By are the horizontal magnetic field components in northward and eastward directions of a local Cartesian coordinate system, respectively. At auroral latitudes, beyond 60° magnetic latitude (MLat), the field lines are almost vertical exhibiting inclination angles larger than 75°. Therefore j_Z represents quite closely the field-aligned current (FAC) density. FACs can be derived more precisely when Mean-Field-Aligned (MFA) coordinates are used in Eq. (6.2). In the MFA frame the z component is aligned with the ambient magnetic field, the y component, pointing eastward, lies within the horizontal plane and is perpendicular to the mean field, and the x component, pointing outward/upward, completes the right-handed triad.

6.2.1 Single-Satellite Field-Aligned Current Estimate

In case magnetic field recordings are available only from a single satellite, data sampled along the track have to be used for estimating j_Z. This requires further simplifications of Eq. (6.2) (e.g. Lühr et al. 1996).

$$jz = \frac{1}{\mu_0}\frac{\Delta By'}{v\Delta t} \tag{6.3}$$

Here By' is the horizontal component perpendicular to the FAC sheet normal, v is the orbital velocity component aligned with the sheet normal and Δt is the time difference between adjacent measurements. Spatial field gradients required for Eq. (6.2) are derived from subsequent readings of By' assuming stationarity of the currents.

Furthermore, FACs are believed to be organised in elongated sheets. The orientation of the current sheet has to be determined by other techniques (e.g. minimum variance analysis, Sonnerup and Cahill 1967). Often it is simply assumed that the satellite crosses the current sheet at a right angle. Any deviation from this geometric assumption causes an underestimation of current density.

In summary, the reliability of FAC density estimates from single-satellite data depends on a number of important assumptions:

1. Stationarity of current density during the crossing
2. Organisation of FACs in elongated sheets
3. Knowledge of the current sheet orientation.

6.2.2 Multi-satellite Current Estimates

Significantly improved current density estimates can be achieved when simultaneous magnetic field measurements at well-spaced distances are available. In the case of multiple data points Ampère's law in integral form is preferably used

$$j = \frac{1}{\mu_0 A} \oint \vec{B} \cdot d\vec{\ell} \tag{6.4}$$

where dl is the path element along the closed contour, A is the encircled area, and j is the mean value of the current density component normal to the plane. Simultaneous measurements from at least three points are needed to calculate a ring integral. For this special case Eq. (6.4) can be written in discrete form

$$j = \frac{1}{2\mu_0 A}\left[\left(\vec{B}_1 + \vec{B}_2\right) \cdot \vec{l}(Q_1, Q_2) + \left(\vec{B}_2 + \vec{B}_3\right) \cdot \vec{l}(Q_2, Q_3) + \left(\vec{B}_3 + \vec{B}_1\right) \cdot \vec{l}(Q_3, Q_1)\right] \tag{6.5}$$

where $\boldsymbol{B}_n$ are the magnetic field vectors at the corners and $l(Qn,Qm)$ are the path elements connecting the points, see Fig. 6.1. The area A can be calculated as

$$A = \frac{1}{2}\left|\vec{l}_a \times \vec{l}_b\right|$$

where l_a and l_b are any two legs $l(Qn, Qm)$ of the triangle. The obtained j represents the mean current density component normal to the encircled plane, and it is assigned in time and space to the barycentre of the triangle.

In cases where simultaneous measurements at four points in space are available (like Cluster or MMS) three such triangles can be formed. In principle this allows to derive current density components in three independent directions and thus to build the complete current density vector. The feasibility of this current estimation

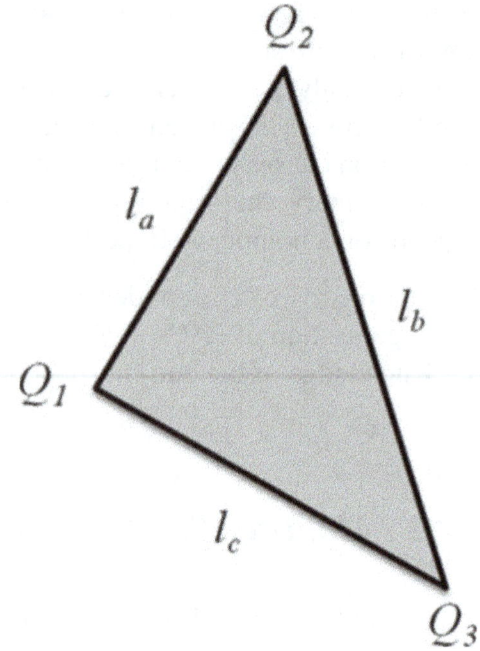

Fig. 6.1 Magnetic field measurements at the three corners of the triangle are used for estimating the mean current density flowing through encircled area

technique was demonstrated successfully in several cases (e.g. Dunlop and Balogh 2005; Zhao et al. 2016). Further details of the so-called 'curlometer technique' can be found e.g. in Dunlop et al. (2002) and references therein.

Occasionally the three Swarm spacecraft pass the auroral region in a close formation. For some of these incidences the 3-point current analysis could be applied, see Dunlop et al. (2015) and Chapter 5. This analysis has also been expanded by including time-shifted measurement points for building up a multi-point virtual constellation, allowing to derive a complete set of current components (Dunlop et al. 2015). Unfortunately, only very few cases are so far available for such an analysis. The orbital plane of Swarm B drifted quickly away from that of the Swarm A & C pair.

However, significant improvements over single-satellite current estimates can already be achieved when magnetic field measurements from a pair of satellites, flying side-by-side, are considered. In general, such a formation is suitable for deriving *radial* currents. In the topside ionosphere, where Swarm satellites are orbiting, the direction of currents is closely controlled by the orientation of the geomagnetic field because of the high electric conductivity along field lines. For that reason the dual-satellite configuration is also well suited for estimating FAC density at high and middle latitudes.

The idea of deriving radial current density estimates from the lower pair of Swarm spacecraft has been intensively tested in preparation for the mission as part of an ESA-sponsored study. With the help of a closed-loop model simulation the reliability of the dual-spacecraft (dual-SC) radial current approach could be demonstrated. The field-aligned currents predicted by a magnetohydrodynamic (MHD) model were compared with the current estimates derived from related magnetic fields sampled along virtual orbits of Swarm. The improvement in reproducing the input FACs by the dual-SC technique as compares to the single-satellite approach was confirmed by this study. More details about this modelling study can be found in the Final Report (Vennerstrøm et al. 2005) or in Ritter and Lühr (2006).

Based on these promising results it was decided by ESA to compute the radial current density routinely from the magnetic field measurements of the two Swarm spacecraft flying side-by-side and to offer the results as a standard Swarm Level 2 product.

6.3 The Swarm Dual-SC Current Estimate Approach

The three Swarm spacecraft (Friis-Christensen et al. 2006) where launched on 22 November 2013 by a single rocket into a high-inclination (~87.5°) orbit. After the commissioning phase was completed an orbit manoeuvre campaign followed for setting up the final constellation. Since 17 April 2014 Swarm A and C are flying side-by-side separated by about 1.4° in longitude at the equator at an initial altitude of about 460 km. Swarm B is cruising somewhat higher at about 510 km and at a larger inclination (~88°). For that reason the orbital plane of Swarm B precesses slightly slower than that of the lower spacecraft pair, gradually building up a local time difference between the orbits.

With Swarm A and C, cruising at the same altitude, there exists the risk of collision at the orbital crossover points near the poles. Therefore the spacecraft are slightly phased along the orbit. Their equator crossing times differ within the range 5-10 s. This time difference has to be taken into account when interpreting the zonal gradient.

Our aim is to derive the radial current density by integrating the magnetic field around a rather regular quad, as depicted in Fig. 6.2. The area does not need to be strictly rectangular, but a large area-to-circumference ration is weighting down the influence of data noise and biases. As a first step, our approach determines the exact times of passing the orbital crossovers of the two spacecraft at high latitudes. From the time difference the orbital phasing (~7 s) is derived. Swarm A is used as reference satellite, and the readings of Swarm C are shifted in time by the phasing. In this way a synchronous side-by-side configuration is achieved. The synchronisation is updated on every passage of the North or South Pole and kept constant over half an orbit. The along-track field gradients, from Point 1 to Point 2 or from Point 4 to Point 3, are obtained by considering measurements 5 s later. This time step is equivalent to a distance of 38 km (7.6 km/s times 5 s). In cross-track direction the distance, equivalent to 1.4° in longitude, varies from 155 km at the equator to zero at the crossover points. An approximately quadratic shape of the quad is achieved around latitudes of 75°, which occurs within the auroral region. Here we can obtain the most accurate results. The position and time stamp of the derived mean radial current density is assigned to the barycentre of the quad (see Fig. 6.2). Here again Swarm A acts as reference.

The basis for the current density estimates are the observed magnetic field deflections, which are caused by the currents. For obtaining them we have to remove the

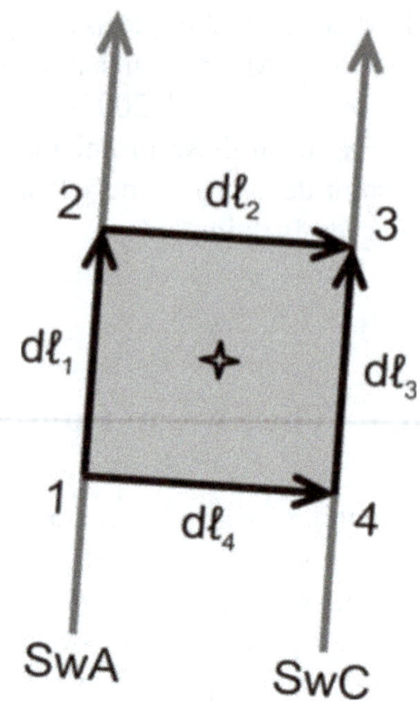

Fig. 6.2 Sketch of the quad of four measurement points used in the Swarm mission for calculating radial currents. The resulting mean current density is assigned to the barycentre (cross)

contributions of the core field, the crustal magnetisation and large-scale magnetic current systems from the original magnetic field readings:

$$B^{RES} = B^{OBS} - B^{COR} - B^{LIT} - B^{EXT} \tag{6.6}$$

where B^{OBS} represent the original readings, B^{COR} is the core field, B^{LIT} is the lithospheric field and B^{EXT} is caused by magnetospheric currents. All these three contributions are represented nowadays reliably by magnetic field models like CHAOS-6 (e.g. Finlay et al. (2016) and Chapter 12). In all subsequent analyses B^{RES} is used. For the sake of clarity we will omit the superfix (RES) in the following.

The radial current density is calculated with the help of Ampére's integral law, as shown in Eq. (6.4). Here we apply a discrete form for solving the ring integral

$$j_r = \frac{-1}{2\mu_0 A}\left[\left(\vec{B}_{AT1} + \vec{B}_{AT2}\right)\cdot d\vec{l}_1 + \left(\vec{B}_{CT2} + \vec{B}_{CT3}\right)\cdot d\vec{l}_2 - \left(\vec{B}_{AT3} + \vec{B}_{AT4}\right)\cdot d\vec{l}_3 - \left(\vec{B}_{CT4} + \vec{B}_{CT1}\right)\cdot d\vec{l}_4\right] \tag{6.7}$$

where B_{ATn} and B_{CTn} are the along-track and cross-track magnetic field components aligned with the respective connections between the quad points, dl_n are the corresponding path elements. The calculation of the area A will be introduced further down.

The magnetic field observations are given in the North-East-Center (NEC) frame. However, for solving Eq. (6.7) we need the field components aligned with the connecting lines between corner points. This requires defined rotations of the horizontal components.

$$B_{ATn} = B_{NECx} \cos\alpha_n + B_{NECy} \sin\alpha_n \quad B_{CTn} = B_{NECx} \cos\beta_n + B_{NECy} \sin\beta_n \tag{6.8}$$

The rotation angles α_n and β_n can be deduced from the position data of the satellites. Here we have to take into account that the spacecraft locations are generally given in Earth-fixed coordinates, but ionospheric currents are better organised in the local time (LT) frame. In order to account for this effect, we introduce an LT-related longitude, λ

$$\lambda = \varphi + (t/86400) * 360^\circ \tag{6.9}$$

where φ is the geographic longitude and t is the time in seconds of a day.

Figure 6.3 shows how the angles α_n and β_n between the path elements and the direction to the pole are defined. The calculation of these angles is based on spherical geometry. The basic equation is the *sine formula* for triangles on a sphere:

Fig. 6.3 Definition of angles α and ß between the four route elements and the connections to the geographic pole (modified after Ritter et al. 2013)

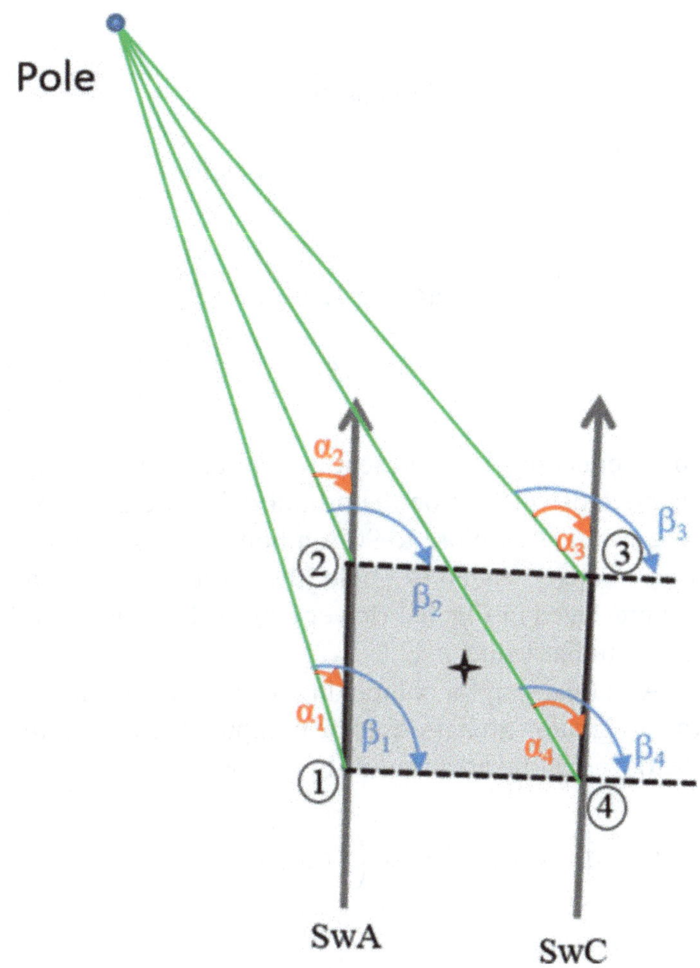

$$\frac{\sin(arc\gamma)}{\sin\gamma} = \frac{\sin(arc\delta)}{\sin\delta} \tag{6.10}$$

where *arc* γ and *arc* δ represent the sides of the triangle opposite to the respective angles γ and δ. For computing *arc* γ and *arc* δ we make use of the *great circle distances*. The angles α_n and β_n are then computed as:

$$\alpha_1 = \arcsin\left[\frac{\sin|\lambda_2 - \lambda_1| \cdot \sin\theta_2}{\sin[\arccos(\cos\theta_1\cos\theta_2 + \sin\theta_1\sin\theta_2\cos(\lambda_2 - \lambda_1))]}\right] \tag{6.11}$$

$$\alpha_2 = \arcsin\left[\frac{\sin\theta_1}{\sin\theta_2}\sin\alpha_1\right] \tag{6.12}$$

$$\alpha_4 = \arcsin\left[\frac{\sin|\lambda_3 - \lambda_4| \cdot \sin\theta_3}{\sin[\arccos(\cos\theta_3\cos\theta_4 + \sin\theta_3\sin\theta_4\cos(\lambda_3 - \lambda_4))]}\right] \tag{6.13}$$

$$\alpha_3 = \arcsin\left[\frac{\sin\theta_4}{\sin\theta_3}\sin\alpha_4\right] \tag{6.14}$$

$$\beta_1 = \arcsin\left[\frac{\sin|\lambda_4 - \lambda_1| \cdot \sin\theta_4}{\sin[\arccos(\cos\theta_1\cos\theta_4 + \sin\theta_1\sin\theta_4\cos(\lambda_4 - \lambda_1))]}\right] \tag{6.15}$$

$$\beta_4 = \arcsin\left[\frac{\sin\theta_1}{\sin\theta_4}\sin\beta_1\right] \tag{6.16}$$

$$\beta_2 = \arcsin\left[\frac{\sin|\lambda_3 - \lambda_2| \cdot \sin\theta_3}{\sin[\arccos(\cos\theta_3\cos\theta_2 + \sin\theta_3\sin\theta_2\cos(\lambda_2 - \lambda_3))]}\right] \tag{6.17}$$

$$\beta_3 = \arcsin\left[\frac{\sin\theta_2}{\sin\theta_3}\sin\beta_2\right] \tag{6.18}$$

In all equations θ_n represent the co-latitude at the quad corner points. The angles α_n are close to zero on the orbital upleg arc and close to 180° on the downleg arc, whereas β_n stay close to 90° on both sides of the orbit. At polar regions the angles vary strongly. As an example, the evolution of the angles α_1 and β_1 over an orbit are presented in Fig. 6.4 (lower panel). The angles at the other corner points exhibit quite similar values.

As a final step for completing the radial current calculation according to Eq. (6.7) the integration area needs to be computed. To be more general we allow for a certain distortion of the area framed by the path elements.

$$A = \frac{1}{2}\left[(dl_2 + dl_4)\left(dl_1\sin\left(\frac{1}{2}(\beta_1 - \alpha_1 + \beta_2 - \alpha_2)\right) + dl_3\sin\left(\frac{1}{2}(\beta_3 - \alpha_3 + \beta_4 - \alpha_4)\right)\right)\right] \tag{6.19}$$

When approaching the poles the cross-track distances, dl_2 and dl_4, become progressively smaller, see Fig. 6.4 (top panel). As a consequence, the vanishing area, A,

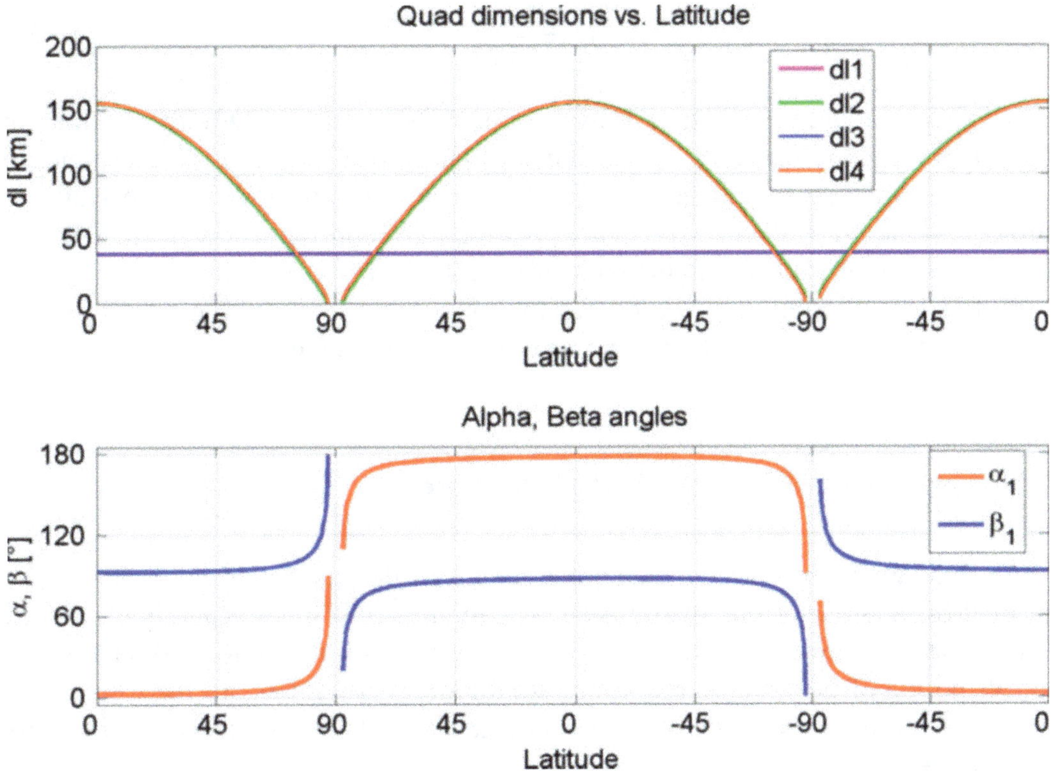

Fig. 6.4 Top: Variation of route elements over an orbit. The along-track elements, *dl1, dl3* stay constant whereas the cross-track ones, *dl2, dl4* become progressively smaller towards the poles. Due to an orbital inclination of less than 90°, the close vicinity of the poles is not sampled. Bottom: Variation of the angles α and ß along the orbit. Here $\alpha 1$ and $\beta 1$ have been chosen as examples for the four angles (modified after Ritter et al. 2013)

make the current estimates unreliable. For that reason no radial currents are calculated from Swarm data at latitudes beyond 86°.

Since the current estimates are based on observations at only four discrete points, the method needs to consider aliasing effects. According to the sample theorem the signal should not contain spatial variations with wavelength shorter than twice the distance between any two corner points. In an attempt to satisfy this requirement we have low-pass filtered the magnetic field residuals with a -3 dB cutoff frequency of 50 mHz. This corresponds to an along-track wavelength of about 150 km. Our assumption is that spatial structures have comparable wavelength in along-track and cross-track directions. With this approach the spatial aliasing effect should be avoided, at least at auroral latitudes.

In particular at auroral latitudes field-aligned currents are of great interest because they transfer energy and momentum from the magnetosphere into the ionosphere-thermosphere system. We derive FAC density estimates by mapping the radial current density, j_r, onto the field direction. This is done by considering the magnetic field inclination, I:

$$j_{\parallel} = \frac{-j_r}{\sin I} \tag{6.20}$$

In order to avoid unrealistic FAC values near the equator, where the inclination angle approaches zero, no FAC values are calculated for $|I| < 30°$. This corresponds to a range of $\pm 15°$ in magnetic latitude MLat. Vertical currents, however, are available through all latitudes. Please note that a positive sign represents downward FACs in the north and upward FACs in the southern hemisphere.

A detailed description of the processing algorithms used for the actual Swarm Level 2 product "FACxTMS_2F" can be found in Ritter et al. (2013).

6.4 Examples of Swarm FAC Estimates

ESA's standard Swarm Level 2 data processing provides current densities derived by the dual-SC approach from Swarm A and C both in radial direction and field-aligned at 1 s cadence. However, due to the low-pass filtering of magnetic field data with a cut-off period of 20 s only large-scale radial current structures are recovered with horizontal wavelengths larger than 150 km when considering the spacecraft velocity of 7.5 km/s.

For completeness, single satellite FAC and radial current densities from all three spacecraft are also calculated and made available. In this case the 1 Hz data are not filtered reflecting all the spatial scales of FAC structures. However, it has to be kept in mind that the reliability of small-scale FAC estimates is questionable because the temporal variability of small-scale FACs may be quite significant and violate one of the basic assumptions in Sect. 6.2.1, as was shown by Lühr et al. (2015a).

Subsequently we will present a few examples of FAC estimates and compare single and dual-SC results. In order to make the curves comparable, the single-satellite results shown here were low-pass filtered in the same way as the dual-SC ones. Generally we find good agreement between the individual FAC estimates from the two spacecraft, Swarm A and C, and also with the dual-SC results. In particular, this is true for the auroral oval region in the northern hemisphere. Figure 6.5 shows an example from a polar passage, crossing first the night-time auroral oval before flying into the pre-noon sector. The current sheet signatures observed are typical for these areas. The large amplitudes at higher latitudes can be related to Region 1 (R1) FACs. Please note that positive values represent downward FACs in the northern hemisphere. At certain locations the dual-SC results reveal larger peak values than the single-satellite estimates. This underestimation of the current density is probably due to a non-perpendicular orientation of the current sheet with respect to the satellite track.

In spite of the generally good agreement there are a number of cases where significant differences appear between FAC estimates from single satellite and dual-SC approach. Two such examples are shown in Fig. 6.6. At auroral latitudes, a good agreement between the two techniques is observed also here, but at higher latitudes,

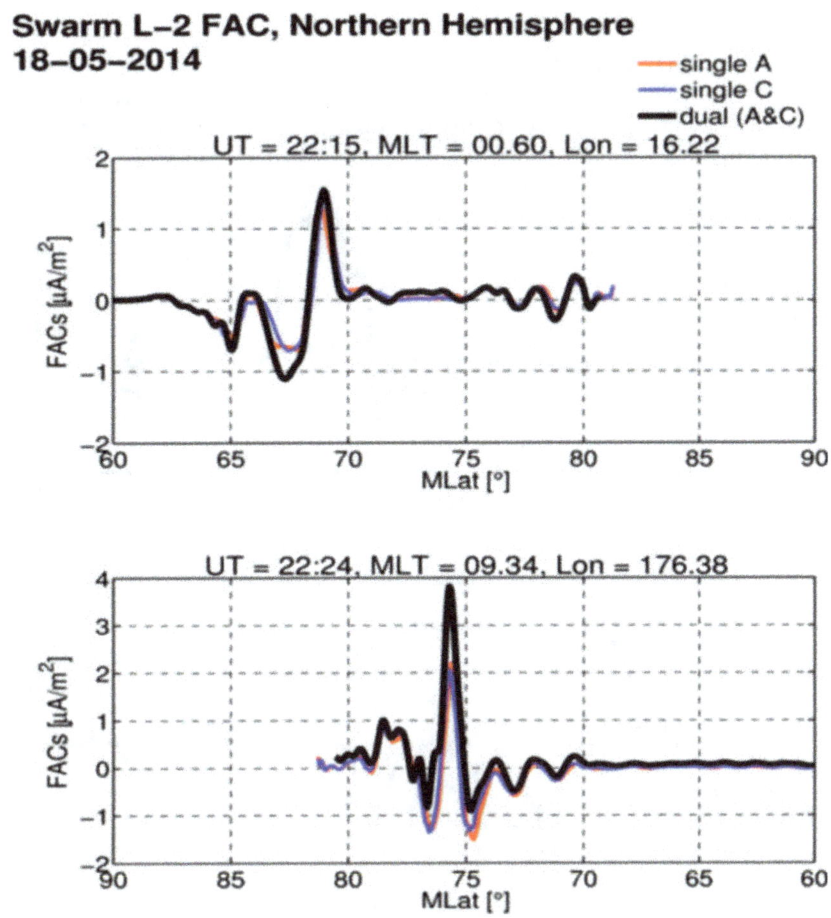

Fig. 6.5 Examples of field-aligned current measurements at auroral latitudes. Individual FAC estimates from the satellites Swarm A and Swarm C are compared with the dual-SC results. In the northern hemisphere positive values represent downward FACs

beyond 75° MLat, clear differences are obvious. Lühr et al. (2016) investigated the properties of such events showing discrepancies in the polar cap region. In some of these cases simultaneous observations by DMSP satellites are available, providing also auroral images. One example from the southern hemisphere is shown in Fig. 6.7. In the top left frame, the differing FAC estimates are plotted. Discrepancies appear poleward of 70° MLat. Below is the auroral image shown, taken by DMSP F16 and mapped onto magnetic coordinates. To this image we added the Swarm orbit track. The satellite pair moves from the early morning sector towards noon. Due to the large offset between geographic and geomagnetic poles in the southern hemisphere, Swarm skims the oval beyond 75° MLat at a small angle. This prevents the single satellite technique to obtain reliable values for FAC densities. The particle instruments on DMSP (see Fig. 6.7, right frames) support the existence of the FAC sheets derived with the dual-SC approach. They show enhanced electron precipitation collocated with the currents. In particular during times of northward interplanetary magnetic field (IMF) orientation, FACs are known to appear frequently in the polar cap (see

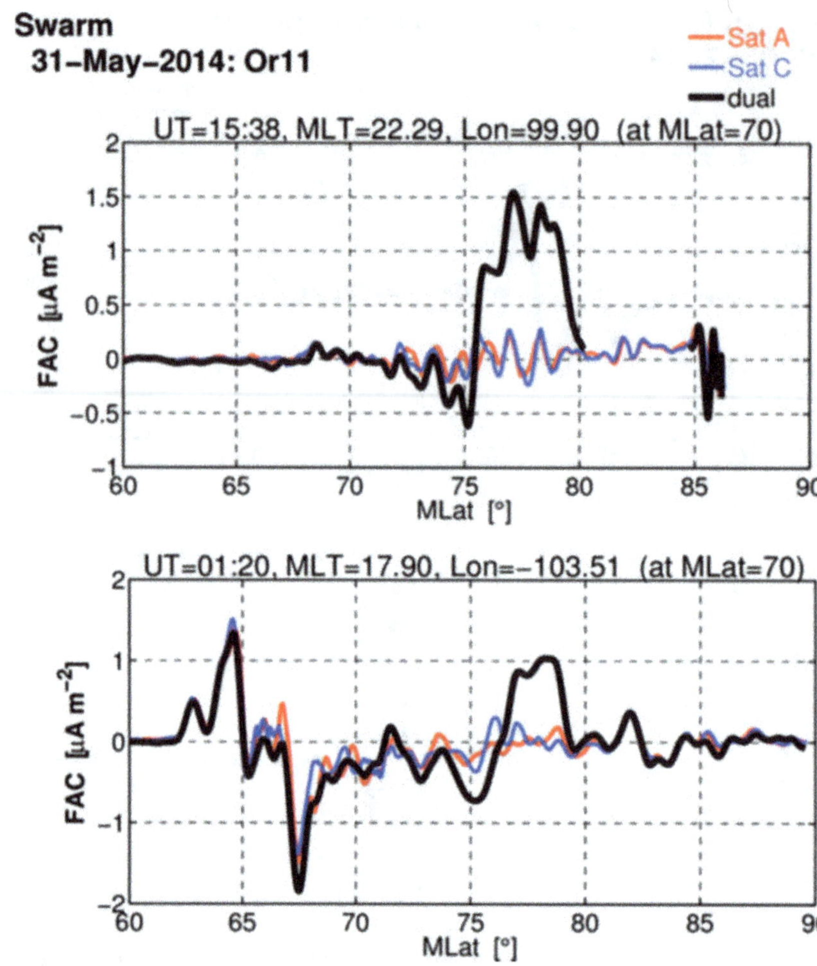

Fig. 6.6 Examples of observations where significant differences are found between FAC estimates from single satellites and the dual-SC approach. At auroral latitudes the agreement is good. Deficits of the single-satellite results appear generally in the polar cap region

Lühr et al. 2016). For those conditions, the dual-SC technique promises to provide new information.

6.5 Assessing the Uncertainties of FAC Estimates

The Swarm dual-SC approach for estimating the radial current density can be regarded as a major progress compared to the traditional way of determining FAC density from single-satellite magnetic field measurements. Still, the dual-SC approach is not perfect, and we will discuss here some of the limitations.

Ambient magnetic field: Before starting with the uncertainties, we'd like to highlight one big advantage of the dual-satellite approach, which overcomes a serious drawback of the single satellite approach. The influence of the various sources of the

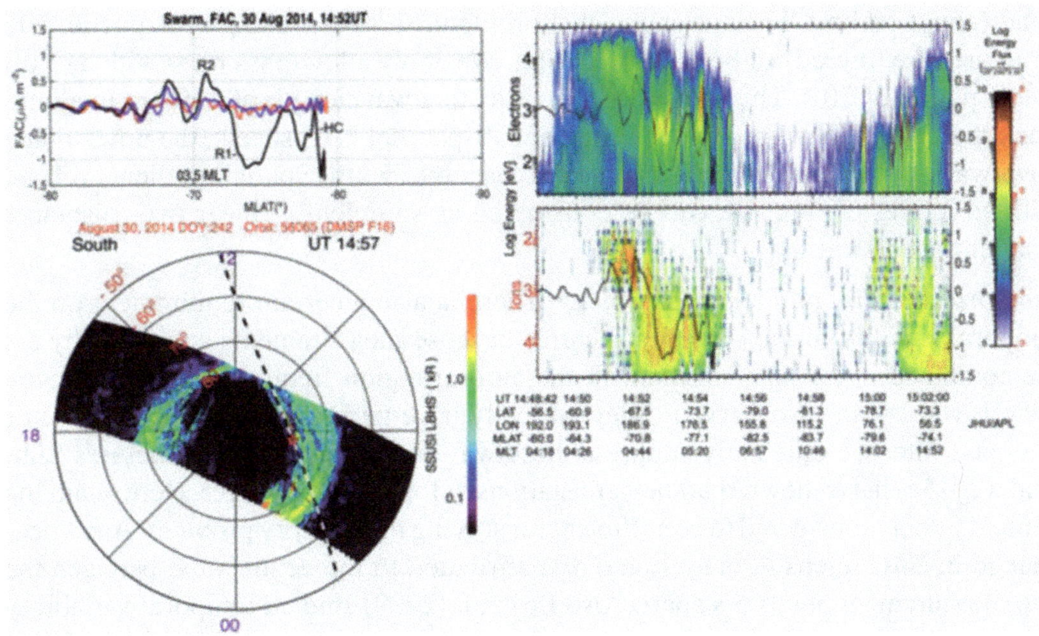

Fig. 6.7 Composite plot of Swarm and DMSP observations of a polar cap event in the Southern Hemisphere. Field-aligned currents along the Swarm track are presented in the top left graph. The black curve shows dual-SC FAC and the red and blue curves present single satellite results of Swarm A and C results, respectively. Auroral images and the Swarm orbit track are illustrated below. The diagrams on the right side show precipitating particle spectrograms with FAC profiles over-plotted (black curve), electrons in the top panel and ions at the bottom (modified after Lühr et al. 2016)

ambient magnetic fields is often ignored when considering error sources. Different from single satellite results, the radial current density derived from a ring integral is hardly influenced by magnetic fields of distant sources: for a field configuration that can be described by a scalar potential, the sum of observations along a closed loop approaches zero. This means that any deficits in main magnetic field removal or in correcting the ring current effect should not compromise the dual-SC current estimates. This holds also for the magnetic effect of the auroral electrojet. Contributions of the electrojets to the magnetic field readings of low-Earth orbiting satellites, such as Swarm, are often underestimated. They can affect significantly the determination of the FAC sheet orientation, for example when employing a minimum variance analysis. In the case of a single-satellite current determination the crossing angle has to be known for a correct estimate. Different from that the skewing angle of the FAC sheet with respect to the flight direction is of no concern for the dual-SC approach. This feature is particularly useful for studies of the southern auroral region, as has been shown above.

Linearity of the magnetic field: One important assumption for the dual-satellite approach is that the magnetic field between the corner points varies linearly. Only in that case the mean value of the two readings at adjacent points represents the average over the entire distance. It is known that the magnetic field at auroral latitudes can vary significantly on scales much shorter than the typical size of our integration quad (about 50 km). For mitigating this problem, short-wavelength magnetic field structures are filtered out before processing. We apply a low-pass filter with a -3 dB cutoff period of 20 s. This period corresponds to a wavelength of 150 km in along-track direction. If Swarm crosses a pair of anti-parallel FAC sheets, the upward and downward FACs detected are separated at least by 75 km. Spatial gradients related with structures of this size can be considered as sufficiently linear over distances smaller than 50 km.

Temporal stability of the current sheets: There is also a certain requirement on the temporal stability of the current structures because measurements separated by 5 s are combined in a single solution. In the auroral region frequently intense kinetic Alfvèn waves can be observed. They typically have periods of 10 s. With our 20 s low-pass filter we efficiently suppress these waves. For larger FAC structures Lühr et al. (2015a) have shown that they are stationary for 1 min or longer. Here again the applied filter is considered to be sufficient for solving the stability problem. Any linear change in FAC intensity is averaged and attributed to the centre time between the two measurement points 5 s apart. Also Le et al. (2009) studied temporal variability based on ST-5 data for meso-scale (~a few 100 km) and large scale (~1000 km) FACs. They confirm a typical stability of about 1 min for FAC of these scales.

Stationarity of the current sheets: The situation is slightly different when the current sheet is moving. In that case the amplitude of current density will not change much, but the current sheet appears broader when the satellites move in the same direction as the sheet and narrower when both move in opposite directions (Doppler Effect). As a consequence, the total current is overestimated or underestimated by the ratio $\frac{v_{sat}+v_{sheet}}{v_{sat}}$ or $\frac{v_{sat}-v_{sheet}}{v_{sat}}$ for parallel or opposite velocities, respectively. A typical sheet velocity may amount to a few hundred meters per second. This has to be compared to 7.6 km/s of the satellite speed in orbit. Therefore resulting uncertainties are generally below 10%. Since independent magnetic field measurements from two spacecraft (Swarm A and C) are available, it is in principle possible to check for the motion of the current sheet. A suitable strategy is to determine first the orientation of the current sheet at the two spacecraft and then check the differences in crossing time. Either they are consistent with passages through a static tilted sheet or a motion along the sheet normal has to be considered to make the times fit. Wang et al. (2009) performed a detailed study on FAC motion at auroral latitudes based on ST-5 measurements. They confirmed that most determined FAC speeds are found in the range 50–200 m/s. Typically the speed increases with higher magnetic activity.

Orientation of the current sheet: In Sect. 6.4 we had claimed that the orientation of the current sheet does not influence the resulting current density, as long as the horizontal scales of the sheet are larger than the dimensions of the measurement quad. In order to demonstrate that, two explicit examples are presented here. For the sake of traceability we make some simplifying assumptions, which however, do not limit the generality of the results.

The current is flowing in an infinitely long vertical sheet of width $2w$. We assume homogenous current distribution within the sheet. Therefore only a transverse magnetic field is induced that varies linearly from B_T to $-B_T$ through the sheet, as shown

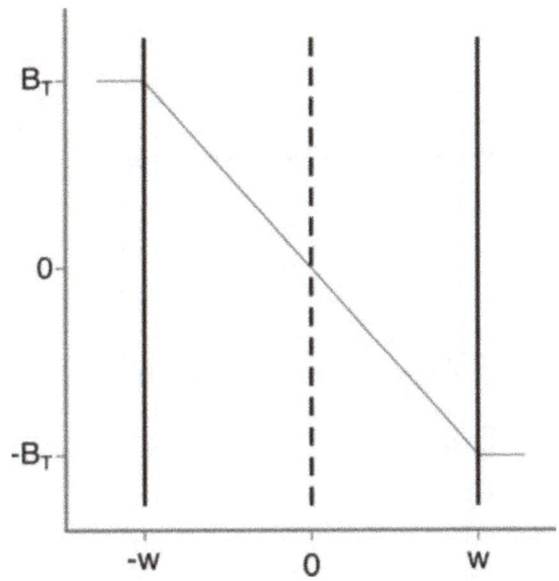

Fig. 6.8 Magnetic field variation across a homogeneous FAC sheet of 2w width. The tangential field component, B_T, varies linearly from one side of the sheet to the other

in Fig. 6.8. In the first example the measurement quad is located well within the current sheet that is oriented at an arbitrary angle φ with respect to the geometry of the quad. Figure 6.9 (top) presents for this case the measurement configuration. Important quantities are the distances u_n ($n = 1, 2, 3, 4$) from the measurement points to the outer border of the current sheet. With some simple geometric analysis we obtain:

$$
\begin{aligned}
u_1 &= (dl_1 + d)\sin\varphi \\
u_2 &= d\sin\varphi \\
u_3 &= d\sin\varphi + dl_2\cos\varphi \\
u_4 &= (dl_1 + d)\sin\varphi + dl_2\cos\varphi
\end{aligned}
$$

where d is the along-track distance from Point 2 to the border of the current sheet. Here it is assumed that $dl_1 = dl_3$, $dl_2 = dl_4$ and the area $A = dl_1 \,.\, dl_2.$ Both conditions are well approximated by the implemented measurement strategy.

The current density is calculated according to Eq. (6.7). For that purpose the magnetic field components B_{AT} and B_{CT} at the corner points, aligned with the path elements, are needed. They can generally be written as:

$$B_{ATn} = B_{Tn}\cos\varphi \quad B_{CTn} = B_{Tn}\sin\varphi \tag{6.21}$$

where $B_{Tn} = B_T(1 - u_n/w)$. When inserting the field components into Eq. (6.7) the current density j results

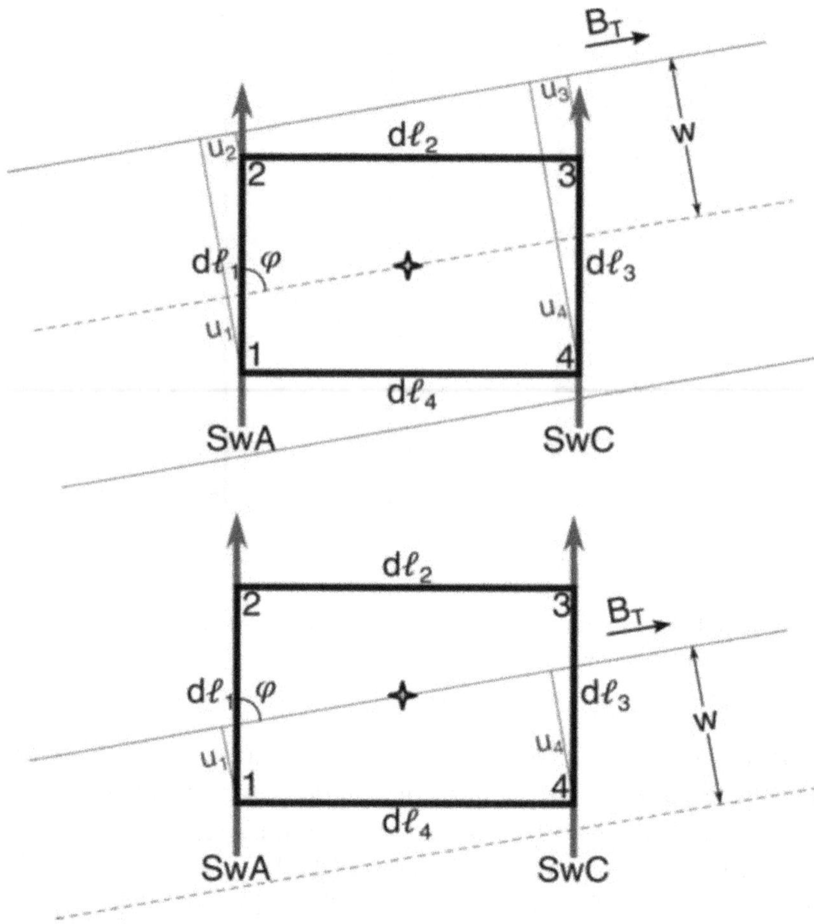

Fig. 6.9 Schematic drawings of two FAC measurement configurations. In the top part the measurement quad is located fully within the current sheet. In the bottom part only half of the quad is engulfed by the FAC sheet

$$j = \frac{B_T}{2w\mu_0 A}[(dl_2 \cos\varphi + dl_2 \cos\varphi)\, dl_1\, \cos\varphi + (dl_1 \sin\varphi + dl_1 \sin\varphi)\, dl_2 \sin\varphi] \tag{6.22}$$

$$j = \frac{B_T}{2w\mu_0 A}[2dl_1 dl_2] \tag{6.23}$$

According to our definition, the term in brackets represents just two times the area A. We thus finally get:

$$j = \frac{B_T}{w\mu_0} \tag{6.24}$$

which is the actual current density within the FAC sheet. The result is independent of the orientation angle, φ, and of the distance, d, to the border.

The second example considers a situation where the measurement quad has been shifted partly out of the current sheet, as shown in Fig. 6.9 (bottom). For convenience, the border runs through the center of the quad, but tilted at an arbitrary angle φ. Now only the distances u_1 and u_4 are of concern

$$u_1 = \frac{dl_1}{2}\sin\varphi - \frac{dl_2}{2}\cos\varphi$$
$$u_4 = \frac{dl_1}{2}\sin\varphi + \frac{dl_2}{2}\cos\varphi$$

Correspondingly, the magnetic field components at the corners amount to:

$$B_{AT1,4} = B_T\left(1 - u_{1,4}/w\right)\cos\varphi \quad B_{CT1,4} = B_T\left(1 - u_{1,4}/w\right)\sin\varphi \tag{6.25}$$

While for the other two corners outside the current sheet we get:

$$B_{AT2,3} = B_T\cos\varphi \quad B_{CT2,3} = B_T\sin\varphi \tag{6.26}$$

When inserting the field components from (6.25) and (6.26) into Eq. (6.7) we obtain for the current density:

$$j = \frac{B_T}{2w\mu_0 A}[dl_2\cos\varphi\, dl_1\cos\varphi + dl_1\sin\varphi\, dl_2\sin\varphi] \tag{6.27}$$

resulting in

$$j = \frac{B_T}{2w\mu_0} \tag{6.28}$$

which is just half the value of (6.24). This makes sense since only half of the quad area is engulfed by the current sheet. Also in this case, no dependence on the tilt angle, φ, results. The independence of our dual-SC current determination technique on the orientation of the sheet can also be shown for more complicated current density profiles. But for them a numerical modelling is recommended, which goes beyond the scope of this chapter.

Measurement accuracies: One important assumption for the dual-SC approach is the accuracy of magnetic field measurements at the two spacecraft. The differences in field readings at two positions are interpreted in terms of currents. Any bias or uncertainty in the scaling factor will cause spurious results. In order to ensure a reliable calibration of the magnetometers, the Swarm spacecraft are carrying Absolute Scalar Magnetometers (ASM). These are used as references for validating the vector field data. Despite these precautions, there may be differences in the magnetic field accuracies. Uncertainties in spacecraft positioning and timing are small. In case of Swarm both quantities are based on GPS navigation solutions and thus do not contribute to the error budget.

Formal uncertainty: For our error estimate we make use of the Swarm specifications. It is required that the difference in magnetic field accuracy between two Swarm spacecraft shall not exceed 1 nT in any component. This value can be inserted in Eq. (6.7), assuming that Swarm A readings are too large by 1 nT with respect to Swarm C. The contributions from the second and fourth term in brackets cancel each other, because the bias can be assumed constant over the 5 s difference between the first and second pair of readings. Only the first and third term in brackets contribute to the uncertainty. Also here we can set, as a reasonable assumption, $dl_1 = dl_3$, $dl_2 = dl_4$ and the area $A = dl_1 \, . \, dl_2$. With that we obtain a resulting error of radial current density, Δj_r, according to Eq. (6.7):

$$\Delta j_r = \frac{2nT \, dl_1}{2\mu_0 dl_1 dl_2} \tag{6.29}$$

Since the zonal separation between the satellite pair varies with the latitude *ß*, $dl_2 = 155$ km cos(ß), the obtained formal uncertainty ranges from 5 nA/m^2 at the equator to 75nA/m^2 at 86° latitude. Beyond 86° no radial currents are determined. The resulting distribution of the uncertainty with changing latitude is shown in Fig. 6.10 for both radial current and field-aligned current estimates.

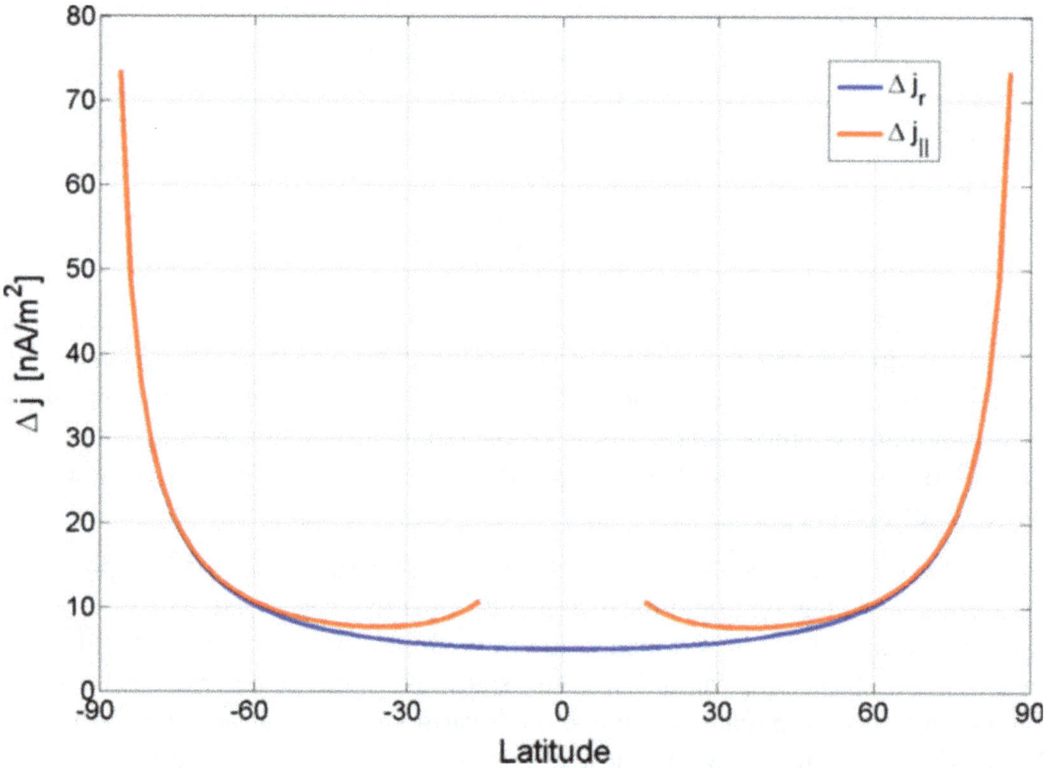

Fig. 6.10 Latitude-dependent variations of the radial current (blue) and FAC (red) estimates. At low latitudes FACs are not calculated due to increasing uncertainties (modified after Ritter et al. 2013)

Divergence of magnetic field: In other missions, such as Cluster or MMS, the divergence of the magnetic field at the four measurement points has frequently been used as an indicator for the quality of the current estimates. In the case of Swarm, only measurements within a plane are available. Therefore it is not obvious what we can learn from the 2D divergence derived from the magnetic fields at the four corner points.

We have tested the response of the 2D divergence to several current configurations. Only for the ideal case of measurements within a homogeneous FAC sheet, as shown in Fig. 6.9 (top), the divergence vanishes. Already when part of the measurement quad is located outside the FAC sheet (e.g. Fig. 6.9, bottom) *div B* attains significant values, and their size is dependent on the orientation angle, φ. In spite of non-vanishing *div B* values the dual-SC technique delivers in this case reliable results, as demonstrated above.

We applied also the 2D divergence calculation to actual measurements, but did not attain convincing relations between FAC quality and *div B* values. In those cases an additional complication arises. Magnetic fields from remote current systems (that can be expressed by scalar potentials) contribute to the 2D divergence. Conversely, they have little influence on the results of our dual-SC FACs. This difference in characteristics further reduces the significance of *div B* for qualifying the FAC results. As a consequence we refrained from using *div B* as a quality indicator.

Low latitudes: Field-aligned currents are estimated reliably at auroral latitudes, as shown above. A different situation is found at low latitudes. Here, the integration quad is quite elongated, as shown in Fig. 6.11. Due to the orbital geometry the east-west separation between Swarm A and C, dl_2, increases to 155 km. This distance is more than 4 times larger than dl_1 (38 km). Such a measurement geometry is not consistent with several FAC types existing at low and middle latitudes. An example of FAC estimates is shown in Fig. 6.12. Slightly poleward of 15° MLat the two spacecraft pass an inter-hemispheric, bipolar FAC structure, with currents directed southward on lower and northward at slightly higher L-shells. In the southern hemisphere this FAC pair is recorded only by Swarm A and 10 min later it is passed by Swarm C in the northern hemisphere. This clearly indicates that the zonal extension of this particular FAC structure is smaller than the satellites separation (155 km). Consequently, the dual-SC

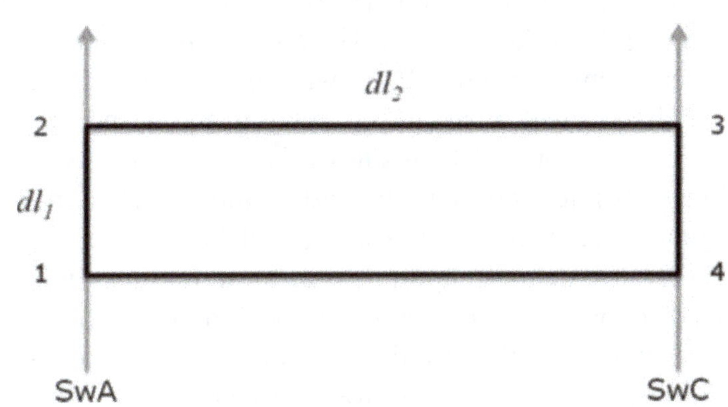

Fig. 6.11 Sketch of measurement quad. At low latitudes the zonal spacecraft separation (dl_2 = 155 km) is 4 times larger than the along-track element ($dl1$ = 38 km). This elongated measurement geometry tends to underestimate FAC densities of certain features

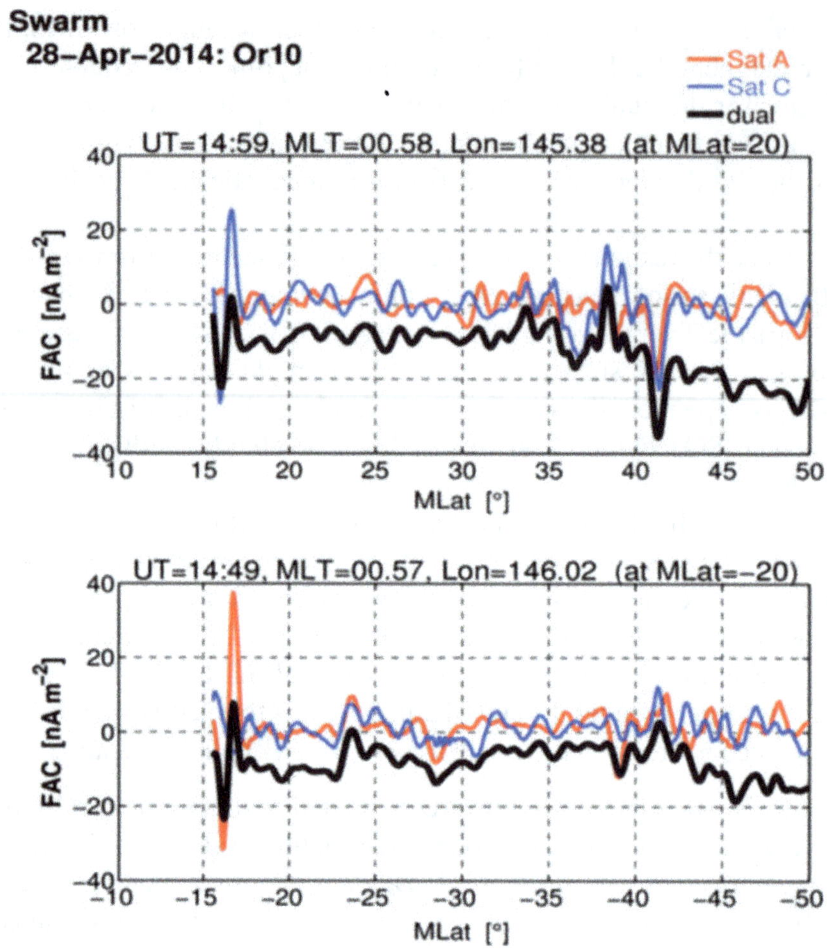

Fig. 6.12 Example of FAC estimates at low and middle latitudes. Near 15° MLat a conjugate FAC feature is observed in the southern hemisphere only by Swarm A and in the northern only by Swarm C. Consequently, the dual-SC method underestimates the current density. Conversely, large-scale inter-hemispheric FACs are sensed by the dual-SC method at mid-latitudes in both hemispheres but missed by the single satellites

approach returns current densities that are too small. From studies of accompanying observations, Park et al. (2015) have identified this event as medium-scale travelling ionospheric disturbance (MSTID). Numerical model simulations of these phenomena, e.g. by Yokoyama and Hysell (2010) predict elongated inter-hemispheric FAC sheets, approximately 50 km wide in zonal direction and several hundred km long, which are tilted by about 30° away from the meridian. These features propagate at a speed of around 100 m/s towards the equator and towards earlier local times. Our low-pass filter does not efficiently remove such elongated FAC sheets. This example shows that, at low latitudes, not all FAC features may be quantified correctly with the Swarm dual-SC approach. In these regions, individual measurements from the satellite pair should be used for checking the dimensions of the features.

At low and middle latitudes there exist also large-scale inter-hemispheric FACs. These can be detected very well with the dual-SC approach. In Fig. 6.12 there is clear evidence for a negative current density at mid-latitude in both hemispheres, which is missed by the single-satellite estimates. This indicates an inter-hemispheric FAC flowing from northern to southern hemisphere. The rather low amplitude, about 10 nA/m^2, is quite typical for the night-time sector, but is well resolvable. Lühr et al. (2015b) performed a systematic study of inter-hemispheric large-scale FACs, based on Swarm measurements, revealing a number of characteristic properties of these low-latitude currents.

6.6 Summary and Conclusions

The constellation mission Swarm can efficiently be used for determining currents in the ionosphere. In particular the magnetic field measurements from the lower pair, Swarm A & C flying side-by-side, provide a good basis for a reliable estimate of the radial current component. The underlying method is Ampère's ring integral. Pairs of field readings from both satellites separated along the track by 5 s are used to define a quad in space. Around this quad the integration is performed to derive the mean value of current density flowing through the encircled area. At auroral latitudes the separation of measurement points is typically 50 km, therefore only large-scale FACs with horizontal scales >150 km are considered. Smaller structures are removed by filtering.

Estimates of the formal uncertainties range from 5 nA/m^2 at the equator to about 100 nA/m^2 near the poles. With this high resolution a whole range of current features can be investigated.

First statistical studies confirm a reliable detection of FAC structures in the auroral oval. Special features like NBZ currents are resolved much clearer than could be achieved previously by single-satellite measurements. In particular in the polar cap region, the dual-SC approach provides a much more detailed picture of FAC distribution.

The high sensitivity of the dual-SC approach also allows for systematic investigations of week inter-hemispheric FACs at middle latitudes. These are closely related to the solar quiet, Sq, current systems and show clear seasonal and longitudinal variations.

The Swarm pair is even able to resolve the wind-driven radial currents over the magnetic equator (see Lühr et al. 2016b). Their intensity and change in direction between noon and evening has clearly been recovered by means of the dual-SC approach.

The presented examples show that the routinely generated radial currents (FACs) densities enable a number of interesting studies. An important constrains to be considered is the size of the current structure. Horizontal scales of the FAC structures interpreted have to be at least twice as large as the gross dimensions of the measurement quad. This requirement can be checked by comparing the magnetic field measurements at satellites Swarm A and C.

All phenomena mentioned here call for further, more detailed investigations. For that purpose more data are necessary. The Swarm constellation needs approximately

5 years for covering all local times evenly during all seasons. The Swarm mission thus offers the opportunity for a number of interesting systematic FAC studies in future when the present Swarm A & C configuration is maintained at least until May 2019.

Acknowledgements The authors are grateful to Jaeheung Park and Ingo Michaelis for fruitful scientific discussions and continuous support during the development of the processing code. The European Space Agency (ESA) is acknowledged for providing the Swarm data and for financially supporting the work on developing the Swarm Level 2 product "FAC". The authors thank the International Space Science Institute in Bern, Switzerland, for supporting the ISSI Working Group: "Multi-Satellite Analysis Tools Ionosphere", in which frame this study was performed. The Editors thanks Guan Le for her assistance in evaluating this chapter.

References

Dunlop, M.W., D.J. Southwood, K.-H. Glassmeier, and F.M. Neubauer. 1988. Analysis of multipoint magnetometer data. *Advances in Space Research* 8: 273–277.

Dunlop, M.W., A. Balogh, K.-H. Glassmeier, and P. Robert. 2002. Four-point Cluster application of magnetic field analysis tools: The Curlometer. *Journal Geophysical Research* 107 (A11): 1384. https://doi.org/10.1029/2001JA005088.

Dunlop, M.W., and A. Balogh. 2005. Magnetopause current as seen by Cluster. *Annales Geophysicae* 23: 901–907.

Dunlop, M.W., Y.-Y. Yang, J.-Y. Yang, H. Lühr, C. Shen, N. Olsen, P. Ritter, Q.-H. Zhang, J.-B. Cao, H.-S. Fu, and R. Haagmans. 2015. Multispacecraft current estimates at Swarm. *Journal of Geophysical Research: Space Physics* 120: 8307–8316. https://doi.org/10.1002/2015JA021707.

Dunlop, M.W., S. Haaland, P.C. Escoubet, and X.-C. Dong. 2016. Commentary on accessing 3-D currents in space: Experiences from cluster. *Journal of Geophysical Research: Space Physics* 121: 7881–7886. https://doi.org/10.1002/2016JA022668.

Eastwood, J.P., et al. 2016. Ion-scale secondary flux ropes generated by magnetopause reconnection as resolved by MMS. *Geophysical Reseach Letters* 43: 4716–4724. https://doi.org/10.1002/2016GL068747.

Finlay, C.C., N. Olsen, S. Kotsiaros, N. Gillet, and L. Tøffner-Clausen. 2016. Recent geomagnetic secular variation from Swarm and ground observatories as estimated in the CHAOS-6 geomagnetic field model. *Earth Planets Space* 68: 112. https://doi.org/10.1186/s40623-016-0486-1.

Gjerloev, J.W., S. Ohtani, T. Iijima, B. Anderson, J. Slavin, and G. Le. 2011. Characteristics of the terrestrial field-aligned current system. *Annales Geophysicae* 29: 1713–1729.

Le, G., Y. Wang, J.A. Slavin, and R.J. Strangeway. 2009. Space Technology 5 multipoint observations of temporal and spatial variability of field-aligned currents. *Journal Geophysical Research* 114: A08206. https://doi.org/10.1029/2009JA014081.

Lühr, H., J. Warnecke, and M.K.A. Rother. 1996. An algorithm for estimating field-aligned currents from single spacecraft magnetic field measurements: A diagnostic tool applied to Freja satellite data. *IEEE Transactions on Geoscience and Remote Sensing* 34: 1369–1376.

Lühr, H., J. Park, J.W. Gjerloev, J. Rauberg, I. Michaelis, J.M.G. Merayo, and P. Brauer. 2015a. Field-aligned currents' scale analysis performed with the swarm constellation. *Geophysical Reseach Letters* 42: 1–8. https://doi.org/10.1002/2014GL062453.

Lühr, H., G. Kervalishvili, I. Michaelis, J. Rauberg, P. Ritter, J. Park, J.M.G. Merayo, and P. Brauer. 2015b. The inter-hemispheric and F-region dynamo currents revisited with the Swarm constellation. *Geophysical Reseach Letters* 42: 3069–3075. https://doi.org/10.1002/2015GL063662.

Lühr, H., T. Huang, S. Wing, G. Kervalishvili, J. Rauberg, and H. Korth. 2016. Filamentary field-aligned currents at polar cap region during northward interplanetary magnetic field derived with the Swarm constellation. *Annales Geophysicae* 34: 901–915. https://doi.org/10.5194/angeo-34-901-2016.

Park, J., H. Lühr, G. Kervalishvili, J. Rauberg, I. Michaelis, C. Stolle, and Y.-S. Kwak. 2015. Nighttime magnetic field fluctuations in the topside ionosphere at midlatitudes and their relation to medium-scale traveling ionospheric disturbances: The spatial structure and scale sizes. Journal of Geophysical Research: Space Physics. 120. https://doi.org/10.1002/2015ja021315.

Ritter, P., and H. Lühr. 2006. Curl-B technique applied to Swarm constellation for determining field-aligned currents. *Earth Planets Space* 58: 463–476.

Ritter, P., H. Lühr, and J. Rauberg. 2013. Determining field-aligned currents with the Swarm constellation mission. *Earth Planets Space* 65: 1285–1294. https://doi.org/10.5047/eps.2013.09.006.

Slavin, J.A., G. Le, R.J. Strangeway, Y. Wang, S.A. Boardsen, M.B. Moldwin, and H.E. Spence. 2008. Space Technology 5 multi-point measurements of near-Earth magnetic fields: Initial results. *Geophysical Reseach Letters* 35: L02107. https://doi.org/10.1029/2007GL031728.

Sonnerup, B.U.O., and L.J. Cahill. 1967. Magnetopause structure and attitude from explorer 12 observations. *Journal Geophysical Research* 72: 171–183.

Vallat, C., I. Dandouras, M. Dunlop, A. Balogh, E. Lucek, G.K. Parks, M. Wilber, E.C. Roelof, G. Chanteur, and H. Rème. 2005. First current density measurements in the ring current region using simultaneous multispacecraft CLUSTER-FGM data. *Annales Geophysicae* 23: 1849–1865.

Vennerstrøm S., E. Friis-Christensen, H. Lühr, T. Moretto, N. Olsen, C. Manoj, P. Ritter, L. Rastätter, A. Kuvshinov, and S. Maus. 2005. The impact of combined magnetic and electric field analysis and of ocean circulation effects on *Swarm* mission performance, *ESA contract No 3-10901/03/NL/CB*, DSRI Report 2/2004.

Vogt, J., E. Sorbalo, M. He, and A. Blagau. 2013. Gradient estimation using configurations of two or three spacecraft. *Annales Geophysicae* 31: 1913–1927. https://doi.org/10.5194/angeo-31-1913-1927-2013.

Wang, Y., G. Le, J.A. Slavin, S.A. Boardsen, and R.J. Strangeway. 2009. Space Technology 5 measurements of auroral field-aligned current sheet motion. *Geophysical Reseach Letters* 36: L02105. https://doi.org/10.1029/2008GL035986.

Yokoyama, T., and D.L. Hysell. 2010. A new midlatitude ionosphere electrodynamics coupling model (MIECO): Latitudinal dependence and propagation of medium-scale traveling ionospheric disturbances. *Geophysical Reseach Letters* 37: L08105. https://doi.org/10.1029/2010GL042598.

Zhao, C., C.T. Russell, R.J. Strangeway, S.M. Petrinec, W.R. Paterson, M. Zhou, B.J. Anderson, W. Baumjohann, K.R. Bromund, M. Chutter, D. Fischer, G. Le, R. Nakamura, F. Plaschke, J.A. Slavin, R.B. Torbert, and H.Y. Wei. 2016. Force balance at the magnetopause determined with MMS: Application to flux transfer events. *Geophysical Reseach Letters* 43: 11941–11947. https://doi.org/10.1002/2016GL071568.

Chapter 7
Science Data Products for AMPERE

Colin L. Waters, B. J. Anderson, D. L. Green, H. Korth, R. J. Barnes and Heikki Vanhamäki

Abstract Birkeland currents that flow in the auroral zones produce perturbation magnetic fields that may be detected using magnetometers onboard low-Earth orbit satellites. The Active Magnetosphere and Planetary Electrodynamics Response Experiment (AMPERE) uses magnetic field data from the attitude control system of each Iridium satellite. These data are processed to obtain the location, intensity and dynamics of the Birkeland currents. The methodology is based on an orthogonal basis function expansion and associated data fitting. The theory of magnetic fields and currents on spherical shells provides the mathematical basis for generating the AMPERE science data products. The application of spherical cap harmonic basis and elementary current system methods to the Iridium data are discussed and the procedures for generating the AMPERE science data products are described.

7.1 Introduction

The electric dynamo comprising the solar wind kinetic energy interacting with Earth's magnetic field in space drives a global electric circuit that couples the magnetosphere with the polar ionospheres through the Birkeland currents. The first averaged spatial configuration and intensity of the region 1 and region 2 Birkeland currents were obtained from the low-Earth orbit Triad satellite around 50 years ago (Iijima and Potemra 1976). Using spatially sparse satellite observations from Triad, MAGSAT, Viking and DMSP, the intensity and location dependence of the currents on the inter-

C. L. Waters (✉)
University of Newcastle, Callaghan, New South Wales, Australia
e-mail: colin.waters@newcastle.edu.au

B. J. Anderson · H. Korth · R. J. Barnes
The Johns Hopkins University Applied Physics Laboratory, Laurel, MD, USA

D. L. Green
Oak Ridge National Labs, Oak Ridge, TN, USA

H. Vanhamäki
University of Oulu, Oulu, Finland

M. W. Dunlop and H. Lühr (eds.), *Ionospheric Multi-Spacecraft Analysis Tools*, ISSI Scientific Report Series 17,
https://doi.org/10.1007/978-3-030-26732-2_7

planetary magnetic field (IMF) was recognised (Bythrow et al. 1982; Iijima et al. 1984; Zanetti et al. 1984; Erlandson et al. 1988). Birkeland currents are typically located between 65°–75° magnetic latitude, expanding to 45° during geomagnetic storms and contracting poleward of 75° during quiet periods with similar excursions for the southern hemisphere. Subsequent statistical and event studies using Iridium satellite data have confirmed the IMF dependence (Anderson et al. 2008; Green et al. 2009; Korth et al. 2010), provided estimates of ionosphere conductance (Green et al. 2007) and energy transfer via the Birkeland currents (Waters et al. 2004; Korth et al. 2008) and revealed the spatial sequence from small to enhanced current flow (Anderson et al. 2014, 2018). These results derived from Iridium satellite data have highlighted limitations of previous, spatially sparse, in situ measurements, which required months of data when using single satellite studies in order to cover all magnetic local times.

The Iridium satellite constellation is a network of about 90 polar orbiting satellites at an altitude of 780 km. Each satellite contains a 3-axis, vector fluxgate magnetometer as part of the attitude control system. The number of satellites and the spatial coverage make the Iridium constellation an excellent sensor system for space physics research. The quantification and validation of the Iridium magnetometer data for studying the Birkeland currents was discussed by Waters et al. (2001) Anderson et al. (2000, 2002, 2008), Green et al. (2009). After removing magnetic field variations longer than 26 min with suitable filtering, the Iridium magnetometer data are processed to give the perturbation magnetic field, $\mathbf{b} = (b_r, b_\theta, b_\phi)$ which are dominated by signatures of the Birkeland currents.

Magnetic field data from the Iridium satellites were first available for space physics research in 1999. These were limited to the cross satellite track components of the perturbation magnetic field. Data were obtained from each satellite at a sample period of 200 s. This provided a large improvement in spatial and temporal studies of the Birkeland currents over previous single satellite measurements. While several months of single satellite data are required to build a global pattern (e.g. Gary et al. 1995; Kosch and Nielsen 1995), the Iridium data allowed the required time frame to be drastically reduced to data averaged over $\approx$1h. These 200 s sampled, cross-track only data will be referred to as *pre-AMPERE* data.

During 2009, enhancements to the Iridium data delivery and processing were introduced under the Active Magnetosphere and Planetary Electrodynamics Response Experiment (AMPERE) project as summarised in Anderson et al. (2014). The data sample period was reduced tenfold from 200 s to a standard 20 s interval from each satellite. Furthermore, a 2 s data sample mode is available for storm case study intervals, enabling higher time resolution studies of active space weather processes. AMPERE data products are available from October 2009 at the web address, http://ampere.jhuapl.edu.

The procedure for estimating the radial current from the pre-AMPERE data was described by Waters et al. (2001). The Iridium, cross-track component data were expanded using spherical cap harmonic basis functions where the expansion coefficients were estimated from a minimum least squared error process. There are two main features of the AMPERE data that required enhancements to the pre-AMPERE data processing described by Waters et al. (2001) and Green et al. (2006). The full vector magnetic field values are available under AMPERE and the higher time cadence provides the ability to resolve smaller spatial structures in latitude. This required high order spherical cap harmonic analyses. In particular, the ability to compute the spatial gradient of the spherical cap harmonic functions for orders greater than 50 over typical spherical cap sizes of $\theta_0 = 50°$.

In this chapter, the theory and data processing used to obtain the AMPERE science data products available from http://ampere.jhuapl.edu are described. We begin in Sect. 7.2 with a relevant summary of the theory of magnetic fields and currents on spherical surfaces, drawing mostly from the work by Backus (1986). This provides the mathematical framework for computing the Birkeland currents from the Iridium satellite data. Spherical harmonic functions are discussed in Sect. 7.3, followed by a description of the expansion of the Iridium data using vector spherical harmonics as orthogonal basis functions in Sect. 7.4. This leads to a discussion of the vector spherical cap harmonics used for confining the data fit over a cap rather than the full sphere at high order for degree-scale spatial resolution.

The underlying theory and application to Iridium data are illustrated in Sect. 7.5 using examples of the preprocessed input magnetic perturbations data (b_r, b_θ, b_ϕ), the estimated perturbation magnetic field data $(b_{r,f}, b_{\theta,f}, b_{\phi,f})$, expanded from the vector spherical cap harmonic basis functions and the derived radial current densities. Properties of the uncertainties in the input and fitted satellite data are discussed in Sect. 7.6 including error statistics and uncertainty estimates in the radial current data product. The spatial resolution of the data is discussed, particularly around the data-dense Iridium satellite track convergence locations in each pole. An alternative to using spherical harmonics is the Spherical Elementary Current System (SECS) basis functions (Amm 1997 and Chap. 2 of this volume). The AMPERE data processed using this approach are discussed in Sect. 7.7. Finally, an example of AMPERE data products combined with other space physics data sets (e.g. SuperDARN radars) is given in Sect. 7.8.

7.2 Magnetic Fields and Currents on Spherical Surfaces

Electric currents have well-defined magnetic fields as specified through the Maxwell equations. The mathematical relationships between magnetic fields and currents on spherical shells were reviewed by Backus (1986). There are three representations that are relevant for magnetic fields in space physics. These are the Gauss, Mie and Helmholtz descriptions. The relevant descriptions, parameters and relationships are summarised in Fig. 7.1. The Gauss form is used for modelling the surface geomagnetic field, where both $\nabla \bullet \mathbf{B}$ and $\nabla \times \mathbf{B}$ are zero. Approximating the ionosphere as a thin current sheet with anisotropic conductance requires the Helmholtz description. The Mie expressions are used to derive the radial current density, given the input perturbation magnetic field data, **b**, from the Iridium satellites.

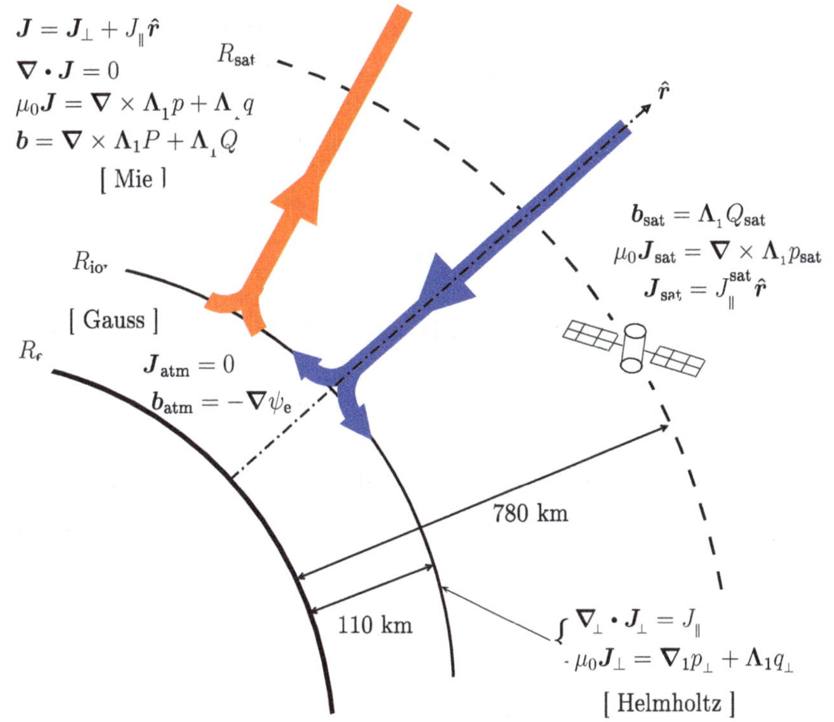

Fig. 7.1 Magnetic fields and currents and relevant mathematics below, within and above the ionosphere. The surface curl operator $\Lambda_1 = \mathbf{r} \times \nabla$ (Green et al. 2006)

Consider a spherical shell denoted $S(a, b)$, with origin at the Earth centre and with inner radius, a and outer radius b at the minimum and maximum satellite altitudes ($a < b$). Since $\nabla \bullet \mathbf{B} = 0$ the vector field, $\mathbf{B}$, is *solenoidal*. The converse is not necessarily true (Backus et al. 1996). A vector field, $\mathbf{Q}$ is *toroidal* in the shell if there is a scalar field, Q, such that

$$\mathbf{Q} = \mathbf{r} \times \nabla Q \tag{7.1}$$

This is the Helmholtz representation of the field, $\mathbf{Q}$ in the surface $S(a, b)$. The toroidal scalar, Q, is uniquely determined from Eq. (7.1) provided that for each r in $a < r < b$, the average is zero, i.e. $\langle Q \rangle_r = 0$.

A vector field, $\mathbf{P}$, is *poloidal* in the shell if there is a scalar field, P, such that

$$\mathbf{P} = \nabla \times (\mathbf{r} \times \nabla P) \tag{7.2}$$

If the current density $\mathbf{J} \neq 0$ in the shell then the Mie representation of the magnetic field, $\mathbf{B}$, is

$$\mathbf{B} = \mathbf{P} + \mathbf{Q}. \tag{7.3}$$

Equation (7.1) was introduced by Lamb (1881). The Gauss representation used in geomagnetism for main field modelling has $\mathbf{Q} = 0$ and $\nabla^2 \mathbf{P} = 0$. The vector fields, $\mathbf{P}$ and $\mathbf{Q}$, are uniquely determined by the field $\mathbf{B}$ in the shell. However, the scalar fields, P and Q are determined by $\mathbf{B}$ only to additive functions of r as discussed by Backus (1986). They can be made unique by the constraint that the average values, $\langle Q\rangle_r = \langle P\rangle_r = 0$.

From Maxwell, with negligible displacement current, the magnetic field, $\mathbf{B}$ and the current density, $\mathbf{J}$ are related by

$$\nabla \times \mathbf{B} = \mu \mathbf{J} \tag{7.4}$$

Taking the curl of Eq. (7.3) and since $\mathbf{J}$ is solenoidal [i.e. $\nabla \bullet \mathbf{J} = 0$] the Mie representation of $\mathbf{J}$ is

$$\mu \mathbf{J} = \nabla \times (\mathbf{r} \times \nabla p) + \mathbf{r} \times \nabla q \tag{7.5}$$

where p and q are the poloidal and toroidal scalars for the current density vector field, $\mathbf{J}$. It follows that these are related to the scalars P and Q describing the magnetic field by

$$Q = p \tag{7.6}$$

$$\nabla^2 P = -q \tag{7.7}$$

Therefore, a toroidal current produces a poloidal magnetic field while a poloidal current gives a toroidal magnetic field. In summary, toroidal fields are solenoidal fields with zero radial component while poloidal fields are solenoidal fields whose curl has no radial component.

We only have information about the horizontal gradients of the magnetic field from the Iridium data. The perturbation magnetic field, $\mathbf{b}$ and current density, $\mathbf{J}$ are solenoidal fields but the horizontal components, $\mathbf{b}_\perp$ and $\mathbf{J}_\perp$ are not. Therefore, the Mie representation cannot be used directly. However, the Helmholtz representation allows the construction of any vector field from curl-free and divergence-free components. The relationships between the Helmholtz and Mie representations for magnetic fields and currents is described by Backus (1986).

The field aligned (radial) currents ($J_\parallel$) are derived from the transverse magnetic field perturbations, $\mathbf{b}_\perp = \mathbf{b}(\theta, \phi)$, where θ is the co-latitude and ϕ is the longitude coordinate. If $\mathbf{v}_s$ is some tangent vector field on the spherical surface, $S(b)$, then there is a unique poloidal field

$$\mathbf{v}_p = \nabla_1 g \tag{7.8}$$

and a unique toroidal field

$$\mathbf{v}_t = \nabla_1 h \tag{7.9}$$

where ∇_1 is the dimensionless surface gradient defined by

$$\nabla_1 = r \nabla_s \tag{7.10}$$

The scalar functions, g and h, are determined by $\mathbf{v}$ if $\langle g\rangle_b = \langle h\rangle_b = 0$ according to

$$\nabla_1^2 g = \nabla_1 \bullet \mathbf{v}_s \tag{7.11}$$

$$\nabla_1^2 h = (\mathbf{r} \times \nabla_s) \bullet \mathbf{v}_s \tag{7.12}$$

for

$$\mathbf{v}_s = \nabla_1 g + \mathbf{r} \times \nabla_s h \tag{7.13}$$

and

$$\mathbf{v} = \mathbf{r}^u f + \mathbf{v}_p + \mathbf{v}_t \tag{7.14}$$

where $\mathbf{r}^u$ is the radial unit vector so $\mathbf{r}^u f$ is the radial part of $\mathbf{v}$.

The derivations for the divergence and curl of the vector field, $\mathbf{v}$, are given in Backus (1986). The results are

$$\nabla \bullet \mathbf{v} = r^{-2}[\nabla_r(r^2 f) + \nabla_1^2(rg)] \tag{7.15}$$

and

$$\nabla \times \mathbf{v} = \mathbf{r}^u f' + \nabla_1 g' + (\mathbf{r} \times \nabla_s)h' \tag{7.16}$$

where

$$f' = r^{-1}\nabla_1^2 h \tag{7.17}$$

$$g' = -r^{-1}\nabla_r(rh) \tag{7.18}$$

$$h' = r^{-1}[\nabla_r(rg) + \langle f \rangle_r - f] \tag{7.19}$$

From (7.14), (7.17) and (7.3) any solenoidal vector field, $\mathbf{b}$, can be written

$$\mathbf{b} = \mathbf{r}^u(r^{-1}\nabla_1^2 P) - \nabla_1[r^{-1}\nabla_r(rP)] + \mathbf{r} \times \nabla_s Q \tag{7.20}$$

and a similar expression can be written for the current density, $\mathbf{J}$

$$\mu\mathbf{J} = \mathbf{r}^u(r^{-1}\nabla_1^2 p) - \nabla_1[r^{-1}\nabla_r(rp)] + \mathbf{r} \times \nabla_s q \tag{7.21}$$

For radial, field aligned current with no horizontal currents in $S(a, c)$, the magnetic perturbations detected by the Iridium constellation form a toroidal magnetic field. From (7.7) and (7.21) the relation used to obtain the radial current density is

$$\mu r J_r = \nabla_1^2 p \tag{7.22}$$

7.3 Spherical Harmonic Basis Functions

The spherical harmonic functions were used by Gauss in 1838 to obtain analytic expressions for the Earth's magnetic field. This is now a core technique in geophysics and the spherical harmonic expansion of Earth's main field is publicised as the International Geomagnetic Reference Field (IGRF) as a set of spherical harmonic coefficients. There is an abundance of literature describing geomagnetic field modelling. An excellent resource is the two volumes by Chapman and Bartels (1940) in addition to any text on mathematical methods (e.g. McQuarrie 2003).

The techniques are very similar for modelling perturbation fields due to the Birkeland currents. This section gives a brief summary of these spherical harmonic methods, providing a foundation for describing a spherical cap harmonic expansion of the Iridium data.

The spherical harmonic functions are solutions to the Laplace equation expressed in spherical coordinates. Electric currents internal to the Earth determine the *main* geomagnetic field, $\mathbf{B}_0$, observed on Earth's surface where $\nabla \bullet \mathbf{B}_0 = \nabla \times \mathbf{B}_0 = 0$. This gives the classic Gauss representation

$$\nabla^2 \psi = 0 \tag{7.23}$$

which is the Laplace equation for some potential function, ψ. In spherical coordinates Eq. (7.23) becomes

$$\nabla^2 \psi = \frac{1}{r^2}\frac{\partial}{\partial r}\left(r^2 \frac{\partial \psi}{\partial r}\right) + \frac{1}{r^2 sin\theta}\frac{\partial}{\partial \theta}\left(sin\theta \frac{\partial \psi}{\partial \theta}\right) + \frac{1}{r^2 sin^2\theta}\frac{\partial^2 \psi}{\partial \phi^2} = 0 \tag{7.24}$$

The solutions to Eq. (7.24) are usually obtained using the separation of variables technique. At constant radial height ($R_{sat} = 780$ km) we are interested in the solutions, $\mathrm{Y}(\theta, \phi)$. The solution in the co-latitude variable, θ involves the associated Legendre differential equation

$$\frac{d}{dx}\left[(1-x^2)\frac{dP(x)}{dx}\right] - \frac{m^2}{1-x^2}P(x) = -\lambda P(x) \tag{7.25}$$

where λ is a constant, $x = cos\theta$ and $m = 0, \pm 1, \pm 2, \ldots.$

For $m = 0$ and $0 \le \theta \le \pi$, the solutions to Eq. (7.25) are the *Legendre polynomials*, $P_n(x)$ which may be generated from Rodrigues' formula

$$P_n(x) = \frac{1}{2^n n!}\frac{d^n}{dx^n}\left(x^2 - 1\right)^n \tag{7.26}$$

for $n = 0, 1, 2, \ldots.$ For example, the Legendre polynomials are $P_0(x) = 1$, $P_1(x) = x$, $P_2(x) = \frac{1}{2}(3x^2 - 1)$ and so on, with $-1 \le P_n(x) \le 1$ and all $P_n(1) = 1$. They satisfy the orthogonality relation

$$\int_{-l}^{+l} P_l(x)P_n(x)dx = \frac{1}{2l+1}\delta_{l,n} \tag{7.27}$$

For $m \neq 0$ we have the *associated Legendre functions*

$$P_{l,|m|}(x) = (1-x^2)^{\frac{|m|}{2}} \frac{d^m P_n(x)}{dx^m} \tag{7.28}$$

which satisfy

$$\int_{-l}^{+l} P_{l,|m|}(x)P_{n,|m|}(x)dx = \frac{2}{2l+1}\frac{(l+|m|)!}{(l-|m|)!}\delta_{l,n} \tag{7.29}$$

for the Kronecker delta function, $\delta_{l,n}$ and $|m| \leq l$. The factorial multiplier in Eq. (7.29) shows that these functions have a large range in magnitude even for small increments in m. Since these will be used to construct a basis set of the experimental data, we would like the coefficients to represent the strength of each basis function in the data. One early choice was to use normalised $P_{l,m}$. However, the standard adopted by the geophysics community in 1939 is *Schmidt semi-normalisation* where

$$P_l^m(\theta) = P_{l,m}(\theta) \quad m = 0 \tag{7.30}$$

$$P_l^m(\theta) = \left[2\frac{(l-m)!}{(l+m)!}\right]^{1/2} P_{l,m}(\theta) \quad m > 0 \tag{7.31}$$

Including the dependence with longitude gives the spherical harmonics

$$Y_l^m(\theta,\phi) = P_l^m(\theta)e^{im\phi} \tag{7.32}$$

and their mean squared value over the spherical surface is

$$\int_0^{\pi}\int_0^{2\pi} \left[Y_l^m(\theta,\phi)\right]^2 sin\theta d\theta d\phi = \frac{4\pi}{2l+1} \tag{7.33}$$

The $e^{im\phi}$ in Eq. (7.32) is implemented by using $P_l^m cos(m\phi)$ for $m > 0$ and $P_l^m sin(m\phi)$ for $m < 0$. The eigenvalues, λ in (7.25) are related to l in Eq. (7.32) by $\lambda = l(l+1)$ so that

$$\nabla^2\psi(\theta,\phi) = -l(l+1)\psi(\theta,\phi) \tag{7.34}$$

This simplifies the computation of J_r in Eq. (7.22). For $0 \leq \theta \leq \pi$ the eigenvalues are integers. This is no longer true for spherical cap harmonics.

7.4 Basis Functions and Data Fitting

The Iridium satellites orbit at 780 km above Earth's surface. At this altitude the magnetic field perturbations detected at high latitudes are assumed to be due to radial currents. Therefore, the perturbation magnetic field is toroidal as defined by Eq. (7.1) and the radial current may be determined using Eq. (7.22). The toroidal scalar, Q, in Eq. (7.1) may be expanded on a basis set of spherical harmonics (Backus 1986)

$$Q = \sum_{l=0}^{K} \sum_{m=-l}^{l} a_{l,m} \; Y_{r,\theta,\phi} \tag{7.35}$$

so that Eq. (7.34) can be used for the second order derivative in Eq. (7.22).

The preprocessing of the magnetic field data from the Iridium satellites for AMPERE provides vector perturbation magnetic field data, $\mathbf{b} = (b_r, b_\theta, b_\phi)$. These are expanded on a basis set of *vector* spherical harmonics

$$Y_r(\theta, \phi) = Y_l^m(\theta, \phi)\mathbf{r}^u \tag{7.36}$$

$$Y_\theta(\theta, \phi) = r\nabla Y_l^m(\theta, \phi) \tag{7.37}$$

$$Y_\phi(\theta, \phi) = \mathbf{r} \times \nabla Y_l^m(\theta, \phi) \tag{7.38}$$

which involves the gradient of the scalar spherical harmonics of Eq. (7.32). The derivative of Y_l^m with respect to ϕ is straightforward, involving a multiplication by m. For $0 \leq \theta \leq \pi$ and Schmidt semi-normalisation, the derivative of P_l^m with respect to θ is Chapman and Bartels (1940)

$$\frac{dP_l^m}{d\theta} = \frac{1}{2}\left[(l+m)(l-m+1)\right]^{1/2} \; P_l^{m-1}(\theta) - \frac{1}{2}\left[(l+m+1)(l-m)\right]^{1/2} \; P_l^{m+1}(\theta) \tag{7.39}$$

and for $m = 0$,

$$\frac{dP_l^0}{d\theta} = -\left[\frac{1}{2}l(l+1)\right]^{1/2} \; P_l^1(\theta) \tag{7.40}$$

The Iridium constellation provides magnetic field data in six longitudinally spaced orbit planes which gives a maximum value, m=6. However, the latitude spatial resolution changes with the data sample rate and values for $l > 60$ are common for the 20 second sampled data. According to Eqs. (7.36)–(7.38) the gradient of the basis functions at each (θ, ϕ) of the recorded magnetic field data is required. Therefore, basis functions and the data fit are calculated in an orthogonal coordinate system related to geographic coordinates. The *design matrix* is a two dimensional array (matrix) containing the basis function values at the locations of the perturbation magnetic field data. If the number of basis functions is *nBF* and the number of data points is *NP* then the design matrix for scalar data would be of size nBF by NP. For AMPERE, each vector component of the perturbation magnetic field is loaded into a column vector and the system is solved for the set of coefficients that minimises the squared difference between the data and the estimated values. Since $NP > nBF$, Singular Value Decomposition (SVD) may be used to solve for the coefficients (Press et al. 1986).

7.5 Practical Considerations

There are a number of reasons for using Spherical Cap Harmonic Analysis (SCHA) to generate AMPERE data products. The region of interest containing the major current systems is generally located within 40° of the poles. Second, AMPERE can provide up to 2 s sampled data resulting in higher spatial resolution which requires a high order fit. For a full sphere spherical harmonic fit, the number of coefficients required to achieve the spatial resolution can become larger than the number of input data points. Third, a basis function fit that employs sine/cosine functions in longitude requires selection of m that satisfies the Nyquist criterion for the largest separation in longitude. Finally, given that the northern and southern hemisphere Iridium satellite track intersection locations are not 180° apart, they may be treated separately by using a spherical cap approach.

The calculation of spherical cap harmonics was described by Haines (1988) and de Santis et al. (1999). The computer program provided by Haines (1988) used a recursion process based on the hypergeometric functions to compute the spherical harmonics over a cap. However, for AMPERE data products these functions need to be evaluated to high order where the recursion relations require the computation of very large numbers, stretching the numerical precision capabilities of computers. In order to circumvent this problem, (Green et al. 2006) used software techniques to increase the computer numerical precision in order to compute these functions. A simpler method is to numerically solve Eq. (7.25) using a finite difference algorithm as an eigen problem, given the two boundary conditions discussed by Haines (1985). In principle, this method allows for the computation of the set of spherical harmonic basis functions over a spherical annulus as used, for example, by Waters and Sciffer (2008).

While the spherical cap harmonic functions may be obtained for any cap size, θ_0, the computation time is decreased if recursion and identity relationships that are well known for the full sphere associated Legendre polynomials and their derivatives are used, (e.g. Eqs. 7.39 and 7.40). The Iridium data covers all latitudes in six orbit planes at 780 km altitude. Therefore, the combined advantages of efficient computation of basis functions and the use of spherical cap harmonics is obtained by selecting $\theta_0 = 90°$. This strategy yields the two basis function sets obtained using the Dirichlet and Neumann boundary conditions from standard math library associated Legendre polynomial routines.

The spherical cap harmonic expansion yields the fitted perturbation magnetic field $(b_{r,f}, b_{\theta,f}, b_{\phi,f})$ and Birkeland current configuration over both the northern and southern hemisphere auroral regions. As an example, the 20 s (b_θ, b_ϕ) data for 1520–1530 UT, 24 Aug 2010 are shown in Fig. 7.2a. The data are shown in the Altitude Adjusted Corrected GeoMagnetic (AACGM) latitude (Baker and Wing 1989) and Magnetic Local Time (MLT) coordinate system with 12 MLT at the top. The vector spherical cap harmonic fit magnetic field data are shown in Fig. 7.2b.

Using Eqs. (7.37)–(7.40) for the basis set, followed by Eqs. (7.22) and (7.35), the radial current is obtained and shown in Fig. 7.2c. The region 1 current system is located near 70° with region 2 located a few degrees poleward of 60° AACGM latitude. The Interplanetary Magnetic Field (IMF) magnitude and orientation modulate the Birkeland current pattern (Green et al. 2009). The time shifted solar wind data from the Advanced Composition Explorer (ACE) spacecraft for this interval shows the IMF was relatively steady with $(B_x, B_y, B_z) = (7, -16, -8)$ nT. The negative B_y of the IMF twists the current pattern pushing current poleward on the dayside.

The southern hemisphere data are represented as looking through the Earth from above the north pole. The Iridium tracks in Fig. 7.2a show the larger offset of the spacecraft intersection location from the AACGM pole in the southern compared with the northern hemisphere. This largely contributed to the difficulty in achieving data fits for the southern hemisphere using the pre-AMPERE, cross-track only Iridium data.

The availability of both b_θ and b_ϕ data from Iridium post-2009 allows improved estimates, particularly for the southern hemisphere. As an example, consider the southern hemisphere data for 2348–2358 UT, 4 April 2010 and the straightforward application of a spherical cap harmonic fit as shown in Fig. 7.3. While the fitted magnetic fields look reasonable, the field aligned current pattern shows 'stripes' of current segments, particularly on the dawn side over $70^0 - 80°$ latitude. This is an effect caused by the mostly longitudinally oriented currents 'slicing' across the finite basis function sum, particularly in longitude where $m \leq 6$ in Eq. (7.35) in order to avoid aliasing.

The 'pole' (or zero co-latitude) location for the basis functions can be placed anywhere on the sphere (with radius of $R_E + 780$ km altitude). In order to maximise longitudinal resolution, zero co-latitude for the data fit is located at the average Iridium track intersection point, which for this case is near 70° and 21 MLT. Given the Iridium satellite altitude of 780 km and the spacing in each orbit plane, it takes about 9 min for full latitude data coverage. Therefore, each fit involves a data collection window of 10 min. For a given 10 min data interval, the average of the satellite track intersections is obtained from the 15 track pair combinations. There are two average intersection locations, one for each hemisphere. For a given hemisphere of data, the 'shifted pole' and the input magnetic perturbation data in (orthogonal) geographic (GEO) coordinates are used for the spherical cap harmonic fit.

An outline of the algorithm to improve the fit is as follows. For a given 10 min interval, the data are separated into northern and southern hemisphere caps. The magnetic field perturbation values midway between the Iridium satellite tracks are estimated and folded into the fit with reduced weighting. The simplest estimation scheme is a linear fit (averaged values) between adjacent Iridium tracks. A more advanced approach is to use Spherical Elementary Currents, as described below. Either way, a set of 'ghost' data tracks are generated between the Iridium satellite

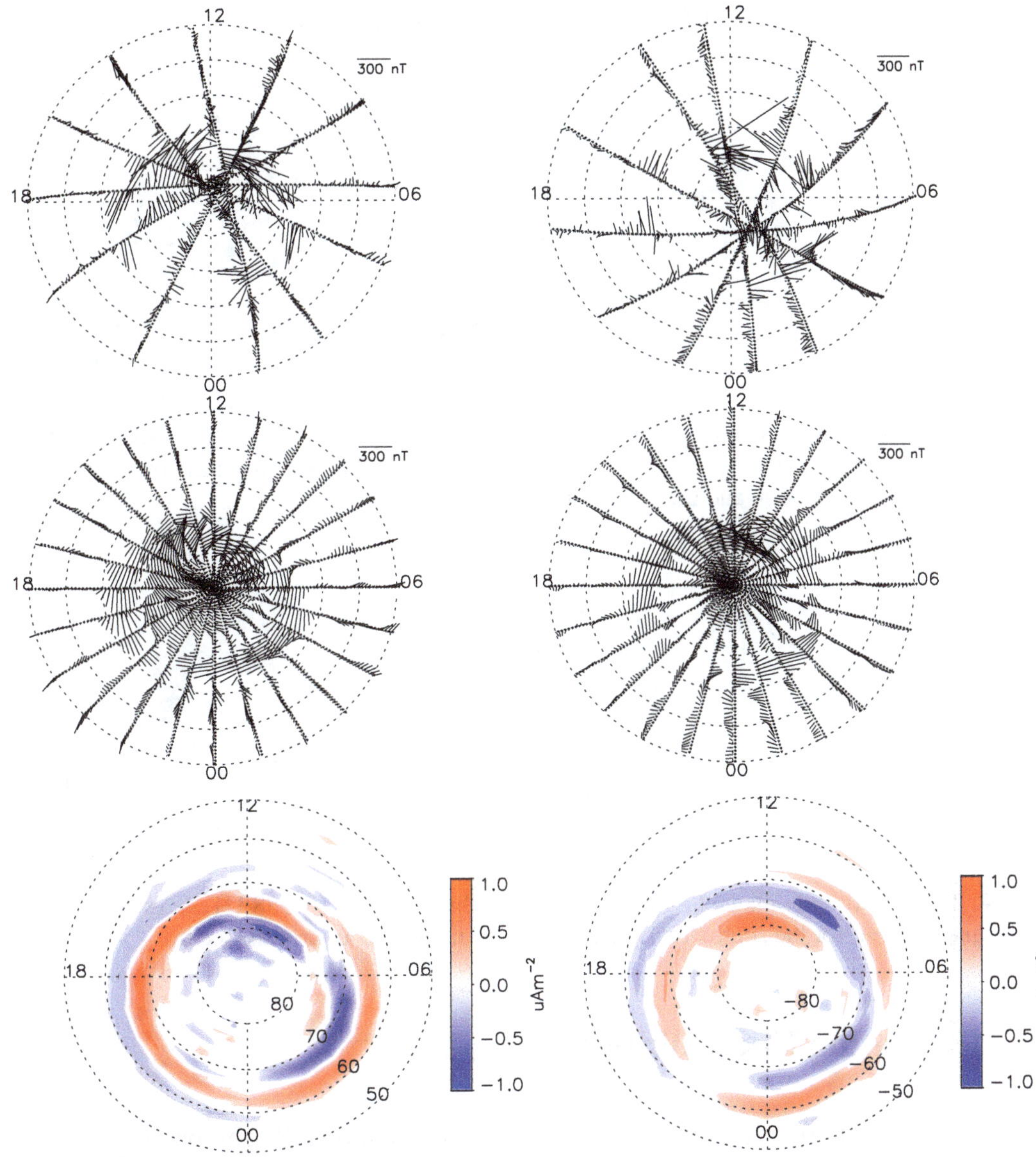

Fig. 7.2 (top) The perturbation magnetic field data from Iridium for 0520–0530 UT, 24 August 2010, (centre) the fitted magnetic field and (bottom) the radial current density for northern (left) and southern (right) hemispheres plotted in AACGM and MLT coordinates (see text). Red identifies outward and blue the inward current

tracks, allowing larger values for m in the fit. The input Iridium data are treated as more important than the 'ghost' data in the weightings for the data fit. The $l = 50$, $m = 8$ spherical cap harmonic fit to the data in Fig. 7.3 is shown in Fig. 7.4.

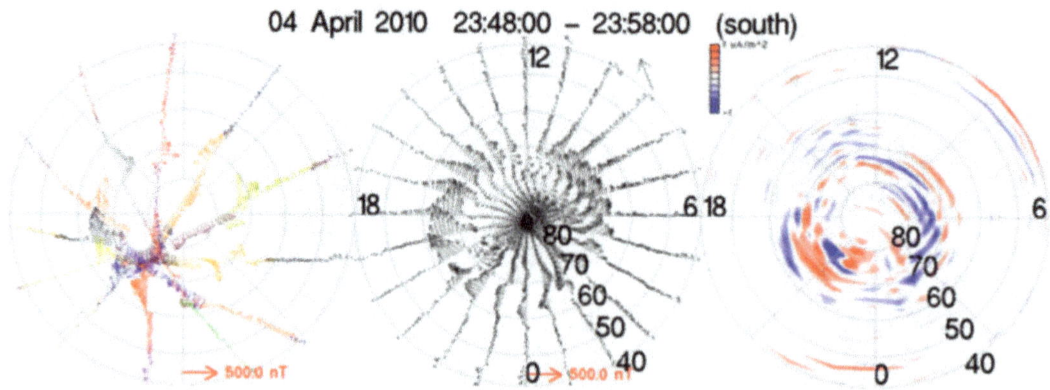

Fig. 7.3 The perturbation magnetic field data from Iridium for 2348–2358 UT, 4 April 2010 for the southern hemisphere. The 'stripes' in current on the dawn side result from the offset between the satellite track intersection and the centre of the current system

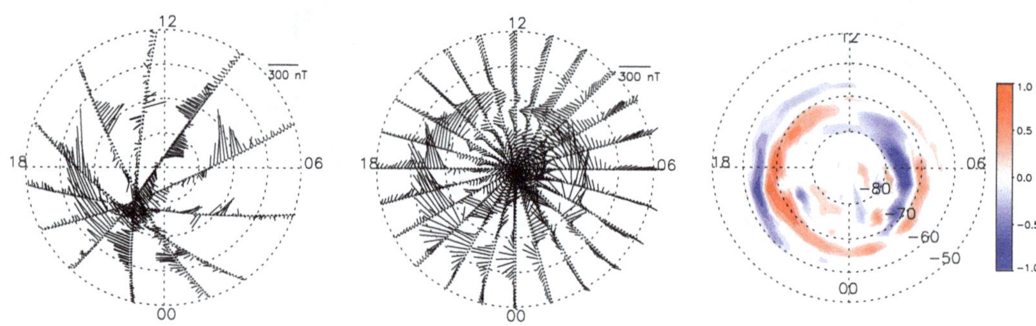

Fig. 7.4 The perturbation magnetic field data from Iridium for 2348–2358 UT, 4 April 2010 and field aligned current for the south hemisphere using the improved algorithm (see text)

7.6 Estimating Uncertainties

Magnetometer data obtained from the Iridium satellite constellation have been used to estimate the field aligned current (Waters et al. 2001; Anderson et al. 2002; Green et al. 2009), high latitude Poynting flux (Waters et al. 2004; Korth et al. 2008) and ionosphere conductance (Green et al. 2007). Since these are key quantities in magnetosphere–ionosphere coupling it is important to estimate the magnitudes and identify sources of uncertainties in the AMPERE data products. In this section, the uncertainties in the magnetic field and the derived current are discussed.

Uncertainties in the measured magnetic field values involves the magnetometer specifications and performance on each Iridium satellite and the associated analog to digital data conversion. The instruments are a fluxgate design which sense the total magnetic field in three orthogonal component directions. These data are reduced to (b_r, b_θ, b_ϕ) by subtraction from a main field model, filtering and adjustments for orthogonality and channel cross-talk as described by Anderson et al. (2000). This process also provides the data quality values which are used as the measurement errors, σ_i.

The perturbation magnetic field data are fit to a set of orthogonal functions with the coefficients determined by minimising the least squared error. The merit function is (Press et al. 1986)

$$\chi^2 = \sum_{i=1}^{N} \left[\frac{b_i(r,\theta,\phi) - Q_i(r,\theta,\phi)}{\sigma_i} \right]^2 \tag{7.41}$$

where $Q_i(r, \theta, \phi)$ are the estimated values obtained from the fit coefficients $a_{l,m}$ via Eq. (7.35) and $b_i(r, \theta, \phi)$ are the experimental data. This is a common data fitting method and discussions of the 'goodness of fit' may be found in many statistical analysis texts such as Johnson and Wichern (2002) and Press et al. (1986). If measurement errors are normally distributed, then the merit function in Eq. (7.41) follows a chi-squared distribution and the least squared and maximum likelihood estimates are equivalent.

The input (b_r, b_θ, b_ϕ) values are located at coordinates (r, θ, ϕ) where the coefficients and basis functions provide the fitted perturbation magnetic field estimates, $(b_{r,f}, b_{\theta,f}, b_{\phi,f})$. For a data sample interval of 20 s, the number of magnetic field values from Iridium for a 10 min interval is $\approx$2000. The residuals are the differences between the input and fitted values and these are included in the AMPERE data products. For estimating the radial current density, we focus on the (b_θ, b_ϕ) data. The input versus the fitted data for 0520–0530 UT, 24 August, 2010 are shown in the two upper panels of Fig. 7.5. A 'good' fit to the data is shown by points close to the line with unity slope. In order to check for anomalies in the residuals and the variance, the residuals are plotted versus the predicted values in the lower panels of Fig. 7.5. There is no clear trend, indicating the basis function model adequately describes most of the data and the variance is independent of the input data.

The minimisation of the merit function also yields a covariance matrix which is related to the data uncertainties if the residuals have a normal distribution. Figure 7.6 shows the residuals as a function of the normal distribution quartile (Q-Q plots) [e.g. see Johnson and Wichern (2002)] for the fitted (b_θ, b_ϕ) data. A straight line indicates the data are from a normal distribution. This is approximately true for the b_θ component and out to one and a half standard deviations from the mean for the b_ϕ component. A measure of the 'straightness' of the line in the Q-Q plot is the correlation coefficient, which for these data are $r_{Q,\theta} = 0.97$ and $r_{Q,\phi} = 0.88$. A test, at some significance level, for rejecting the hypothesis that the residuals are from a normal distribution may be formulated using these values of r_Q. The change in slope in the tails of the b_ϕ component is the reason for the smaller $r_{Q,\phi}$ and this corresponds to the larger values of b_ϕ which are often confined to a few degrees in latitude and would be better estimated using a higher order fit.

Given estimates of the uncertainties in (b_θ, b_ϕ), the associated uncertainties in the radial current may be determined. The procedure for taking the perturbation magnetic field values obtained from Iridium to estimate the radial current density involves the derivatives of potential functions according to Eq. (7.22). An uncertainty estimate

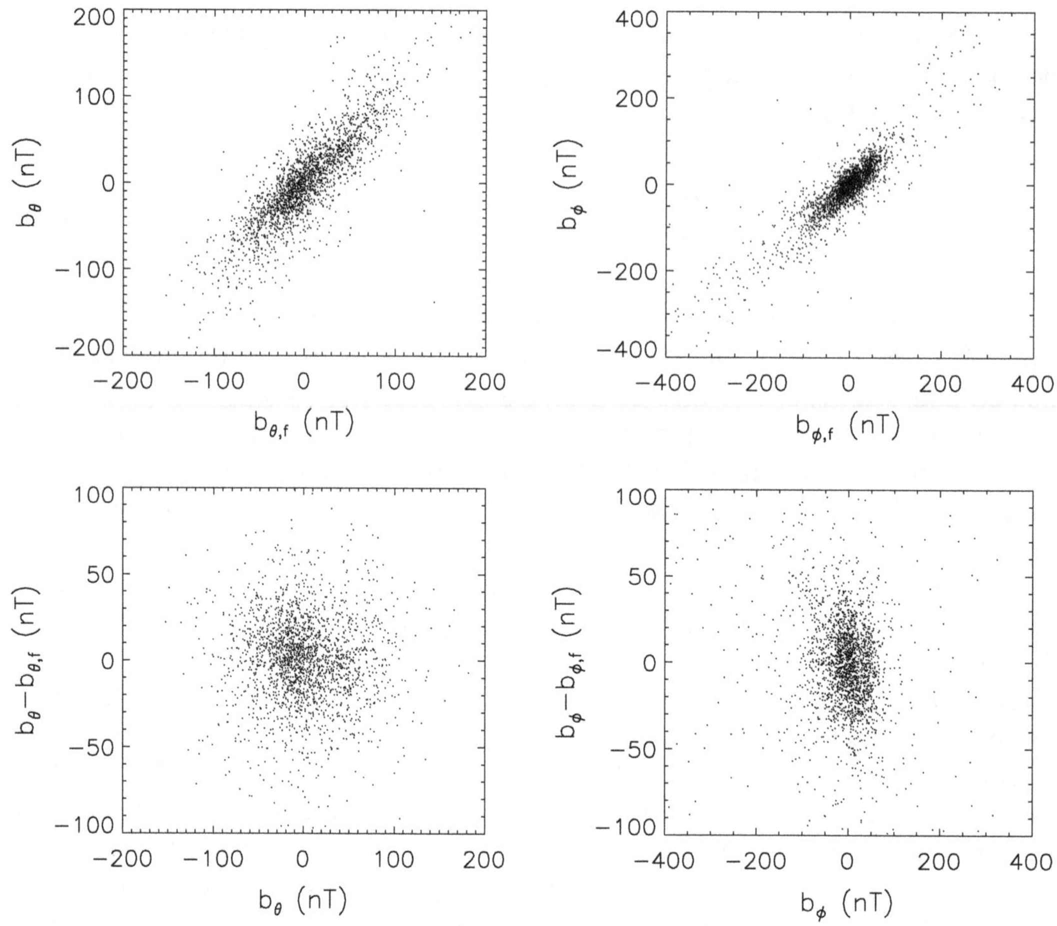

Fig. 7.5 Top panels: Input versus fitted magnetic perturbation data for 0520–0530 UT, 24 August 2010. Bottom panels: Residuals versus spherical cap harmonic fitted magnetic perturbation data

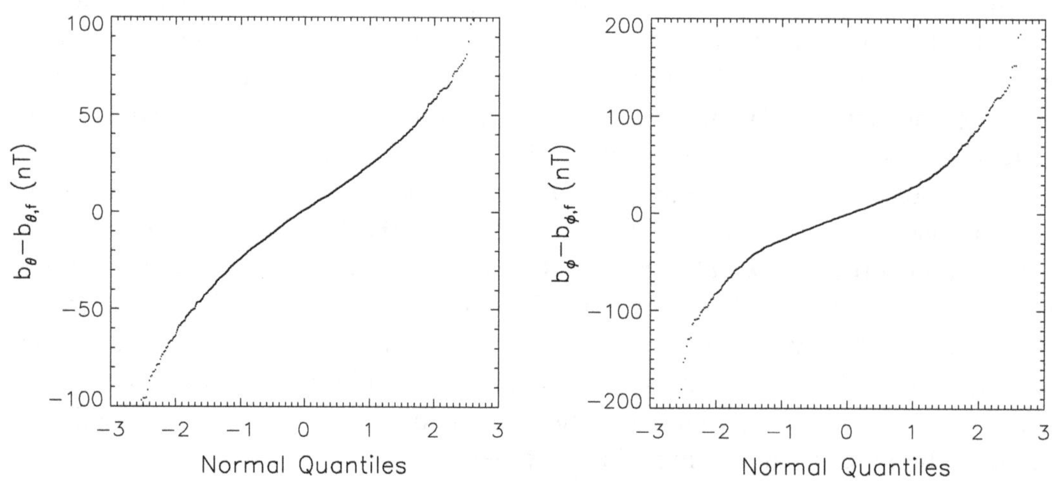

Fig. 7.6 Normal distribution quartiles versus residuals for the Iridium data and spherical harmonic fit of the data in Fig. 7.2

may be obtained by assuming a sheet current structure that flows radially and extends along the ϕ coordinate. Integration of Ampere's law (Eq. 7.4) gives $b = \mu_0 K/2$ where K is the current density in Am^{-1}. The field aligned current density, J (Am^{-2}) is obtained through division by the current sheet width. Therefore, the proportional error in J is the sum of the proportional errors in the perturbation magnetic field and the resolution in latitude (θ).

An estimate for the uncertainty in the radial current may also be obtained using the statistics of the perturbation magnetic field values. The radial component of Eq. (7.4) is

$$J_r = \frac{1}{\mu_0 r sin\theta}\left[\frac{\partial}{\partial\theta}(sin\theta b_\phi) - \frac{\partial b_\theta}{\partial\phi}\right] \tag{7.42}$$

Therefore, $J_r = j_{r,1} + j_{r,2} + j_{r,3}$ where

$$j_{r,1} = \frac{1}{\mu_0\ r}\frac{\partial b_\phi}{\partial\theta} \tag{7.43}$$

$$j_{r,2} = \frac{1}{\mu_0\ r\ tan\theta} b_\phi \tag{7.44}$$

$$j_{r,3} = -\frac{1}{\mu_0\ r\ sin\theta}\frac{\partial b_\theta}{\partial\phi} \tag{7.45}$$

The measurement resolution in the θ coordinate is related to the spherical harmonic fit order, l (or n_k for cap harmonics). The highest order, associated Legendre function has a minimum wavelength, λ_{min} so the uncertainty in θ is $\delta\theta = \lambda_{min}/2$. Similarly, the separation of the Iridium orbit planes gives the resolution in ϕ as $\delta\phi = \pi/6$. The uncertainties in b_θ and b_ϕ are δb_θ and δb_ϕ and may be estimated from the statistics of the residuals. From Eqs. (7.43)–(7.45), the uncertainties in the field aligned current density are

$$\delta j_{r,1} = \frac{1}{\mu_0\ r}\frac{2}{\lambda_{min}}\sqrt{2\ (\delta b_\phi)^2} \tag{7.46}$$

$$\delta j_{r,2} = \frac{1}{\mu_0\ r\ tan\theta}\delta b_\phi \tag{7.47}$$

$$\delta j_{r,3} = \frac{1}{\mu_0\ r\ sin\theta}\frac{6}{\pi}\sqrt{2\ (\delta b_\theta)^2} \tag{7.48}$$

with the total uncertainty in the field aligned current, $\delta J = \sqrt{\delta j_{r,1}^2 + \delta j_{r,2}^2 + \delta j_{r,3}^2}$. The mass production of data products for AMPERE involves a fixed set of analysis and fit parameters. Users of these data products should communicate with the AMPERE science data team for advice on the quality and uncertainties in the data for specific intervals.

7.7 Spherical Elementary Currents and Iridium Data

The orbit configuration of the Iridium satellites provides data at varying spatial separations in longitude. The spatial density of data samples is greater around the Iridium satellite track intersection locations providing increased spatial resolution of the currents where the satellite tracks are close. The average location of pairs of Iridium track intersections moves relative to geomagnetic coordinates and the currents form different spatial patterns depending on the IMF and magnetic activity. Therefore, an average Iridium track intersection location in a given hemisphere might occur between the main current systems. This represents a difficult situation for data fitting using a finite number of SCHA basis functions. One alternative approach to SCHA is the Spherical Elementary Currents Systems (SECS).

The SECS basis functions were described by Amm (1997, 2001). This approach has many similarities with the multiple multipole method used for solving the Maxwell equations (e.g. Ballisti and Hafner 1983). The SECS basis functions have been applied to both ground and satellite magnetometer data. For the latter, with data confined along single satellite tracks, a one-dimensional (1D) version of the method has been developed (Vanhamaki et al. 2003; Juusola et al. 2006). In this section, application of the 2D, curl-free SECS basis functions to Iridium data is described. These are used for the AMPERE *regional fit* data products with the aim of providing enhanced spatial resolution around the Iridium satellite track intersection regions.

As discussed in Chap. 2 (this volume), the SECS expansion of spatial data are based on a divergence-free and a curl-free basis set to give the total field. They relate the field aligned and (horizontal) ionospheric currents with the perturbation magnetic fields. The curl-free current basis is

$$J_{cf} = \frac{I_0}{4\pi R} cot \frac{\theta'}{2} \mathbf{e}_{\theta'} \tag{7.49}$$

where the dashed coordinates reference the local SECS coordinate system with the pole at $\theta' = 0$, R is the radius of the sphere on which the poles are placed and $\mathbf{e}_{\theta'}$ is the unit vector. The magnetic field from the curl-free current basis is

$$b_{cf} = -\frac{\mu_0 I_0}{4\pi R} cot \frac{\theta'}{2} \mathbf{e}_{\phi'} \tag{7.50}$$

for r > R and zero for r < R . It is straightforward to show that

$$\nabla \bullet J_{cf} = -\frac{I_0}{4\pi R^2} \tag{7.51}$$

$$\nabla \times b_{cf} = \mu_0 J_r \tag{7.52}$$

There are a number of parameters to adjust in the SECS approach to data fitting. The main considerations are the number and location of the poles and solution grid

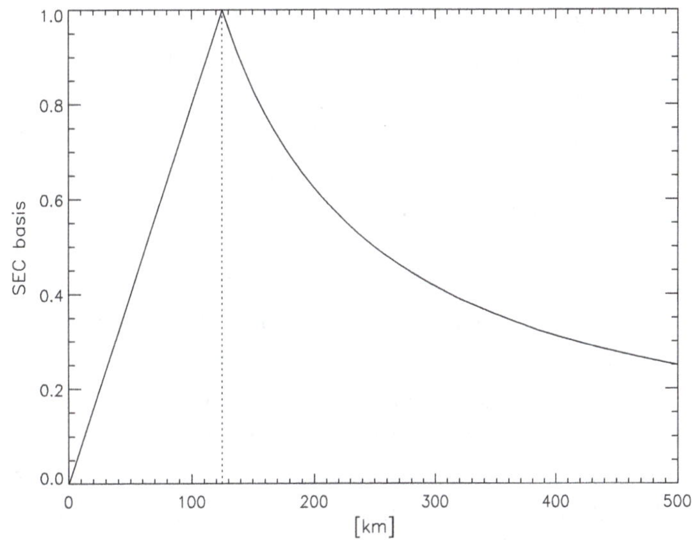

Fig. 7.7 Normalised basis function used in the 2D spherical elementary current system. The dashed line shows the limit angle where the tangent function is used to avoid infinity at the pole

points. For dense spatial grids compared to the experimental data locations and over limited spatial extent, the modelled magnetic field and currents appear reasonable as illustrated in examples provided by Amm (1997) and Amm and Viljanen (1999). Constraints on the number of pole and solution locations include computation time required to solve the matrix equation for the fit coefficients and the spatial information afforded by the input experimental data. Depending on the number of poles and the spatial separation between the input data and the pole locations, the current pattern can become less smooth with the appearance of patches of localised current.

A direct application of the SECS basis to Iridium magnetic field perturbation data proved unsatisfactory for a number of reasons that are related to the spatial properties of the basis functions and the Iridium data. Equations (7.49) and (7.50) show an infinity at $\theta' = 0$. This was eliminated by changing the cotangent function to a tangent function within a 'limit angle' that spans a SECS grid cell (Vanhamaki et al. 2003). The modified basis function is shown in Fig. 7.7 with the limit angle shown by a dashed line. The horizontal axis scale is $R\theta'$ where R is 6370 + 780 km, the radius of the Iridium constellation.

Equation (7.49) and Fig. 7.7 show that there is no flexibility for the spatial extent of the basis function. However, the Iridium data have finer spatial resolution around the satellite track intersection locations compared with data at lower latitudes. In practice, this results in a mismatch between the spatial properties of the basis function and the data. In fact, locating the SEC poles midway between the satellite tracks gave very small root mean squared error values between the input magnetic field and fitted values. However, the currents became very localised and did not span the space between the satellite tracks. This is an illustration of the care required when interpreting fitted data using a root mean squared error metric.

In order to design a more robust metric for fitting the 2D data from Iridium, a region 1 and 2 model current system was constructed. The magnetic perturbations from this model current system were calculated using both a Biot–Savart integration and a high order spherical cap harmonic expansion. This provided model radial current and horizontal magnetic field values with an improved data fit metric based on the root mean squared values for both magnetic field and current. Using these model data, a number of spatial data fit methods and their variations were assessed.

The spatial information needs to be more flexible for the 2D SECS method applied to Iridium data. One approach was to extend the design matrix by including a constraint on the longitudinal derivative of the combined basis function solution. While this modification smoothed the resulting radial current, it required an estimate of a 'reasonable' derivative to be applied a-priori, and the current magnitudes were reduced compared with the input values.

Another approach was to consider a different basis function formulation with spatial extent flexibility. For the curl-free magnetic field and radial currents in AMPERE, the source current is assumed to be radial but the horizontal spatial variation may be different to a cotangent function. In principle, a Biot–Savart integration may be coded in order to obtain the magnetic field from any current distribution. However, analytic expressions have advantages for computational speed. A Gaussian basis function was trialed, with different widths used to adjust the spatial extent. A disadvantage with this approach was found to be related to the symmetric nature of the basis function, while the input Iridium data have less data in longitude compared with latitude.

Before describing the approach chosen, a comment or two on the spatial grids are in order. The SECS algorithm involves three coordinate systems. For AMPERE these are the input data in GEO, the number and location of the SECS basis function poles and the grid on which the solution is desired. The input data locations are determined by the Iridium satellite locations. The SECS pole and solution spatial grids were constrained by being interlaced as described in Chap. 2 (this volume). The desired data product is the radial current, in μAm^{-2}, which requires calculation of the fit coefficients, I_0 and horizontal area. Therefore, a quasi-equal area grid was chosen, similar to that used by Ruohoniemi and Baker (1998) for HF radar data. The latitudinal separation is fixed and the longitudinal separation varies with latitude in order to ensure an integer number of grid cells with the same area in each latitudinal ring.

The method chosen for obtaining *regional data* fits around the Iridium satellite track intersection locations was a combination of the SECS and spherical cap harmonic bases. The input, preprocessed Iridium magnetometer data are used to compute the SCHA solution, $(b_{\theta,f}, b_{\phi,f})$ over the quasi-equal area grid. This grid is then searched for locations where the input data are located and the SCHA model values are replaced by the input (b_θ, b_ϕ). The combined input and SCHA data are then used to compute the SECS solution, weighting the SCHA values as less important. The weighting was determined by using the model current and magnetic field data, computing data fits with different weighting in order to minimise the residuals. Once the SECS coefficients are obtained, values for the model magnetic field may be computed on any spatial grid encompassed by the input data. Vanhamaki (2007) recommend that the SECS grid exceed the spatial extent of the solution grid. It is straightforward to also obtain values for the equivalent horizontal current density.

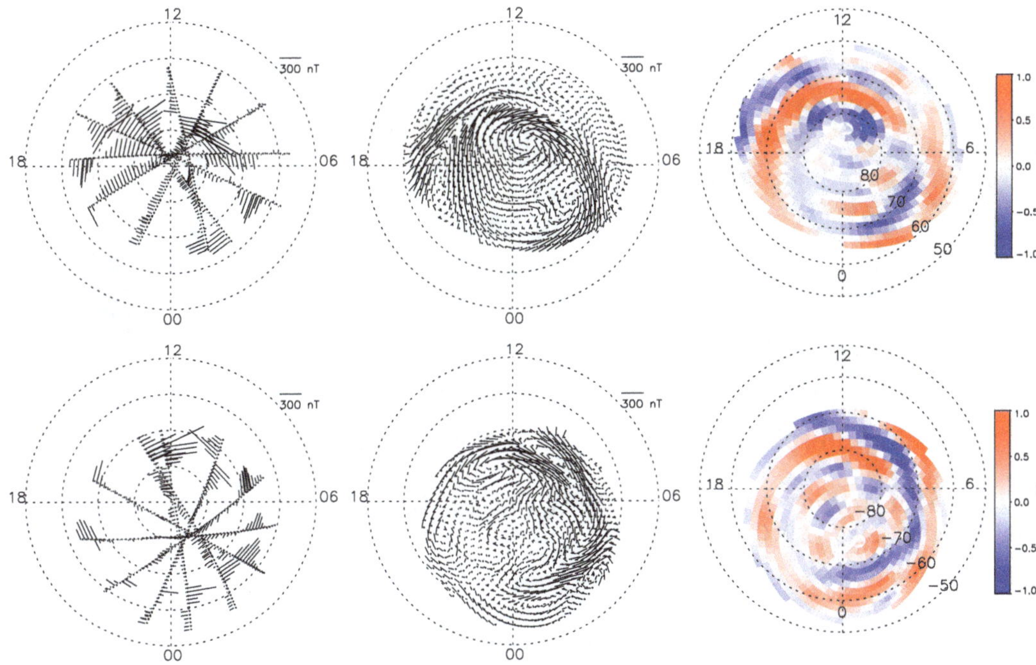

Fig. 7.8 Regional fit data for 0520–0530 UT, 24 August 2010 processed using the SECS method; (top) northern and (bottom) southern hemisphere data. Red identifies outward and blue the inward current

The AMPERE *regional data* method applied to the Iridium data for 0520–0530 UT 24 Aug 2010 is shown in Fig. 7.8.

7.8 AMPERE and Other Data Sets

The temporal and spatial resolution of the AMPERE data provides unique opportunities to investigate details of the electrodynamics of magnetosphere–ionosphere energy exchange and coupling. The previous, cross-track only, Iridium data have been combined with other comprehensive data sets to provide estimates of the input Poynting flux (Waters et al. 2004; Korth et al. 2008) and ionosphere conductivity (Green et al. 2007). In addition to improved temporal resolution, the AMPERE data may yield improved estimates of these parameters through the introduction of the full vector perturbation magnetic field values. The integration of AMPERE data with the electric field measurements obtained from the Super Dual Auroral Radar Network (SuperDARN) is an example.

The SuperDARN is an international consortium that operate 30 high frequency (8–20 MHz) over-the-horizon research radars located primarily to study the auroral and high latitude regions (http://vt.superdarn.org). Using a 16 antenna broadside array, the signal is phased to form a broad vertical but narrow azimuth (≈3°) radiation pattern that is steered over 52° in 16 equi-spaced azimuth directions. A multi-pulse

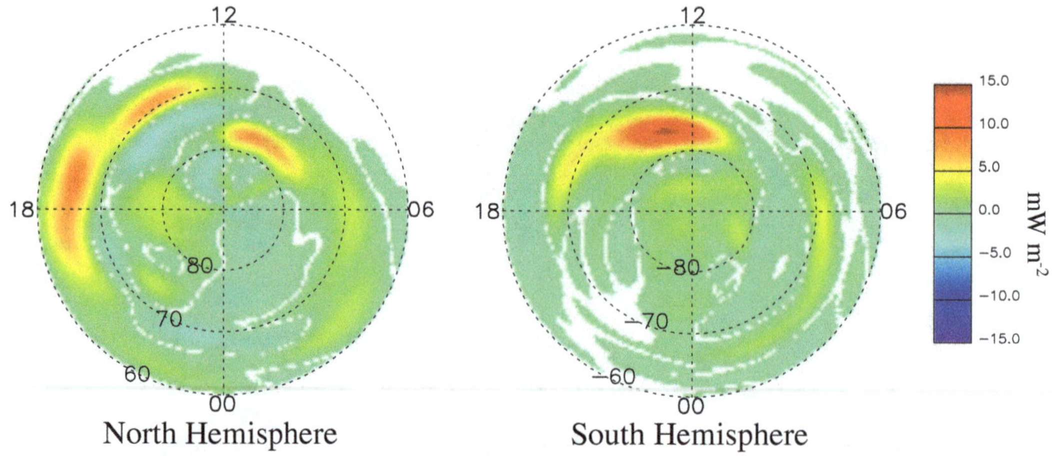

Fig. 7.9 Radial Poynting flux calculated from combined Iridium and SuperDARN data for 0520–0530 UT on 24 August 2010

transmit pattern and autocorrelation data processing allow ranges out to ≈3000 km and Doppler shifts from target velocities up to ≈1 km/s to be resolved. The dual instrumentation features overlapping radar fields of view in order to resolve the horizontal plasma velocity vectors, **v**. The electric field vector, **E** in the ionosphere is then estimated from the geomagnetic field, **B** using $\mathbf{E} = -\mathbf{v} \times \mathbf{B}$. The location of the radars limits the magnetic latitude extent of the data to ≈60°, although expansion of the network to lower latitudes is progressing.

The electric field estimates from the radars are available from both hemispheres over latitudes extending from the poles to a spherical cap size of 30°. The AMPERE data provide the perturbation magnetic field estimates at the same locations as the radar data. The combined data are used to calculate the radial component of the Poynting flux, $\mathbf{E} \times (b_{\theta,f}, b_{\phi,f})$, as shown in Fig. 7.9. The latitude extent is limited by the radar data. Comparing with the Birkeland current pattern in Fig. 7.2c shows that the downward Poynting flux is largest at latitudes between the region 1 and 2 current systems. The spatial distribution is often quite different in the northern compared with the southern hemisphere and this is thought to be controlled by details of the ionosphere conductance. The total power is obtained by integrating the radial Poynting flux over the area of the caps. For the 24 August 2010, 0520–0530 UT interval the power estimate is ≈30 GW for the north and ≈20 GW for the southern hemisphere. These are most likely underestimates due to factors such as the limited spatial coverage and that the estimated electric and magnetic fields are probably smaller due to the choice of latitude resolution (spherical harmonic fit order) in both data sets as discussed by Korth et al. (2008).

7.9 Conclusion

The AMPERE project provides estimates of the radial current and the full vector perturbation magnetic field at any location over a sphere with origin at Earth's centre and radius 780 km above Earth's surface. The magnetic field may be adjusted for other altitudes by multiplication of the data by a factor that has an $r^{3/2}$ dependence. For AMPERE, the magnetometer data from the Iridium satellites are sampled at 20 s intervals, a factor of 10 increase over previous data obtained from this constellation. For storm time case study intervals, the sample interval may be reduced to 2 s. This improves the spatial resolution in latitude, requiring a high order in the spherical harmonic fitting process. The time interval required to obtained full latitude coverage is several minutes which corresponds with the time taken for each Iridium satellite to move (in their orbit plane) the distance equal to their latitude spacing. These two factors determine the minimum number of data points required to calculate the spherical harmonic expansion coefficients. The full vector perturbation magnetic field and the field aligned current are then estimated over a suitable spatial grid and made available to the scientific community.

Alternative basis functions, such as the SECS, may also be used. While the spherical cap harmonics have a more global reach due to the domain of the basis functions, the SECS basis is more localised and may be used for regional data fits. AMPERE data products are available for download from http://ampere.jhuapl.edu. Although every effort is made to provide the highest quality, as with all experimental data, the Iridium data are not perfect. There are a number of parameters that are involved in the data processing, starting from receipt of the data from Iridium Communications through to the final AMPERE data products. Through experience, various parameters have been chosen for bulk data processing and web display. The AMPERE web site provides researchers with sufficient information to identify intervals of interest. As specified on the AMPERE web site, in order to ensure high quality and the use of optimum parameters for the processed data, any publication of AMPERE data products should involve consultation with the AMPERE science team.

Acknowledgements Support for AMPERE has been provided under NSF sponsorship under grants ATM-0739864 and AGS-1420184. We are greatly indebted to the contribution of Iridium Communications for providing the data for AMPERE. We thank the International Space Science Institute (ISSI) in Bern, Switzerland for supporting the Working Group Multi Satellite Analysis Tools—Ionosphere from which this chapter resulted. The editors thank Ryan Mcgranaghan for his assistance in evaluating this chapter.

References

Amm, O. 1997. Ionospheric elementary current systems in spherical coordinates and their application. *Journal of Geomagnetism and Geoelectricity* 49: 947–955.

Amm, O. 2001. The elementary current method for calculating ionospheric current systems from multisatellite and ground data. *Journal of Geophysical Research* 106: 24,843–24,855.

Amm, O., and A. Viljanen. 1999. Ionospheric disturbance magnetic field continuation from the ground to the ionosphere using spherical elementary current systems. *Earth Planet Space* 51: 431–440.

Anderson, B.J., K. Takahashi, and B.A. Toth. 2000. Sensing global Birkeland currents with Iridium engineering magnetometer data. *Geophysical Research Letters* 27: 4045–4048.

Anderson, B.J., K. Takahashi , T. Kamei, C.L. Waters, and B.A. Toth. 2002. Birkeland current system key parameters derived from Iridium observations: Method and initial validation results. *Journal of Geophysical Research* 107 (A6). https://doi.org/10.1029/2001JA000080.

Anderson, B.J., H. Korth, C.L. Waters, D.L. Green, and P. Stauning. 2008. Statistical Birkeland current distributions from magnetic field observations by the Iridium constellation. *Annales Geophysicae* 26: 671–687.

Anderson, B.J., H. Korth, C.L. Waters, D.L. Green, V.G. Merkin, R.J. Barnes, and L.P. Dyrud. 2014. Development of large-scale Birkeland currents determined from the active magnetosphere and planetary electrodynamics experiment. *Geophysical Research Letters* 41: 3017–3025. https://doi.org/10.1002/2014GL059941.

Anderson, B.J., Olson H.C.N. Korth, R.J. Barnes, C.L. Waters, and S.K. Vines. 2018. Temporal and spatial development of global Birkeland currents. *Journal of Geophysical Research* 123: 4785–4808. https://doi.org/10.1029/2018JA025254.

Backus, G. 1986. Poloidal and toroidal fields in geomagnetic field modeling. *Reviews of Geophysics* 24: 75–109.

Backus, G., R. Parker, and C. Constable. 1996. *Foundations of geomagnetism*. Cambridge University Press.

Baker, K., and S. Wing. 1989. A new magnetic coordinate system for conjugate studies at high latitudes. *Journal of Geophysical Research* 94: 9139–9243.

Ballisti, R., and C. Hafner. 1983. The multiple multipole method in electro- and magnetostatic problems. *IEEE Transactions on Magnetics* 19: 2367–2370. https://doi.org/10.1109/TMAG.1983.1062871.

Bythrow, P., T. Potemra, and R. Hoffman. 1982. Observations of field-aligned currents, particles, and plasma drift in the polar cusps near solstice. *Journal of Geophysical Research* 87: 5131–5139. https://doi.org/10.1029/JA087iA07p05131.

Chapman, S., and J. Bartels. 1940. *Geomagnetism*, vol. 2. Oxford: Clarendon Press.

de Santis, A., J.M. Torta, and F.J. Lowes. 1999. Spherical harmonic caps revisited and their relationship to ordinary spherical harmonics. *Physics and Chemistry of the Earth* 24: 935–941.

Erlandson, R.E., L.J. Zanetti, T.A. Potemra, P.F. Bythrow, and R. Lundin. 1988. IMF B(y) dependence of region 1 Birkeland currents near noon. *Journal of Geophysical Research* 93: 9804–9814. https://doi.org/10.1029/JA093iA09p09804.

Gary, J.B., R.A. Heelis, and J.P. Thayer. 1995. Summary of field aligned Poynting flux observations from DE 2. *Geophysical Research Letters* 22: 1861.

Green, D.L., C.L. Waters, B.J. Anderson, H. Korth, and R.J. Barnes. 2006. Comparison of large-scale Birkeland currents determined from Iridium and SuperDARN data. *Annales Geophysicae* 24: 941–959.

Green, D.L., C.L. Waters, H. Korth, B.J. Anderson, A.J. Ridley, and R.J. Barnes. 2007. Technique: Large-scale ionospheric conductance estimated from combined satellite and ground-based electromagnetic data. *Journal of Geophysical Research* 112(A05303). https://doi.org/10.1029/2006JA012069.

Green, D.L., C.L. Waters, B.J. Anderson, and H. Korth. 2009. Seasonal and interplanetary magnetic field dependence of the field-aligned currents for both northern and southern hemispheres. *Annales Geophysicae* 27: 1701–1715.

Haines, G.V. 1985. Spherical cap harmonic analysis. *Journal of Geophysical Research* 90: 3195–3216.

Haines, G.V. 1988. Computer programs for spherical cap harmonic analysis of potential and general fields. *Computers and Geosciences* 14: 413–447.

Iijima, T., and T.A. Potemra. 1976. The amplitude distribution of field-aligned currents at northern high latitudes observed by Triad. *Journal of Geophysical Research* 81: 2165–2174.

Iijima, T., T.A. Potemra, L.J. Zanetti, and P.F. Bythrow. 1984. Large-scale Birkeland currents in the dayside polar region during strongly northward IMF: A new Birkeland current system. *Journal of Geophysical Research* 89: 7441–7452.

Johnson, R.A., and D.W. Wichern. 2002. *Applied multivariate statistical analysis*. Prentice-Hall.

Juusola, L., O. Amm, and A. Viljanen. 2006. One-dimensional spherical elementary current systems and their use for determining ionospheric currents from satellite measurements. *Earth Planets and Space* 58: 667–678.

Korth, H., B.J. Anderson, J.M. Ruohoniemi, H.U. Frey, C.L. Waters, T.J. Immel, and D.L. Green. 2008. Global observations of electromagnetic and particle energy flux for an event during northern winter with southward interplanetary magnetic field. *Annales Geophysicae* 26: 1415–1430.

Korth, H., B.J. Anderson, and C.L. Waters. 2010. Statistical analysis of the dependence of large-scale Birkeland currents on solar wind parameters. *Annales Geophysicae* 28: 515–530.

Kosch, M.J., and E. Nielsen. 1995. Coherent radar estimates of average high-latitude ionospheric Joule heating. *Journal of Geophysical Research* 100:12,201–12,215.

Lamb, H. 1881. On the oscillations of the viscous spheroid. *Proceedings of the London Mathematical Society* 13: 51.

McQuarrie, D.A. 2003. *Mathematical methods for scientists and engineers*. Sausalito, California: University Science Books.

Press, W.H., S.A. Teukolsky, W.T. Vetterling, and B.P. Flannery. 1986. *Numerical recipes in Fortran 77: The art of scientific computing*, 2nd ed. Cambridge University Press.

Ruohoniemi, J.M., and K.B. Baker. 1998. Large scale imaging of high latitude convection with Super Dual Auroral Radar Network HF radar observations. *Journal of Geophysical Research* 103:20,797–20,811.

Vanhamaki, H. 2007. Theoretical modeling of ionospheric electrodynamics including induction effects. Finnish Meteorological Institute Contributions, electronic version available at http://ethesishelsinkifi/66.

Vanhamaki, H., O. Amm, and A. Viljanen. 2003. One-dimensional upward continuation of the ground magnetic field disturbance using spherical elementary current systems. *Earth Planets Space* 55: 613–625.

Waters, C.L., M.D. Sciffer. 2008. Field line resonant frequencies and ionospheric conductance: Results from a 2-D MHD model. *Journal of Geophysical Research* 113(A05219), https://doi.org/10.1029/2007JA012822.

Waters, C.L., B.J. Anderson, and K. Liou. 2001. Estimation of global field aligned currents using the Iridium$^{riptsize®}$ system magnetometer data. *Geophysical Research Letters* 28 (11): 2165–2168.

Waters, C.L., B.J. Anderson, R.A. Greenwald, R.J. Barnes, and J.M. Ruohoniemi. 2004. High-latitude Poynting flux from combined Iridium and SuperDARN data. *Annales Geophysicae* 22: 2861–2875.

Zanetti, L.J., T.A. Potemra, T. Iijima, W. Baumjohann, and P.F. Bythrow. 1984. Ionospheric and Birkeland current distributions for northward interplanetary magnetic field: Inferred polar convection. *Journal of Geophysical Research* 89: 7453–7458. https://doi.org/10.1029/JA089iA09p07453.

Chapter 8
ESA Field-Aligned Currents—Methodology Inter-comparison Exercise

Lorenzo Trenchi and The FAC-MICE Team

Abstract Various ESA projects and several proposals to first Swarm DISC Call for Ideas (May 2016) suggested possible evolution for the current Swarm Level 2 FAC products, and the implementation of quality flags for the FAC products. The Field-Aligned Currents—Methodology Inter-Comparison Exercise (FAC-MICE) consisted in comparison of the various methods to determine the FAC from Swarm data, with a test dataset of 28 Swarm auroral crossings delivered to participants last June. Eight groups performed the FAC-MICE analysis. The results of this exercise, discussed in the dedicated 'Swarm Ionospheric Currents Products workshop' in ESTEC on September 2017, highlighted the strengths of the various methods/approaches. Following discussion with the participants to this workshop, we are now working to develop an open source platform for user-definable FAC calculation.

8.1 Introduction

Field-aligned currents (FACs), also called Birkeland currents, are electric currents flowing along geomagnetic field lines at high magnetic latitudes, where they couple the magnetosphere and the ionosphere.

The presence of FACs was first proposed by the Norwegian explorer Kristian Birkeland, who reported magnetic perturbations on the ground near the Artic circle associated with auroras (Birkeland 1908, 1913). The existence of FACs was then confirmed by magnetic field perturbations detected in space, when the polar orbiting satellite 1962 38C made the first observations (Zmuda et al. 1966). On a global scale, FACs are organized along a persistent pattern around both magnetic poles, first highlighted by the statistical studies based on Triad data (Iijima and Potemra 1976a, 1976b, 1978). The FAC pattern consists of two concentric rings in the auroral

For the full author list of the FAC-MICE Team, please see the last page of this chapter.

L. Trenchi (✉)
Serco for ESA—European Space Agency, ESA/ESRIN, Directorate of Earth Observation Programmes, Frascati (Roma), Italy
e-mail: Lorenzo.Trenchi@ESA.int

M. W. Dunlop and H. Lühr (eds.), *Ionospheric Multi-Spacecraft Analysis Tools*, ISSI Scientific Report Series 17,
https://doi.org/10.1007/978-3-030-26732-2_8

oval, with opposite polarities at dawn and dusk: the poleward ring (Region 1) connects with the magnetopause current, and the equatorward ring (Region 2) connects with the magnetospheric ring current (e.g. Cowley 2000). Both Region 1 and Region 2 currents are closed via horizontal currents (primarily Pedersen currents) within the ionosphere current layer, at altitudes between approximately 100–150 km. The Region 1–2 pattern is an almost permanent feature, while the shape, the size and the intensity of these currents are related to the amount of energy transferred from the solar wind into the magnetosphere, which in turn is determined by the orientation of the interplanetary magnetic field (IMF), and the solar wind parameters.

Swarm is a three-satellite constellation mission developed from the European Space Agency (ESA) to measure, as a primary goal, the Earth's magnetic field with unprecedented accuracy. The three identical spacecraft, launched on November 22, 2013, are flying in polar orbits: the lower pair (Swarm A and C) at 460 km altitude, while Swarm B is flying at 510 km. Each spacecraft carries two magnetometers that measure the vector magnetic field and its absolute value, an electric field instrument, an accelerometer, and the instruments to determine the attitude and the position of the spacecraft, which are the GPS receiver, the startracker and a laser retroreflector.

By using magnetic field models, it is possible to separate the various contributions of the magnetic field signals measured by Swarm, i.e. the contribution from the core, the lithosphere and the magnetosphere components. This allows to use Swarm data to study several different phenomena. One of the main objectives of Swarm regards the study of the Ionospheric currents, and in particular FACs. The 'in situ' determination of current densities in space plasmas is generally obtained from spacecraft magnetic field data using the Ampere's law, under various assumptions regarding the stationarity and the orientation of the current sheet, and the uniform spatial variation of the magnetic field. Less strict assumptions are needed when multi-spacecraft data are used. For this reason, the multipoint measurements provided by the Swarm constellation, together with the unprecedented accuracy of Swarm magnetic field measurements, represent an optimal asset for the determination of FACs. Currently, ESA provides two different estimates of FAC, based on a single or multi-spacecraft approach (Ritter et al. 2013, see also Chap. 6 of this book):

- single spacecraft FAC products, based on 1 Hz magnetic field data, which provide three individual products for each of the 3 satellites
- dual spacecraft FAC product, based on 1 Hz magnetic field data collected by the lower pair (Swarm A and Swarm C), low-pass filtered at 20 s time scale in order to meet the time stationarity assumption (Ritter et al. 2013, see Chap. 6 of this book).

Various ESA projects are developing new methods to compute FAC and the ionospheric currents with Swarm data. In particular the Swarm data quality Investigation of Field-Aligned Current products, Ionoshepere, and Thermosphere systems (SIFACIT) is developing new algorithms to estimate FAC from single, dual and three spacecraft methods, together with quality indicators about the assumptions underlying FAC estimates (http://gpsm.spacescience.ro/sifacit/, see also Chap. 4 of this book).

Swarm-SECS is instead applying a two-dimensional version of the Spherical Elementary Current Systems (SECS) method to recover the Ionospheric currents and conductivities from Swarm magnetic and electric field data (http://space.fmi.fi/MIRACLE/Swarm_SECS/, see also Chaps. 2 and 3 of this book).

Also several proposals in response to the first ESA Swarm Call for Ideas for new data products and services (May 2016, https://earth.esa.int/web/guest/news/-/article/esa-swarm-call-for-ideas-for-new-data-products-and-services) focussed on FACs, suggesting possible new approaches to estimate FAC densities and quality indicators.

In order to identify a possible evolution of the present FAC products, and/or potential new FAC products and FAC quality indicators, ESA organized a FAC Methodology Inter-comparison Exercise (FAC-MICE), which consisted in comparison of the different FAC methods, based on a test dataset of 28 Swarm auroral crossings.

8.2 The FAC-MICE Test Dataset

The events in the FAC-MICE test dataset have been selected according to the following criteria:

- events uniformly distributed in local time
- events sampled during both quiet and active geomagnetic conditions, identified according to the values of the AE index
- some of the events were sampled when the orbit of Swarm B was close enough to the one of the lower pair, to allow FAC estimates based on the data collected by all the three Swarm spacecraft
- for some of the events, also the high time resolution magnetic field data (50 Hz) were provided
- some events were selected near the geographic poles in the exclusion zone (latitude>86°), where the longitudinal separation between Swarm A and C is small, and therefore FAC estimation from the dual spacecraft method is more difficult.

According to these criteria, ESA identified the 28 events sampled by Swarm in years 2014–2015 listed in Table 8.1, which are part of the FAC-MICE test dataset.

Moreover, in order to estimate the possible influences of the choice of magnetic residuals on FAC results, as input for FAC calculation, ESA provided, for each of these events, the magnetic field residuals obtained from two different field models: the same residuals as L2 products (AUX_CORE, Lithosphere-MF7, magnetosphere-POMME-6) and also CHAOS-6 residuals (Finlay et al. 2016; see also Chap. 12 of this book).

Table 8.1 The 28 events sampled by Swarm in years 2014–2015, which are part of the FAC-MICE test dataset

#	Date (YYYY-MM-DD)	Start	Stop	Swarm s/c	1 Hz data	50 Hz data	3 s/c conjunction
1	2014-04-22	04:05	04:20	A-B-C	YES	NO	YES
2	2014-04-22	04:40	05:00	A-B-C	YES	NO	YES
3	2014-04-28	09:45	10:00	A-B-C	YES	YES	YES
4	2014-04-28	10:30	10:45	A-B-C	YES	YES	YES
5	2014-05-04	15:20	15:35	A-B-C	YES	NO	YES
6	2014-05-10	21:30	21:45	A-B-C	YES	NO	YES
7	2014-05-17	03:05	03:15	A-B-C	YES	NO	YES
8	2014-05-29	13:35	13:45	A-B-C	YES	YES	YES
9	2014-06-04	18:10	18:25	A-B-C	YES	NO	YES
10	2014-06-04	18:20	18:30	A-B-C	YES	NO	YES
11	2014-06-10	23:40	23:50	A-B-C	YES	NO	YES
12	2014-06-17	03:50	04:00	A-B-C	YES	NO	YES
13	2014-07-11	19:30	19:45	A-B-C	YES	NO	YES
14	2014-07-30	06:05	06:15	A-B-C	YES	NO	YES
15	2014-08-11	10:00	10:10	A-B-C	YES	NO	YES
16	2015-03-16	22:35	23:00	A-C	YES	NO	NO
17	2015-03-17	08:50	09:15	A-C	YES	NO	NO
18	2015-03-21	08:55	09:20	A-B-C	YES	NO	NO
19	2015-08-15	04:30	04:50	A-C	YES	YES	NO
20	2015-08-15	12:20	12:45	A-C	YES	YES	NO
21	2015-09-20	02:00	02:25	A-B-C	YES	YES	NO
22	2015-09-20	11:20	11:45	A-B-C	YES	YES	NO
23	2015-10-07	02:10	02:30	A-C	YES	YES	NO
24	2015-10-08	17:10	17:35	A-B-C	YES	YES	NO
25	2014-04-22	05:40	05:55	A-C	YES	NO	NO
26	2014-05-04	19:05	19:30	A-C	YES	NO	NO
27	2014-06-29	12:25	12:45	A-B-C	YES	NO	NO
28	2014-07-11	20:20	20:40	A-B-C	YES	NO	NO

8.3 The Active Participants to FAC-MICE

The FAC-MICE test dataset has been delivered by ESA on June 2016 to a list of 38 potential participants, identified among the researchers who already have developed algorithms to compute FACs, or other researchers who have a good experience with high latitude currents. This list included also researchers who have developed new FAC algorithms, as part of ESA projects. The various FAC algorithms have been tested based on this dataset, and the first results of this exercise have been discussed in ESA—ESTEC on 07th—8th September 2017, during a dedicated meeting called Swarm Ionospheric Currents Products Workshop (https://earth.esa.int/web/guest/missions/esa-eo-missions/swarm/activities/conferences/swarm-ionospheric-currents).

Eight different groups performed the FAC-MICE analysis:

- Finnish Meteorological Institute (FMI)
- Institute of Space Science (ISS)—Jacobs University Bremen (JUB)
- Rutherford Appleton Laboratory (RAL)
- Deutsches GeoForschungsZentrum (GFZ)
- University of Alberta (UofA)
- Mullard Space Science Laboratory (UCL-MSSL)
- University of Calgary (UC)
- Johns Hopkins University (JHU)

Some of the new approaches adopted by these groups for FAC-MICE are illustrated in other Chapters of this book, while other approaches are documented by scientific papers.

FMI adopted a new two-dimensional version of the Spherical Elementary Current Systems (SECS) developed in the context of Swarm-SECS ESA project (http://space.fmi.fi/MIRACLE/Swarm_SECS/). This approach is applied to magnetic and electric field data measured by Swarm A and C, and allows the determination of the ionospheric currents, i.e. the Hall and Pedersen horizontal components and FAC, the ionospheric conductances and the plasma convection in the Ionospheric current layer. For FAC-MICE, FMI delivered three FAC density products: two single-spacecraft FAC products based on data measured by Swarm A and C, and one dual-spacecraft FAC product based on data acquired by the lower pair (Swarm A and C). The theory of SECS approach is described in Chap. 2, and the application of this method to Swarm data is described in Chap. 3.

ISS-JUB adopted the new methods to estimate FAC based on the data collected by single, dual or three Swarm spacecraft developed in the context of SIFACIT project (http://gpsm.spacescience.ro/sifacit/). For FAC-MICE, ISS-JUB delivered several different new FAC estimates and FAC quality indicators. Particularly interesting, FAC product based on the 3 spacecraft data, provided for the conjunction events, that does not require the hypothesis of time stationarity, being based on simultaneous Swarm measurements. These conjunction events occurred more frequently at the beginning of the mission, and will be again very frequent in year 2021 when the orbit of Swarm B will be coplanar with the one of the lower pair. ISS-JUB also provided a number of quality indices that can be used to verify the assumptions needed for the calculation of FAC, based on Minimum Variance Analysis (MVA, see e.g. Sonnerup and Scheible 1998) and cross-correlation analysis of time lagged magnetic field time series from Swarm A and C. They also provided the time intervals corresponding to the location of the auroral oval, selected with an automatic identification procedure based on FAC density, estimated from single spacecraft method. All the ISS-JUB approaches are described in Chap. 4 of this book.

RAL adopted the new multi-spacecraft methods based on 2, 3, 4 (or more) point measurements, partly obtained by time-shifting the positions of Swarm spacecraft. They delivered for FAC-MICE these multi-point FAC estimates, based on all the three spacecraft data during the conjunction events. The comparison among these products can be used to verify the time stationarity assumption, and validate the interpretation of the current components flowing predominantly along the field-aligned direction. These methods are illustrated in (Dunlop et al. 2015), and also in Chap. 5 of this book.

GFZ provided the single and the dual spacecraft FAC method, adopting the same algorithms illustrated by Ritter et al. (2013). Therefore, FAC products provided by GFZ are equivalent to the official ESA Level 2 single and dual spacecraft FAC products. They applied these algorithms also to 50 Hz data, not implemented in the Level 2 products, and provided also the orientation of the current sheet obtained from the relation between the two transverse magnetic components, Bx and By, in Mean-Field-Aligned (MFA) coordinates. These methods are illustrated in Chap. 6 of this book.

UofA provided estimates of dual and single spacecraft FAC density, for both Swarm A and C. The single spacecraft FAC estimates are very similar to the ESA Level 2 single-spacecraft FAC products, calculated using Ampere's law/curlometer (Lühr et al. 1996). The dual spacecraft FAC estimate is also similar to the ESA Level 2 dual-spacecraft FAC product (Ritter et al. 2013), but it is obtained without the low-pass filtering applied to the magnetic field time series. UofA also provided a quality parameter, based on cross-correlation analysis of time lagged magnetic field time series from Swarm A and C, similar to the parameter provided by ISS.

UCL-MSSL provided new quality indicators, based on cross-correlation of time lagged single spacecraft FAC estimates from Swarm A and C. Differently from the cross correlation analysis performed by ISS-JUB and UofA, UCL-MSSL performed the cross-correlation analysis separating the signals using five different band-pass filters. Under the assumption that the currents are temporally and spatially stationary, these bands correspond to the different spatial scales of FAC structure. Therefore, the indices provided by UCL-MSSL allow to assess the time stationarity assumption of FAC at different spatial scales.

UnCa provided both single and dual S/C FAC estimates, adopting the same algorithm as the Level 2 products (Ritter et al. 2013), using both the magnetic field residuals from the standard Level 2 model, but also the residuals from the CHAOS-6 model.

JHU provided a number of different FAC products, based on multi-point measurements obtained from two or three Swarm S/C. They used only high resolution magnetic field data (50 Hz), and were able to recover FAC at various spatial scales.

8.4 FAC-MICE Comparison Results

In order to compare the various FAC products provided by FAC-MICE participants, we decided to start from the events characterized by higher values of the quality parameters based on correlation analysis of magnetic field time series measured by Swarm A and C, provided by ISS-JUB and UofA. In theory, higher correlations are expected when the two assumptions needed for FAC calculation, i.e. time stationarity of the current sheet and linear variation of the magnetic signatures measured by the two spacecraft (i.e. constant gradient), are satisfied. This latter condition implies that when the correlation coefficients are very high, localized FAC structures, detected by one spacecraft but not the other, are absent. In this case, the various FAC estimates based on different Swarm spacecraft should agree well with each other.

The correlation analyses performed by ISS-JUB and UofA are similar: ISS-JUB performed the correlation analysis on the magnetic perturbations along the direction of maximum magnetic field variance, obtained from MVA, in the auroral time intervals obtained from their automatic identification procedure, using both filtered (with the low-pass filter) and unfiltered magnetic field time series. The correlation analysis is computed as a function of the time lag between Swarm A and C field data, the optimal correlation is selected, and the time lag corresponding to optimal correlation is also provided. Therefore, ISS-JUB provided the average value of the correlation coefficients, for each FAC-MICE event.

On the contrary, UofA computed the correlation of 2nd NEC magnetic field component (East-West component) of the magnetic residuals obtained by removing the provided core, external and lithospheric fields from the CHAOS model. They performed the analysis in a 4 min sliding window. The maximum correlation is achieved by first calculating the along-track latitude difference between Swarm A and C, and then fine-tuning this time lag corresponding to optimal correlation. Therefore, UofA provided a time series of the correlation coefficient for each of the events. In order to obtain an average correlation coefficient also for the analysis from UofA, we averaged the correlation time series provided by UofA in the auroral time intervals provided by ISS.

The values of the correlation parameters provided by ISS-JUB, based on filtered and unfiltered data, and the values of the averaged correlation parameter from UofA, for all the events of FAC-MICE are shown in Fig. 8.1. In this graph, we also reported the average values of the AE index, downloaded from NASA cdaweb (https://cdaweb.sci.gsfc.nasa.gov/index.html/). Along the X axis is reported the number of the event, as listed in Table 8.2, identified through the auroral automatic identification procedure from ISS-JUB. The number of events is now 36, because some of the intervals in the FAC-MICE list (Table 8.1) contained two auroral crossings, corresponding to the ascending and descending branches of the orbit. In Fig. 8.1, we can note that the values of the three correlation coefficients for each event are quite similar to each other, and no relation with the geomagnetic activity is observed.

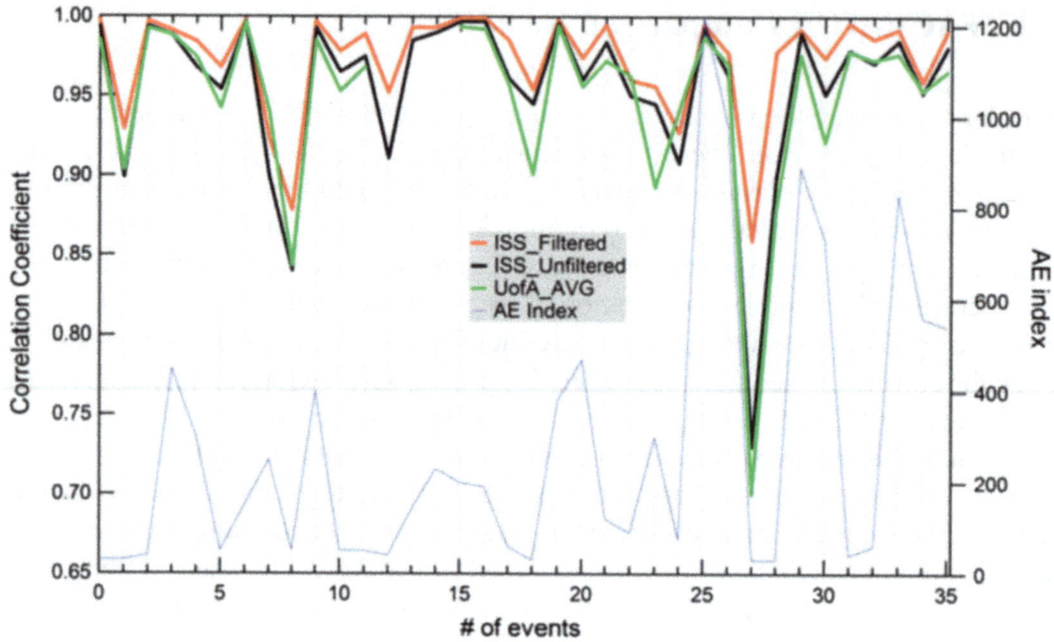

Fig. 8.1 FAC quality parameters based on cross-correlation analysis of magnetic field data measured by Swarm A and C, for the 36 events of the FAC-MICE dataset listed in Table 8.2. In this figure, also reported, is the average value of the AE index for each of the events

8.4.1 Comparison for the 'High Correlation' Events

As a first event for the comparison of various FAC estimates, let's consider the event # 15, observed on 2014-07-11 in the interval 19:30–19:45 UT, characterized by very high correlation coefficients and low AE index. This is a conjunction event, when all the three Swarm spacecraft are close together during the auroral crossing, and it was observed in the northern hemisphere at 18 MLT.

The magnetic field residuals from the standard magnetic field models (AUX_CORE, Lithosphere-MF7, magnetosphere-POMME-6) detected by the three Swarm spacecraft are illustrated in Fig. 8.2, in NEC reference frame. We can see that the magnetic signatures observed by the three spacecraft appear very similar, and are nearly simultaneous, apart for the time delay among Swarm A and C, which is approximately 8 s for this event.

Figure 8.3 shows the comparison among some of the FAC products delivered to ESA for FAC-MICE, for the event # 15. In panel (a) are displayed the single and dual S/C products provided by GFZ, which are equivalent to ESA Level 2 FAC products. The filtered single S/C products, the three S/C products and the unfiltered dual S/C products provided by ISS-JUB are displayed in panels (b), (c), (d) respectively. In panel (e) the products from FMI based on SECS method, in panel (f) the multi-point estimates from RAL, and in panel (g) the single S/C products from UnCA. In all these panels, the dual S/C product from GFZ is also shown for comparison.

The single and dual FAC estimates provided by GFZ, which are equivalent to ESA Level 2 FAC products, are displayed in panel a. The single spacecraft FAC product

Table 8.2 The FAC-MICE auroral events, as identified through the auroral automatic identification procedure from ISS-JUB. The number of events is now 36, because some of the intervals in the FAC-MICE list (Table 8.1) contain two auroral crossings, corresponding to the ascending and descending branches of the orbit

#	Year	Month	Day	Time start	Time stop
0	2014	4	22	04:05:00	04:20:00
1	2014	4	22	04:40:00	05:00:00
2	2014	4	22	05:40:00	05:55:00
3	2014	4	28	09:45:00	10:00:00
4	2014	4	28	10:30:00	10:45:00
5	2014	5	4	15:20:00	15:35:00
6	2014	5	4	19:05:00	19:21:00
7	2014	5	10	21:30:00	21:45:00
8	2014	5	17	03:05:00	03:15:00
9	2014	5	29	13:35:00	13:45:00
10	2014	6	4	18:10:00	18:22:00
11	2014	6	4	18:20:00	18:30:00
12	2014	6	10	23:40:00	23:50:00
13	2014	6	17	03:50:00	04:00:00
14	2014	6	29	12:25:00	12:40:00
15	2014	7	11	19:30:00	19:45:00
16	2014	7	11	20:27:00	20:40:00
17	2014	7	30	06:05:00	06:15:00
18	2014	8	11	10:00:00	10:10:00
19	2015	3	21	08:55:00	09:04:00
20	2015	3	21	09:04:00	09:20:00
21	2015	9	20	02:00:00	02:14:00
22	2015	9	20	02:13:00	02:25:00
23	2015	9	20	11:20:00	11:35:00
24	2015	9	20	11:35:00	11:45:00
25	2015	10	8	17:10:00	17:24:00
26	2015	10	8	17:24:00	17:35:00
27	2015	3	16	22:35:00	22:49:00
28	2015	3	16	22:48:00	23:00:00
29	2015	3	17	08:50:00	09:03:00
30	2015	3	17	09:03:00	09:15:00
31	2015	8	15	04:30:00	04:39:00
32	2015	8	15	04:39:00	04:50:00
33	2015	8	15	12:20:00	12:45:00
34	2015	10	7	02:10:00	02:18:00
35	2015	10	7	02:18:00	02:30:00

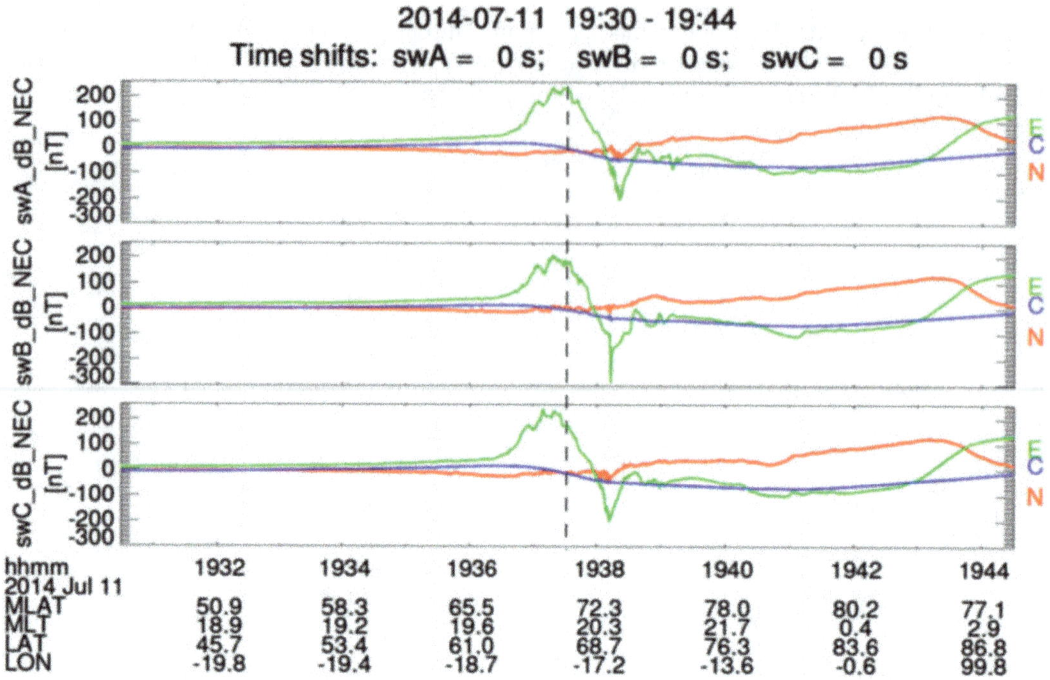

Fig. 8.2 The magnetic field residuals detected by the three Swarm spacecraft, in NEC reference frame, for the auroral event on 2014-07-11, 19:30–19:44 UT. Note that this is a conjunction event, when the three spacecraft are close together during the auroral crossing

obtained from Swarm C has been delayed by 7.9 s to take into account the spacecraft separation along the orbit.

The single spacecraft FAC products show several high frequency fluctuations, at time scales of few seconds, which sometimes are similar among the three spacecraft.

The dual spacecraft FAC product appears much smoother with respect to the single spacecraft FAC products. This difference is related to the low-pass band filter adopted for the dual spacecraft FAC product, at 20 s time scale (Ritter et al. 2013). This filter is necessary in the dual spacecraft FAC product to meet the time stationarity assumption, suppressing the FACs carried by kinetic Alfven waves and also to filter out FAC structures smaller than the separation among the 4 point measurements used for FAC calculation (quad size), which is approximately 50 km. Both these conditions are equally important.

In order to better compare the single and dual spacecraft FAC products, ISS-JUB provided filtered single spacecraft FAC products, obtained using the same the low-pass band filter adopted for the dual FAC product, which are shown in Fig. 8.3 panel (b). The agreement of the filtered single spacecraft FAC products with the GFZ dual spacecraft FAC product is now remarkable. This confirms that the current sheet is time stationary and invariant in longitude during this event since all the three spacecraft, at different times, recover the same FAC structure.

ISS-JUB also provided a FAC product based on data measured from all the three Swarm spacecraft, which can be obtained only during the conjunction events. Since this three-spacecraft ISS-JUB FAC product is based on simultaneous multipoint

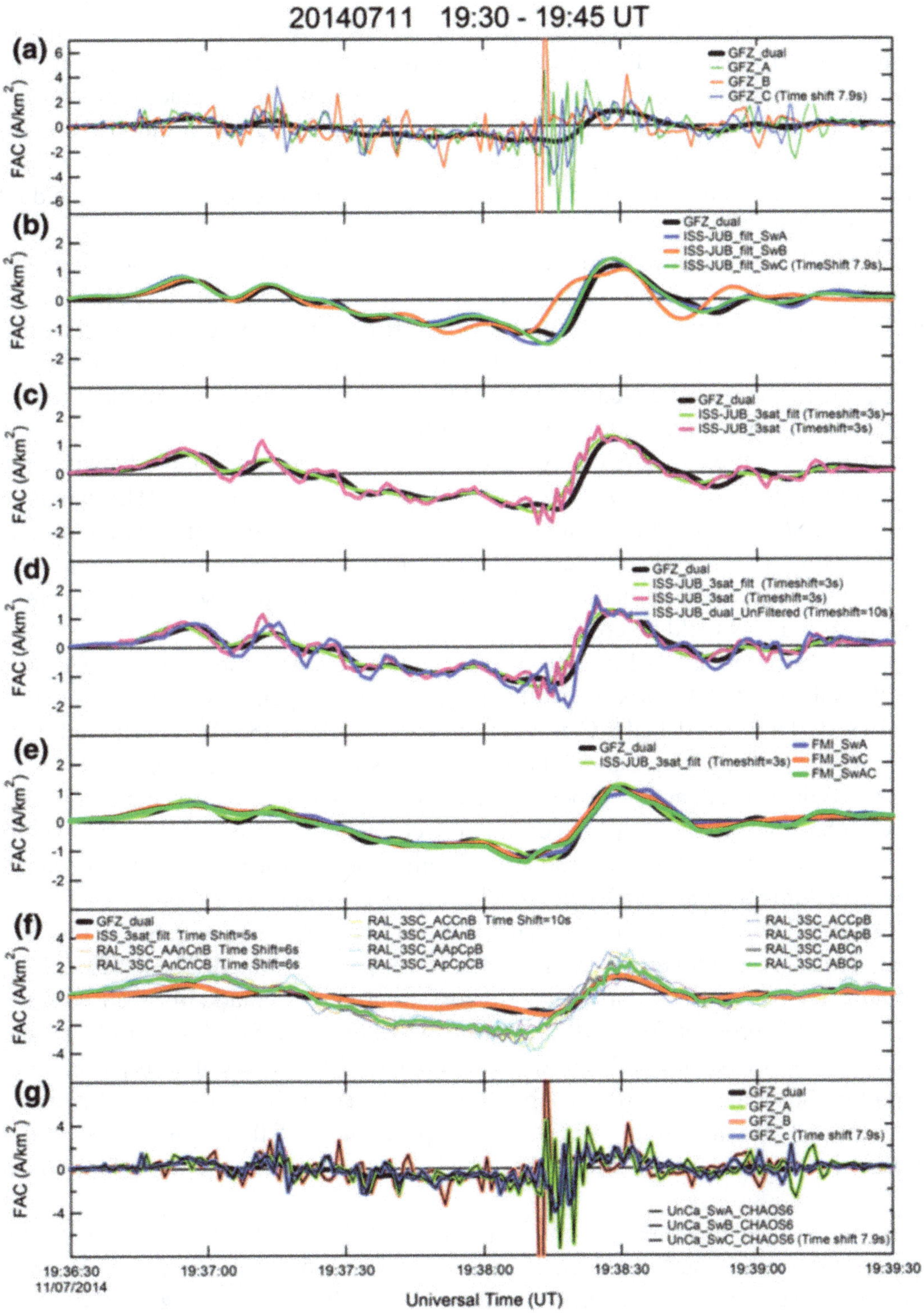

Fig. 8.3 Some of FAC products delivered to ESA for FAC-MICE, for the auroral event observed on 2014-07-11, 19:30–19:44 UT. Panel **a** shows the single and dual s/c products provided by GFZ, which are equivalent to ESA Level 2 FAC products. The filtered single S/C products, the three S/C products and the unfiltered dual S/C products provided by ISS-JUB are displayed in panels **b**, **c**, **d** respectively. In panel **e** the products from FMI based on SECS method, in panel **f** the multi-point estimates from RAL, and in panel **g** the single S/C products from UnCA. In all these panels, the dual S/C product from GFZ is shown for comparison

measurements, it does not require the filtering in order to satisfy the time-stationarity assumption. However, as for the dual spacecraft Level 2 FAC product, also in this case only current structures with dimensions larger than the separation among the three spacecraft can be recovered, otherwise spatial aliasing can occur. This three-spacecraft ISS-JUB FAC product is computed both from the unfiltered 1 Hz data, and from the filtered magnetic field data, using the same low-pass filter as the dual S/C Level 2 product.

Figure 8.3 panel (c) shows the comparison of dual GFZ FAC with the three ISS-JUB FAC products. This comparison can be considered as a sort of validation for the dual S/C product, given that the three spacecraft product is based on simultaneous measurements, not requiring the stationarity assumption. The general trend of these FAC products agree quite well and the unfiltered ISS-JUB three s/c product provide more details about small scales currents features.

For the events characterized by very high correlation coefficients between the unfiltered magnetic field series (higher than 0.99, like this event), ISS-JUB delivered also a dual S/C FAC product obtained from 1 Hz unfiltered magnetic field data. The justification for this product relies on the fact that, when the correlation coefficient is very high, local FAC structures with dimension smaller than separation between Swarm A and C, should be absent.

The comparison of this ISS-JUB dual S/C unfiltered, with the GFZ dual S/C, and ISS-JUB three S/C FAC product is illustrated in Fig. 8.3, panel (d). We can note that this ISS-JUB unfiltered dual S/C product agree well with the ISS-JUB unfiltered three S/C product, and provide more details about the small scale structure of FAC with respect to the GFZ dual S/C product. This suggests that, at least for the 'stationary events' characterized by very high correlation coefficient, it is possible to remove the low-pass filter in the dual S/C product, obtaining more information about the small scale structures with respect to the Level 2 dual S/C product.

In Fig. 8.3 panel (e) are displayed the FAC products obtained with SECS by FMI, using Swarm A and Swarm C alone, or the lower pair combined. These SECS FAC products recover only the large scale features, but agree remarkably well with the dual S/C FAC product from GFZ and also with the three S/C FAC product from ISS-JUB, also displayed in Panel (e) for comparison. This agreement is particularly important, since the FAC estimates from FMI are based on the expansion in Spherical Elementary Current System (SECS, see Chaps. 2 and 3 of this book), which differ substantially with respect to the methods based on the Ampere's law, used for all the other FAC products.

In Fig. 8.3 panel (f) are shown the various RAL FAC estimates based on 4 non-planar points using the full curlometer technique (Dunlop et al. 1988, 2002). These 4-point measurements are obtained from 3 S/C data by time-shifting the data from one S/C. The various FAC products displayed in panel (f) differ each other according to which S/C is time-shifted to obtain the fourth point measurement, and the names in the caption adopt the same convention of Dunlop et al. (2015), where p refers to positive time-shift and n to negative time-shift. The various FAC products provided by RAL show a similar behaviour, with a similar trend also with respect to the Level 2 dual S/C product. The time-shift among the various RAL products is related to

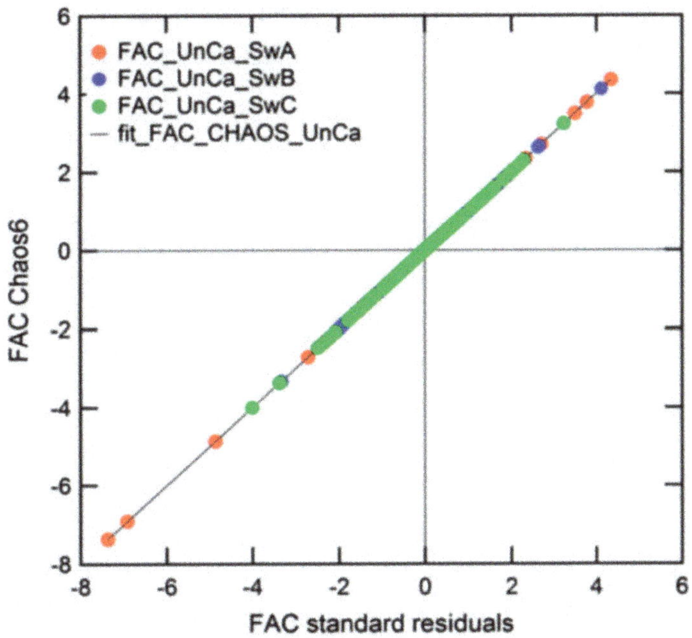

Fig. 8.4 A comparison among the single S/C FAC products from UnCA based on residuals obtained from CHAOS-6 model, with the single S/C Level 2 FAC product, for the auroral event displayed in Fig. 8.3

the change in effective barycentre and different tetrahedral shapes formed by each configuration. All the RAL products recover a significantly larger current density than the standard Level 2 dual S/C product, also larger with respect to the other FAC products from ISS. This can be due to the fact that RAL estimated FAC from the three components of the current density vector, while the other FAC products used the radial component only, which can omit some of the actual FAC. Dunlop et al. (2015) suggested that the degree of stationarity of the current sheet can be obtained from the variance over all different FAC estimates.

The single s/c products obtained from UnCa, based on magnetic residuals from CHAOS-6 model are compared to single s/c Level 2 products provided by GFZ in Fig. 8.3, panel (g). These FAC products from UnCa are perfectly coincident with the Level 2 products, which are instead obtained from the residuals from different magnetic field models (AUX_CORE, Lithosphere-MF7, magnetosphere-POMME-6, see Chap. 6 of this book). This perfect agreement is also confirmed in the scatter plot in Fig. 8.4, where the single s/c FAC products obtained from UnCa using the CHAOS-6 models are displayed as a function of the Level 2 single s/c FAC products. This agreement demonstrates that the choice of the model used to compute the residuals does not affect FAC calculation.

Other examples of events characterized by high correlation coefficients are the events numbered 25 and 26 in Table 8.2, observed on 20151008 in the intervals 17:10–17:24 UT and 17:24–17:35 UT.

In Fig. 8.5, the magnetic field residuals are reported, in NEC reference frame measured by the three Swarm spacecraft, with the same format as Fig. 8.2. Note that

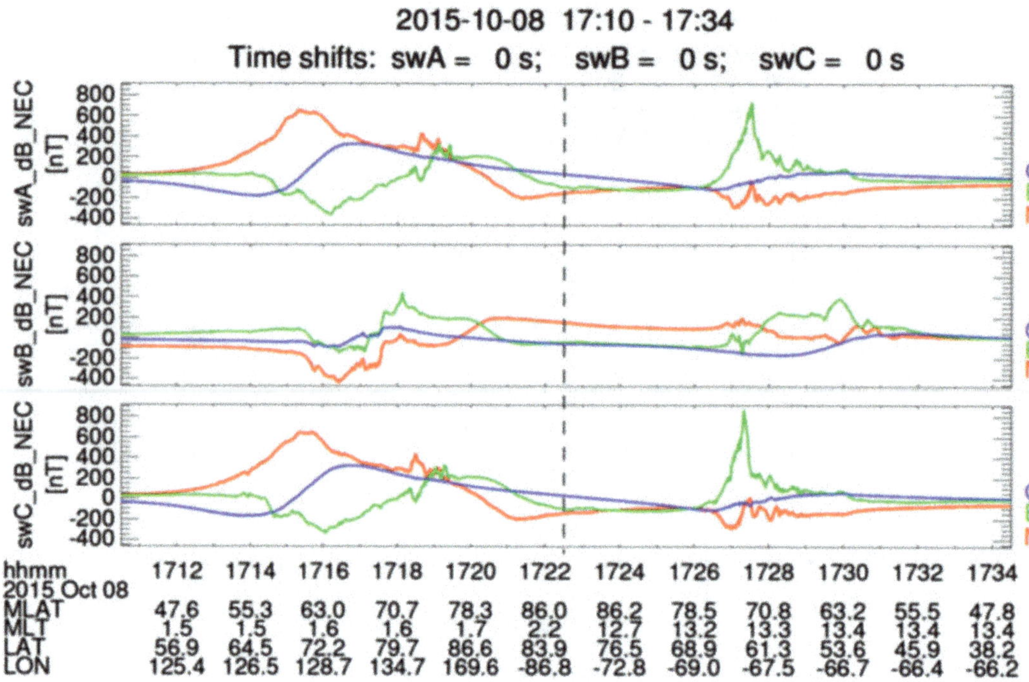

Fig. 8.5 The magnetic field residuals detected by the three Swarm spacecraft, in NEC reference frame, for the auroral event on 2015-10-08, 17:10–17:34 UT. Note that this is not a conjunction event among the three Swarm spacecraft. Indeed, Swarm A and C were in the northern hemisphere, and the magnetic signatures detected around 17:16 and 17:28 UT correspond to the entry and exit from the polar cap. Swarm B instead was flying in the southern hemisphere

this is not a conjunction event among the three Swarm spacecraft. Indeed, Swarm A and C were in the northern hemisphere, and the magnetic signatures detected around 17:16 and 17:28 UT correspond to the entry and exit from the polar cap. Swarm B instead, even if it detected similar magnetic signatures approximately at the same times as Swarm A and C, was flying in the southern hemisphere.

In Fig. 8.6 are reported some of the FAC estimates provided for FAC-MICE: In panel (a) are displayed the single and dual S/C products provided by GFZ, which are equivalent to ESA Level 2 FAC products. In panel (b) the filtered single S/C products provided by ISS-JUB and in panel (c) the products from FMI based on SECS method, together with the dual S/C product from GFZ for comparison.

With respect to the previous event (20140711), this event is characterized by higher geomagnetic activity, with AE $\approx$ 1000, and FACs are more intense. The comparison illustrated in Fig. 8.6 confirms the features highlighted in the previous event: in panel (a) the single and dual S/C products provided by GFZ, which are equivalent to the ESA Level 2 FAC products, show large discrepancies related to the high frequency fluctuations characterizing the single S/C FAC products. This discrepancy is related to the low-pass filters used for the dual S/C FAC product. Indeed, the comparison in panel (b) the dual S/C products from GFZ show a reasonably good agreement with the filtered single S/C FAC products delivered by ISS, obtained using the same low-pass filter.

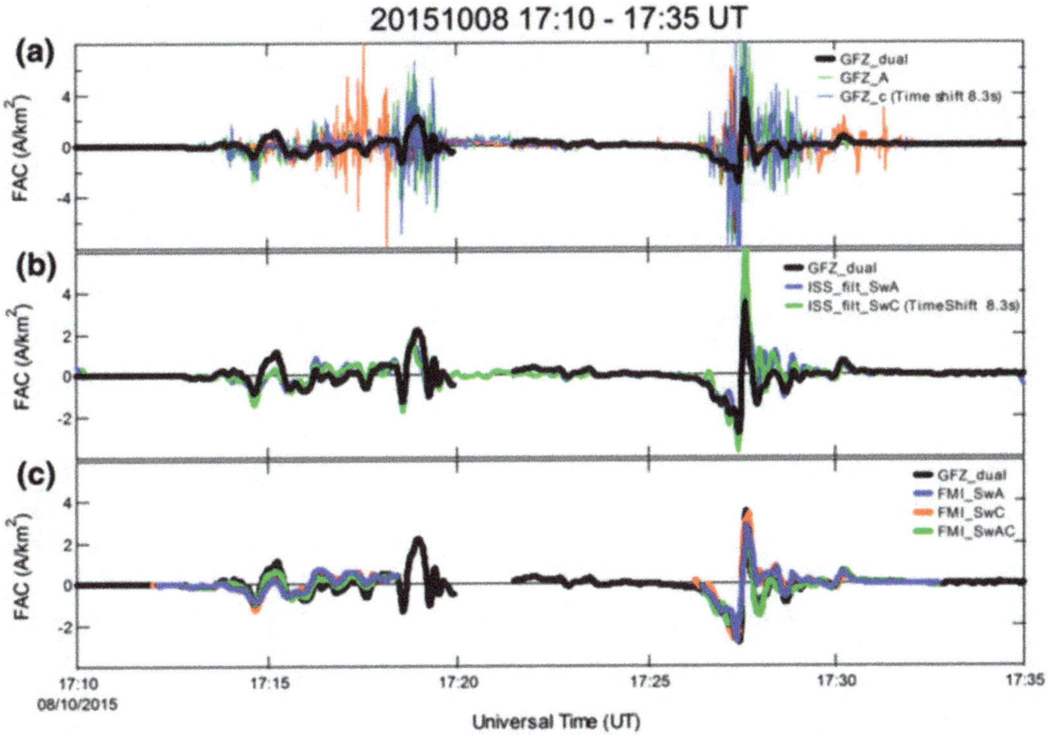

Fig. 8.6 Some of the FAC products delivered to ESA for FAC-MICE, for the auroral event observed on 2015-10-08, 17:10–17:34 UT. In panel **a** the single and dual S/C products provided by GFZ, which are equivalent to ESA Level 2 FAC products. In panel **b** the filtered single S/C products provided by ISS-JUB and in panel **c** the products from FMI based on SECS method, together with the dual S/C product from GFZ for comparison

Also in this event, the FAC products obtained from SECS method agree reasonably well with the dual S/C product from GFZ (see panel c).

8.4.2 *Comparison for the 'Low Correlation' Events*

The comparison among the various FAC products delivered for FAC-MICE has been performed also for a number of events characterized by lower values of the correlation parameters (Fig. 8.1), to see how the various FAC estimates compare with each other during these 'low-correlation' events.

As an example, we considered the event observed on 2014-05-17 in the interval 03:05–03:15 UT in the southern hemisphere. This is a conjunction event, during which the three Swarm spacecraft were close together, and the magnetic field residuals measured by the three spacecraft are shown in Fig. 8.7, in the NEC reference frame, with the same format as Fig. 8.2. The current sheet is observed approximately at the same time by the three spacecraft, but the differences among the magnetic signatures measured by the three spacecraft are now noticeable.

In Fig. 8.8 are reported some of the FAC estimates provided for FAC-MICE: in

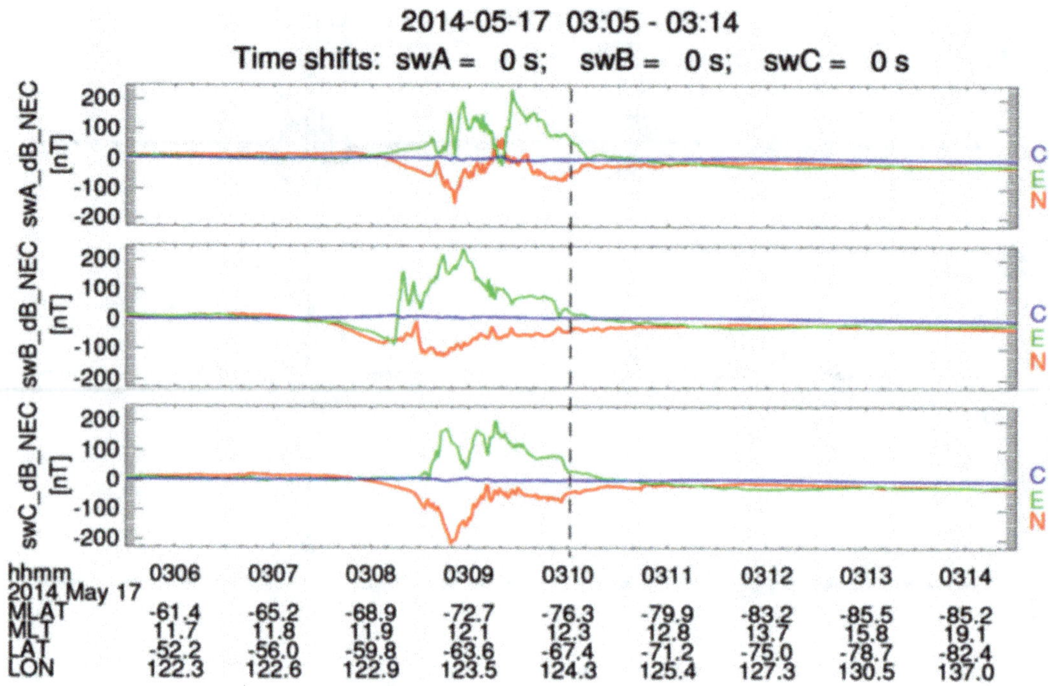

Fig. 8.7 The magnetic field residuals detected by the three Swarm spacecraft, in NEC reference frame, for the auroral event on 2014-05-17, 03:05–03:14 UT. Note that this is a conjunction event, when the three spacecraft are close together during the auroral crossing

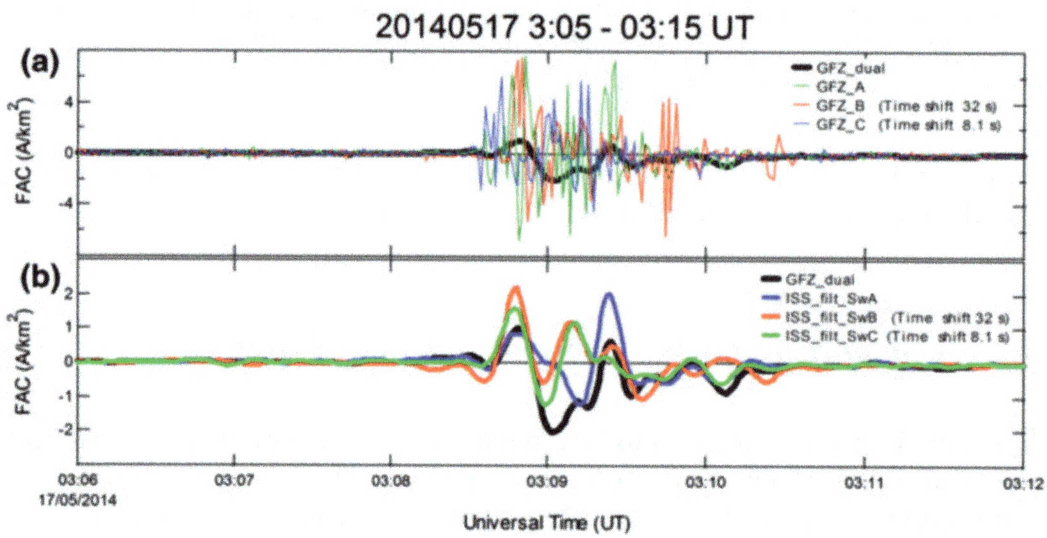

Fig. 8.8 Some of the FAC products delivered to ESA for FAC-MICE, for the auroral event observed on 2014-05-17, 03:05–03:14 UT. In panel **a** the single and dual S/C products provided by GFZ, which are equivalent to ESA Level 2 FAC products. In panel **b** the filtered single S/C products provided by ISS-JUB, together with the dual S/C product from GFZ for comparison

panel (a) are displayed the single and dual S/C products provided by GFZ, which are equivalent to the ESA Level 2 FAC products, and in panel (b) the filtered single S/C products provided by ISS.

Again, the single and dual S/C products provided by GFZ in panel (a) show large discrepancies, related to the high frequency fluctuations characterizing the single S/C FAC products. However, in this case, also the filtered single S/C FAC products provided by ISS-JUB differ substantially from each other, and also with respect the dual S/C L2 product. It can be noted that the larger deviation is shown by Swarm A single S/C filtered FAC product, while the filtered FAC products obtained from Swarm B and C data agree better to each other, at least until 03:09:20 UT.

This substantial difference is probably related to the presence of local FAC structures, with dimension smaller than the spacecraft separation, rather than non-stationarity of the current sheet. Indeed, only the FAC from Swarm A differ much, while FACs recovered by Swarm B and Swarm C agree quite well to each other, even if the relative time-shift between the two spacecraft is approximately 24 s. The presence of these local FAC structures affect the dual spacecraft FAC product, which is therefore not reliable for this event.

For this event, the three-S/C ISS-JUB FAC product is not available because the configuration of the Swarm constellation was too elongated. Moreover, also the FAC products based on SECS method are not available, since Swarm spacecraft were beyond 76° in magnetic latitude, the threshold beyond which the SECS analysis is not performed. Therefore, for this event, we cannot verify the large scale behaviour of the FAC structure.

As a further example of a 'low correlation' event, we considered the event observed on 2015-03-16, in the interval 22:35–23:00 UT in the northern hemisphere. This is not a conjunction event, and the magnetic field residuals measured by the lower pair are shown in Fig. 8.9, in the NEC reference frame, with the same format as Fig. 8.2.

In Fig. 8.10 are reported some of the FAC estimates provided for FAC-MICE: In panel (a) are displayed the single and dual S/C products provided by GFZ, which are equivalent to ESA Level 2 FAC products. In panel (b) the filtered single S/C products provided by ISS-JUB and in panel (c) the products from FMI based on SECS method, together with the dual S/C product from GFZ for comparison.

The single S/C products in panel (a) show high frequency fluctuations, while the dual S/C product is more stable. The comparison in panel (b) with the filtered single S/C products from ISS-JUB shows a better agreement in the first part of the event, while later on, i.e. after 22:43 UT, also the filtered single S/C FAC estimates differ noticeable from the dual S/C product. This can again be explained in terms of non-stationarity of the current sheet, and/or to strong longitudinal gradients. The signatures in the dual S/C product around 22:45, at $\approx$80°MLAT, could instead be due to high latitude filamentary FAC, which cannot be detected by single S/C methods because these currents are not planar, and therefore violate the orientation assumption needed for the single S/C FAC calculation (Lühr et al. 2016).

The FAC products based on SECS illustrated in panel (c) follow reasonably well the L2 FAC in the first part of the event (especially the dual spacecraft SECS). After

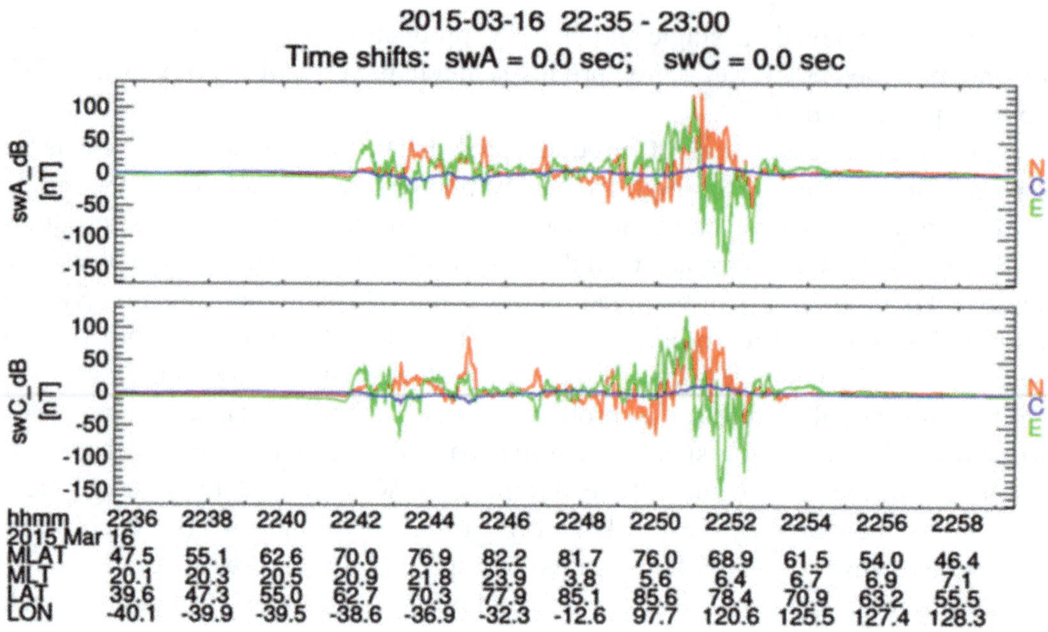

Fig. 8.9 The magnetic field residuals detected by Swarm A and C, in NEC reference frame, for the auroral event on 2015-03-16, 22:35–23:00 UT

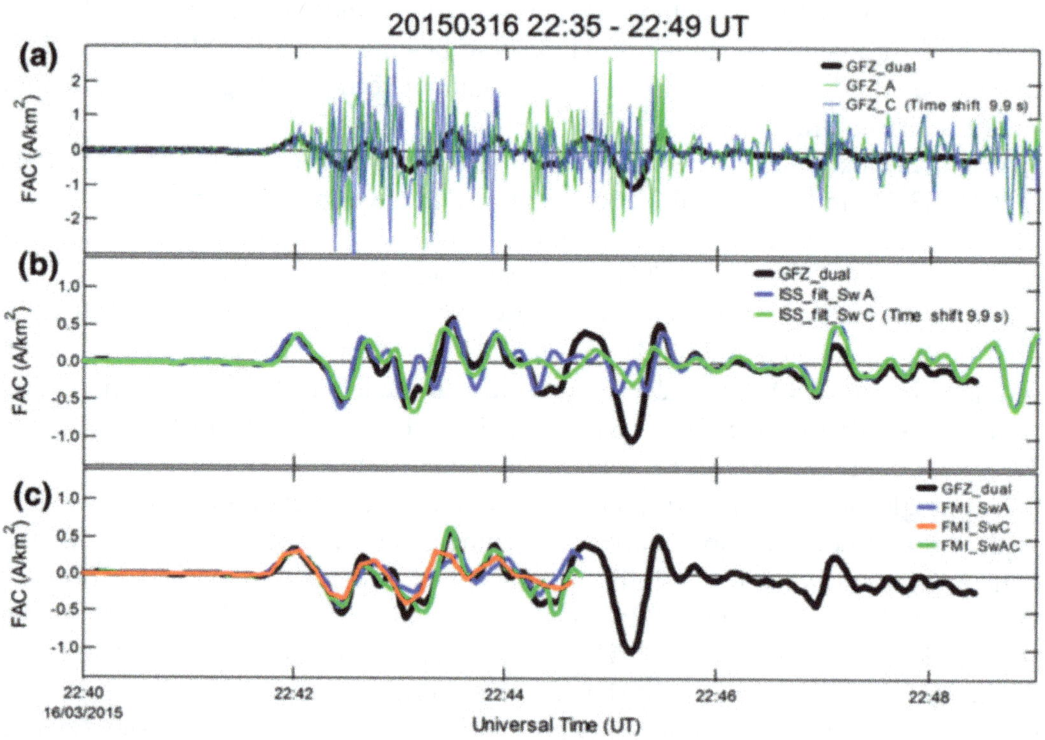

Fig. 8.10 Some of the FAC products delivered to ESA for FAC-MICE, for the auroral event observed on 2015-03-16, 22:35–23:00 UT. In panel **a** the single and dual S/C products provided by GFZ, which are equivalent to ESA Level 2 FAC products. In panel **b** the filtered single S/C products provided by ISS-JUB and in panel **c** the products from FMI based on SECS method, together with the dual S/C product from GFZ for comparison'

22:44:40 the SECS product are not available because the magnetic latitude is larger than 76°. Therefore, it's not possible to verify the signature at 22:45 UT.

8.5 FAC-MICE Comparison Summary

Three different groups participating to FAC-MICE (ISS, UofA, UCL-MSSL) provided quality parameters based on cross-correlation of lagged magnetic field time series (or FAC single spacecraft time series) measured by Swam A and C. These parameters agree well to each other for the various FAC—MICE events, and suggest the presence of two classes of events: the 'high-correlation' events characterized by the higher correlation coefficients and the 'low-correlation' events.

For the 'high-correlation' events, the various FAC estimates provided by the various groups show a reasonably good agreement. In particular, the noticeable discrepancies between the single and dual spacecraft Level 2 FAC products is highly reduced when the same low-pass filter used for dual S/C product is adopted also for the single S/C products. This suggests that the current sheet is both stationary and approximately invariant in longitude during these events. Indeed, the three spacecraft cross the current sheet at different times and in different locations, and recover approximately the same FAC structure.

Another important validation for FAC structure is given by the three spacecraft FAC products, provided for the conjunction events when Swarm B was near the lower pair (Swarm A and C), computed both from filtered and unfiltered 1 Hz data. Indeed, the three S/C products are based on simultaneous measurements from the three spacecraft, and do not rely on the stationarity assumption or orientation assumptions which are instead needed for the single or dual S/C FAC products. However, like the other FAC estimations based on multipoint measurements, these three spacecraft FAC products can recover only FAC with spatial scales larger than the spacecraft separations. During these 'high-correlation' events, also these three S/C products agree well with the other FAC estimates.

During these 'high-correlation' events, since the large scale FAC structure is reliably recovered by these filtered products, it seems possible to look at the smaller scale features of the FAC sheet, provided by various FAC estimates based on unfiltered 1 Hz magnetic field data. However, also during these 'high correlation' events, some of the high frequency fluctuations in the single S/C Level 2 FAC products differ among the various spacecraft. In particular, the differences observed between Swarm A and C, which are close together, suggesting the presence of local structures that do not satisfy the one-dimensional current sheet assumption needed for single spacecraft FAC calculation (Lühr et al. 2015).

During the low correlation events, the comparison among the various FAC products looks instead quite different: the large discrepancies among single and dual S/C FAC estimates remain, even when considering the filtered single S/C products. This suggests that the stationarity assumption is violated, and therefore FAC densities provided by the various products can be affected by non-negligible errors. Some of the large-scale features of FAC structure given by the filtered dual S/C FAC products are confirmed by the SECS analysis also during these low-correlation events.

Other FAC estimates can be used in synergy with Swarm data to investigate the high latitude current system. For example, measurements from incoherent scatter radars can be very useful for this purpose. These radars provide various properties of the ionosphere, like electron density, ion velocity, ion and electron temperatures, ion composition, and collision frequencies, which can be used to infer the other important ionospheric parameters. However, these radars have a limited field of view, and can be used in conjunction with Swarm data only during Swarm overflights. Other methods, like AMPERE or the AMIE procedure, illustrated in Chaps. 7 and 10 of this book respectively, can provide the configuration of FACs on a global scale. Even if the resolution is not comparable with FAC obtained from Swarm, these methods can provide the global context for the more detailed Swarm measurements.

8.6 FAC-MICE Round Table Discussion and Way Forward

During the second day of the workshop in ESA-ESTEC, a Round Table discussion with all the participants debated the possible evolution of the current Level 2 FAC products, and/or potential new FAC product and FAC quality indicators.

During the discussion, it was pointed out that the various new FAC products developed for FAC-MICE could be useful to examine different features of the FAC structure:

- the small scale FAC products (single S/C, possibly also evaluated from 50 Hz mag data) correlate better with images from optical cameras and can be used to study stable arcs. Nice correspondences of the auroral structures with the single S/C level 2 FAC products have been reported in Gillies et al. (2015). The dual S/C FAC product was not used in these studies, since it is not able to recover the smaller scales structures of the arcs, which have a typical size of 10–20 km.
- the multi- S/C products (dual or three S/C) appear to be more suitable for recovering the large-scale behaviour of the FAC structure, and can be used in combination with ground based radar observation (e.g. SuperDARN Chisham et al. 2007) or incoherent scatter radar)
- a quality flag, based on cross-correlation analysis of magnetic field data measured by Swarm A and C, can give important indications about the stationarity of the FAC structure, and therefore about the reliability of the FAC products. However, it has been pointed out that a flag 'good' or 'bad' value could be dangerous, and instead a decimal value between 0 and 1, indicating the validity of stationarity assumption could be preferable.
- the inclination of the FAC sheet estimated from the MVA or from the correlation between the Bx and By components.

It has been suggested to develop a toolbox for user definable FAC calculation, where the users can choose among the various algorithms to compute FAC. The suggested solution is to implement an open source platform, where the various codes for FAC calculation can be uploaded and shared among the users. In order to guarantee the reproducibility of the results, all the algorithms, filters and documentation used in this FAC toolbox need to be published.

Acknowledgements The FAC-MICE activity has been supported by ESA. The authors thank the International Space Science Institute in Bern, Switzerland, for supporting the ISSI Working Group: "Multi-Satellite Analysis Tools, Ionosphere", in which frame this study was performed. The Editors thanks an anonymous reviewer for his assistance in evaluating this chapter.

References

Birkeland, K. 1908. *The Norwegian aurora polaris expedition 1902–1903*, vol. 1. Christiania, Norway: H. Aschelhoug & Co.

Birkeland, K. 1913. *The Norwegian aurora polaris expedition 1902–1903*, vol. 2. Christiania, Norway: H. Aschelhoug & Co.

Cowley, S.W.H. 2000. Magnetosphere-ionosphere interactions: A tutorial review. *Magnetospheric Current Systems, Geophysical Monograph Series* 118: 91–106, AGU, Washington, D.C.

Chisham, G., M. Lester, S.E. Milan, M.P. Freeman, W.A. Bristow, A. Grocott, K.A. McWilliams, J.M. Ruohoniemi, T.K. Yeoman, P.L. Dyson, R.A. Greenwald, T. Kikuchi, M. Pinnock, J.P.S. Rash, N. Sato, G.J. Sofko, J.-P. Villain, and A.D.M. Walker. 2007. A decade of the Super Dual Auroral Radar Network (SuperDARN): Scientific achievements, new techniques and future directions. *Surveys of Geophysics* 28: 33–109. https://doi.org/10.1007/s10712-007-9017-8.

Dunlop, M.W., D.J. Southwood, K.-H. Glassmeier, and F.M. Neubauer. 1988. Analysis of multipoint magnetometer data. *Advances in Space Research* 8: 273.

Dunlop, M.W., A. Balogh, K.-H. Glassmeier, and the FGM team. 2002. Four-point cluster application of magnetic field analysis tools: The curlometer. Journal of Geophysical Research 107 (A11): 1385. https://doi.org/10.1029/2001ja005088.

Dunlop, M.W., et al. 2015. Multispacecraft current estimates at Swarm. *Journal of Geophysical Research Space Physics* 120. https://doi.org/10.1002/2015ja021707.

Finlay, C.C., N. Olsen, S. Kotsiaros, N. Gillet, and L. Tøffner-Clausen. 2016. Recent geomagnetic secular variation from Swarm and ground observatories as estimated in the CHAOS-6 geomagnetic field model. *Earth, Planets and Space* 68: 112. https://doi.org/10.1186/s40623-016-0486-1.

Gillies, D.M., D. Knudsen, E. Spanswick, E. Donovan, J. Burchill, and M. Patrick. 2015. Swarm observations of field-aligned currents associated with pulsating auroral patches. *Journal of Geophysical Research: Space Physics* 120: 9484–9499. https://doi.org/10.1002/2015JA021416.

Iijima, T., and T.A. Potemra. 1976a. The amplitude distribution of field-aligned currents at northern high latitudes observed by Triad. *Journal Geophysical Research* 81: 2165–2174.

Iijima, T., and T.A. Potemra. 1976b. Field-aligned currents in the dayside cusp observed by Triad. *Journal Geophysical Research* 81: 5971–5979.

Iijima, T., and T.A. Potemra. 1978. Large-scale characteristics of field-aligned currents associated with substorms. *Journal Geophysical Research* 83: 599–615.

Lühr, H., J.J. Warnecke, and M. Rother. 1996. An algorithm for estimating field-aligned currents from single-spacecraft magnetic field measurements: A diagnostic tool applied to Freja satellite data. *IEEE Transactions on Geoscience and Remote Sensing* 34: 1369–1376.

Lühr, H., J. Park, J.W. Gjerloev, J. Rauberg, I. Michaelis, J.M.G. Merayo, and P. Brauer. 2015. Field-aligned currents' scale analysis performed with the swarm constellation. *Geophysical Reseach Letters* 42: 1–8. https://doi.org/10.1002/2014GL062453.

Lühr, H., T. Huang, S. Wing, G. Kervalishvili, J. Rauberg, and H. Korth. 2016. Filamentary field-aligned currents at the polar cap region during northward interplanetary magnetic field derived with the Swarm constellation. *Annales Geophysicae* 34: 901–915. https://doi.org/10.5194/angeo-34-901-2016.

Ritter, P., H. Lühr, and J. Rauberg. 2013. Determining field-aligned currents with the Swarm constellation mission. *Earth, Planets and Space* 65: 1285–1294. https://doi.org/10.5047/eps.2013.09.006.

Sonnerup, B.U.Ö., and M. Scheible. 1998. Minimum and maximum variance analysis. In *Analysis methods for multi-spacecraft data*, ed. G. Paschmann and P.W. Daly, 185–220. Int. Space Sci. Inst.: Bern, Switzerland.

Zmuda, A.J., J.H. Martin, and F.T. Heuring. 1966. Transverse magnetic disturbances at 1100 kilometers in the auroral region. *Journal Geophysical Research* 71: 5033–5045.

The FAC-MICE Team

K. Kauristie[1], S. Käki[1], Heikki Vanhamäki[2], L. Juusola[1], Adrian Blagau[3,4], Joachim Vogt[4], Octav Marghitu[3], M. W. Dunlop[5,6], Y.-Y. Yang[7], J.-Y. Yang[5], Hermann Lühr[8], Guram Kervalishvili[8], Jan Rauberg[8], Claudia Stolle[8], Ivan P. Pakhotin[9], Ian R. Mann[9], C. Forsyth[10], I. J. Rae[10], Jiashu Wu[11], J. Gjerloev[12], S. Ohtani[12], M. Friel[12]

[1] Finnish Meteorological Institute, Helsinki, Finland
[2] University of Oulu, Oulu, Finland
[3] Institute of Space Science, Bucharest, Romania
[4] Jacobs University Bremen, Bremen, Germany
[5] Beihang University, Beijing, China
[6] Rutherford Appleton Laboratory, Chilton, UK
[7] The Institute of Crustal Dynamics, CEA, Beijing 100085, China
[8] Deutsches GeoForschungsZentrum, Potsdam, Germany
[9] University of Alberta, Edmonton, AB, Canada
[10] Mullard Space Science Laboratory, Holmbury Saint Mary, UK
[11] University of Calgary, H., Calgary, AB, Canada
[12] Johns Hopkins University, Baltimore, MD, USA

Chapter 9
Spherical Cap Harmonic Analysis Techniques for Mapping High-Latitude Ionospheric Plasma Flow—Application to the Swarm Satellite Mission

Robyn A. D. Fiori

Abstract This chapter describes spherical cap harmonic analysis (SCHA) for mapping ionospheric plasma flows measured by the Swarm satellites. In Sect. 9.1, SCHA is introduced as a tool for mapping a variety of one, two, and three-dimensional parameters. Section 9.2 provides a detailed summary of the theory pertaining to SCHA including a discussion of the spherical cap coordinate system, boundary conditions and basis functions, calculation of non-integer degree, and practical considerations. Section 9.3 provides a practical example of SCHA mapping of ionospheric plasma flow for a ground-based data set, and Sect. 9.4 focuses on two-dimensional SCHA mapping of Swarm ion drift measurements both independently and in conjunction with measurements from other instruments.

9.1 Introduction

A wealth of data exists to describe physical parameters in the near-Earth environment with data sets taken from instruments that are either discretely located or in constant motion taking measurements that range in spatial and temporal resolution. Although examination of these individual data sets provides important information on the evolution of the parameter at a single location over time, or at several discrete locations at a single time, it is often desirable to ascertain what is happening over a specific spatial region at a given time through interpolation or mapping techniques. The technique examined in this Chapter is a form of harmonic analysis, which involves the series expansion of an infinite set of appropriately chosen basis functions. The choice of which basis functions to use is dependent on the distribution of data, the area of interest, and the nature of the function being mapped. For data distributed about the entire spherical Earth, the normal approach is spherical harmonic analysis with basis functions that are comprised of the Associated Legendre polynomials in co-latitude and trigonometric functions in longitude. However, when either the data or region of

R. A. D. Fiori (✉)
Geomagnetic Laboratory, Natural Resources Canada, Ottawa, ON, Canada
e-mail: robyn.fiori@canada.ca

M. W. Dunlop and H. Lühr (eds.), *Ionospheric Multi-Spacecraft Analysis Tools*, ISSI Scientific Report Series 17,
https://doi.org/10.1007/978-3-030-26732-2_9

interest are confined to only a portion of the sphere, a spherical harmonic expansion cannot be applied because it is not defined across the entire natural interval (i.e. the sphere), and new basis functions must be defined for the relevant portion of the sphere. Although the new basis functions are less convenient because the function is not likely periodic over the subinterval, they are mathematically correct and allow the expansion to be applied. In the case of mapping data over a portion of a sphere, a spherical cap harmonic analysis (SCHA) technique has been developed.

Spherical cap harmonic analysis was originally developed by G. V. Haines for mapping regional magnetic fields, and has been applied to create regional maps across the globe (e.g. Haines 1985a, 1988, 2007; An 1993; An et al. 1998; De Santis et al. 1997b; Feng et al. 2015; Ji et al. 2006; Kotze, 2014; Pavón-Carrasco et al. 2008; Pothier et al. 2015; Stening et al. 2008; Tozzi et al. 2013; Weimer 2013; Waters et al. 2015). As a practical example, the Canadian Geomagnetic Reference Field (CGRF) is a regional magnetic field map that was produced every 5 years from 1985 to 2005 over Canada and adjacent regions (Haines and Newitt 1997).

Spherical cap harmonic analysis has been used for mapping a wide variety of ionospheric and magnetospheric parameters including secular variation over Canada (Haines 1985c, 1993), China (Chen et al. 2011), Europe (Korte and Haak 2000), Italy (De Santis et al. 1997a), the North Atlantic (Pavón-Carrasco et al. 2013), South Africa (Nahavo et al. 2011) and Spain (Garcia et al. 1991; Torta et al. 1992); vertical magnetic field anomalies poleward of 40° (Haines 1985b); magnetic field intensity (Kotzé 2002); Birkeland current systems (Green et al. 2006); equivalent current systems (Gaya-Pique et al. 2008; Haines and Torta 1994); ionospheric total electron content (TEC) (Liu et al. 2011, 2014; Ghoddousi-Fard et al. 2011; Otsuki et al. 2011); electric fields (Fiori et al. 2010), ionospheric conductance estimates (Green et al. 2007); and the critical frequency of the F_2 layer (f_oF_2 peak) (De Santis et al. 1991, 1992, 1994; Lazo et al. 2004). Notably, SCHA algorithms have been incorporated into the Assimilative Mapping of Ionospheric Electrodynamics (AMIE) mapping algorithm (Richmond and Kamide 1988). In addition to ionospheric and magnetospheric parameters, SCHA has been applied for mapping other quantities. One example is regional gravity field modelling on the moon (Han 2008), and on the Earth to which SCHA was originally applied by Jiancheng et al. (1995) with clarifications and corrections on its use by De Santis and Torta (1997) and additional work by Hwang et al. (2012). Another example is modelling sea level data (Hwang and Chen 1997).

This Chapter discusses the application of SCHA for mapping plasma flow in the high-latitude ionosphere based on data from multi-satellite missions such as is available from the Swarm mission.

9.2 Theory

The theory of SCHA was originally described in Haines (1985a) for mapping regional magnetic fields. It is assumed that a data set or region of interest is described on a limited portion of the spherical Earth described by a spherical cap. Data may be mapped in one, two or three dimensions. For the purpose of the electrostatic potential and plasma flow, two-dimensional mapping is considered. The necessary theory is summarized here with discussions on the relevant boundary conditions and recommendations on the appropriate choice of mapping parameters.

9.2.1 Spherical Cap Geometry

The spherical cap is constructed by extracting a conical section of angular radius θ_c from a sphere and considering the outer slab of the section which extends from radius r_1 to r_2, see Fig. 9.1. A spherical cap coordinate system is defined with respect to the centre of the outer shell of the spherical cap with θ and ϕ representing co-latitude and longitude, respectively, in the spherical cap coordinate system. Co-latitude spans from 0° to θ_c whereas φ spans the full 0°–360° range. For the two-dimensional mapping considered here, $r_1 = r_2$ and the spherical cap reduces to a thin spherical cap shell. In general, the parameter being mapped is given in a spherical coordinate system (i.e. geographic or geomagnetic) and it is necessary to first transform both coordinate locations and the vector-pointing directions to the spherical cap coordinate system to simplify calculations.

Figure 9.2 illustrates the transformation from an arbitrary spherical coordinate system to the spherical cap coordinate system. Consider an arbitrary point *P* in the spherical coordinate system having coordinates (θ', ϕ') measured with respect to the North Pole (NP). Point P has corresponding coordinates of (θ, ϕ) in the spherical cap

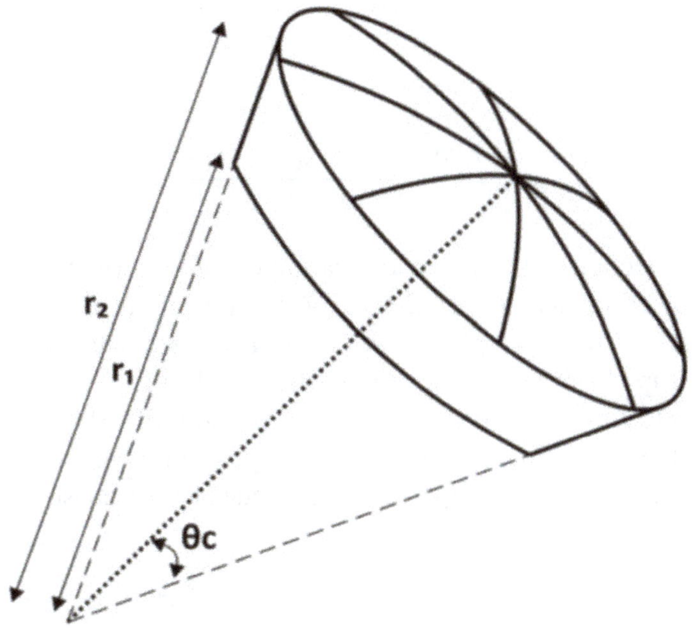

Fig. 9.1 Three-dimensional spherical cap

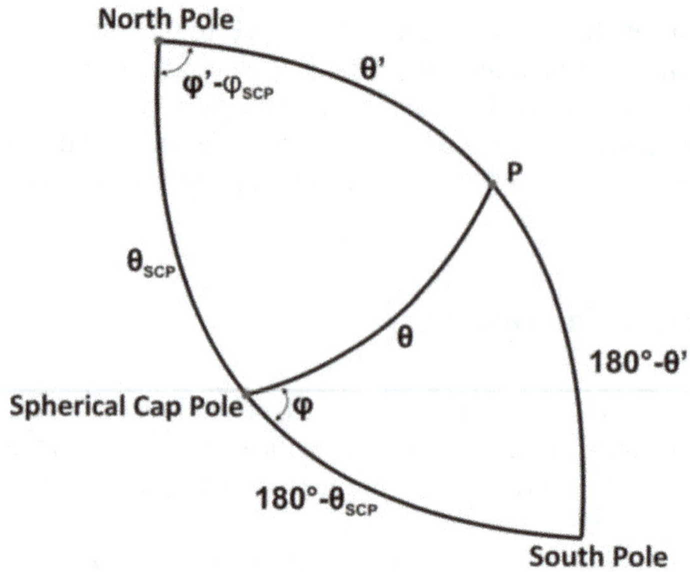

Fig. 9.2 Cartoon illustrating the transformation of the north-pole centred coordinate system to the spherical-cap coordinate system

coordinate system which is centred about the spherical cap pole (SCP), which does not necessarily coincide with NP. If the location of SCP corresponds to (θ_{SCP}, φ_{SCP}) in the spherical coordinate system, then (θ, ϕ) can be determined from (θ', ϕ') using spherical trigonometry as follows:

$$\cos\theta = \cos\theta'\cos\theta_{SCP} + \sin\theta'\sin\theta_{SCP}\cos(\varphi' - \varphi_{SCP}) \tag{9.1}$$

$$\cos(180^\circ - \theta') = \cos\theta\cos(180^\circ - \theta_{SCP}) + \sin\theta\sin(180^\circ - \theta_{SCP})\cos\varphi, \tag{9.2}$$

where Eq. (9.2) can be rearranged to solve for ϕ

$$\cos\varphi = \frac{\cos\theta\cos\theta_{SCP} - cos\theta'}{\sin\theta\sin\theta_{SCP}}. \tag{9.3}$$

Appropriate selection of the size and location of the spherical cap is necessary to ensure optimal performance of the SCHA algorithm. Both the central location of the spherical cap and θ_c should be chosen to ensure adequate data coverage across the entire spherical cap to support the desired spatial resolution of the resultant mapped data (e.g. see Sect. 9.5). When data are not distributed uniformly techniques can be applied to describe the region of the spherical cap which is adequately constrained by data to produce an accurate map. Such techniques are discussed in Sect. 9.5.

9.2.2 *Mapping Electrostatic Potential and Ion Flow by Series Expansion*

It is often desirable to represent the electrostatic potential (Φ_E) over a global or discrete region. For regional mapping, SCHA may be applied. Consider a spherical cap located on a thin spherical shell within the ionosphere on which the potential is to be mapped. Regional mapping of Φ_E is performed by representing Φ_E by series expansion (e.g. Fiori et al. 2010)

$$\Phi_E(\theta, \varphi) = \sum_{k=0}^{K_{max}} \sum_{m=0}^{M_{max}} [A_{km}\cos(m\varphi) + B_{km}\sin(m\varphi)]P^m_{n_k(m)}(cos\theta) \tag{9.4}$$

where θ and φ represent co-latitude and longitude in the spherical cap coordinate system, $P^m_{n_k(m)}(cos\theta)$ are the Associated Legendre functions of the first kind having a non-integer degree $n_k(m)$ and order m, k is an integer degree-indexing term, and K_{max} and M_{max} are the maximum degree-index and order which truncate the series expansion. $M_{max} \leq K_{max}$ and the choice of both M_{max} and K_{max} is dependent on the desired latitudinal and longitudinal mapping resolution and the distribution of data within the spherical cap. In the following notation, M_{max} is taken to be k. In Eq. (9.4), coefficients A_{km} and B_{km} are constants for each combination of k and m which represent the amplitude of the harmonics. These coefficients must be determined to describe Φ_E at any point within the spherical cap. Equations for calculating the Associated Legendre functions with non-integer degree follow in Sect. 9.2.3.

Electrostatic potential is related to the electric field (**E**) and velocity (**v**) on the spherical shell through

$$\boldsymbol{E} = -\nabla\Phi_E, \tag{9.5}$$

and

$$\begin{aligned} \mathbf{v} &= \frac{\mathbf{E}\times\mathbf{B}}{\mathrm{B}^2} \\ &= \frac{E_\varphi}{B}\hat{\theta} + \frac{E_\theta}{B}\hat{\varphi} \end{aligned} \tag{9.6}$$

where **B** is magnetic field and is commonly taken to be the radial component of either the dipole magnetic field or of the International Geomagnetic Reference Field (IGRF), although measurements could be used. Breaking Eq. (9.5) into components yields

$$E_\varphi = -\frac{1}{r\sin\theta}\frac{d\Phi_E}{d\varphi} = \frac{1}{r\sin\theta}\sum_{k=0}^{K_{max}}\sum_{m=0}^{k}[A_{km}\sin(m\varphi) - B_{km}\cos(m\varphi)]mP^m_{n_k(m)}(cos\theta) \tag{9.7}$$

$$E_\theta = -\frac{1}{r}\frac{d\Phi_E}{d\theta} = -\frac{1}{r}\sum_{k=0}^{K_{max}}\sum_{m=0}^{k}[A_{km}\cos(m\varphi) - B_{km}\sin(m\varphi)]\frac{dP^m_{n_k(m)}(cos\theta)}{d\theta} \tag{9.8}$$

which can be substituted into Eq. (9.6) to describe velocity as

$$v_\theta = \sum_{k=0}^{K_{max}}\sum_{m=0}^{k}[A_{km}\sin(m\varphi) - B_{km}\cos(m\varphi)]\frac{m\,P^m_{n_k(m)}(cos\theta)}{r\,\mathrm{B}\sin\theta} \tag{9.9}$$

$$v_\varphi = \sum_{k=0}^{K_{max}}\sum_{m=0}^{k}[A_{km}\cos(m\varphi) + B_{km}\sin(m\varphi)]\frac{1}{rB}\frac{dP^m_{n_k(m)}(cos\theta)}{d\theta}. \tag{9.10}$$

In Eqs. (9.7)–(9.10) r represents the radius of the spherical shell on which the surface is mapped. Typically this is taken to be the radius of measurement. For representing ionospheric plasma flow this is more commonly the radius of the F-region ionosphere (typically ~300 km altitude), and it is sometimes necessary to map measurements to this altitude.

Coefficients A_{km} and B_{km} are determined by minimizing the difference between parameters that are measured by the instrument and represented using Eqs. (9.4), (9.7) and (9.8), or (9.9) and (9.10), depending on the parameter being measured. Minimization is performed using a method such as linear regression or singular value decomposition (SVD) (Press et al. 1992; Brandt 1998). To minimize anomalies due, for example, to a poorly constrained solution over a portion of the spherical cap, the coefficient rejection procedure described in Haines and Fiori (2013), which eliminates non-significant fitting coefficients, can be applied.

To better understand how the fitting coefficients are determined regardless of whether Φ_E, $\boldsymbol{E}$, or $\boldsymbol{v}$ is known, consider observations of some arbitrary parameter $f(r, \theta, \varphi)$ scattered across a spherical cap which can be represented as

$$\mathrm{f}(r, \theta, \varphi) = \sum_{k=0}^{K_{max}}\sum_{m=0}^{k} A_{km}\mathrm{c}_{km}(r, \theta, \varphi) + B_{km}\mathrm{d}_{km}(r, \theta, \varphi) \tag{9.11}$$

where A_{km} and B_{km} are fitting coefficients as before, and $\mathrm{c}_{km}(r, \theta, \varphi)$ and $\mathrm{d}_{km}(r, \theta, \varphi)$ are basis functions. Here there are $(K_{max} + 1)^2$ fitting coefficients. To model $\mathrm{f}(r, \theta, \varphi)$ anywhere within the spherical cap, physical measurements f_i must be substituted into Eq. (9.11) to solve for the fitting coefficients where

$$\mathrm{f}_i(r_i, \theta_i, \varphi_i) = \sum_{k=0}^{K_{max}}\sum_{m=0}^{k} A_{km}\mathrm{c}_{km}(r_i, \theta_i, \varphi_i) + B_{km}\mathrm{d}_{km}(r_i, \theta_i, \varphi_i) \tag{9.12}$$

Consider N observations of some measured quantity $\mathrm{f}_i(r_i, \theta_i, \varphi_i)$. An array of N equations may be written as follows:

$$
\begin{aligned}
f_1 &= A_{00}c_{00_1} + B_{00}d_{00_1} + A_{10}c_{10_1} + B_{10}d_{10_1} + \cdots + A_{KK}c_{KK_1} + B_{KK}d_{KK_1} \\
f_2 &= A_{00}c_{00_2} + B_{00}d_{00_2} + A_{10}c_{10_2} + B_{10}d_{10_2} + \cdots + A_{KK}c_{KK_2} + B_{KK}d_{KK_2} \\
&\vdots \\
f_N &= A_{00}c_{00_N} + B_{00}d_{00_N} + A_{10}c_{10_N} + B_{10}d_{10_N} + \cdots + A_{KK}c_{KK_N} + B_{KK}d_{KK_N}
\end{aligned}
$$

where, for simplification, the notation $K = K_{max}$, $\mathrm{c}_{km_i} = \mathrm{c}_{km}(r_i, \theta_i, \varphi_i)$, and $\mathrm{d}_{km_i} = \mathrm{d}_{km}(r_i, \theta_i, \varphi_i)$ has been used. Such equations may be written in matrix form as

$$
\begin{pmatrix} f_1 \\ f_2 \\ \vdots \\ f_N \end{pmatrix} = \begin{pmatrix} c_{00_1} & d_{00_1} & c_{10_1} & d_{10_1} & \cdots & c_{KK_1} & d_{KK_1} \\ c_{00_2} & d_{00_2} & c_{10_2} & d_{10_2} & \cdots & c_{KK_2} & d_{KK_2} \\ \vdots & \vdots & \vdots & \vdots & \vdots & \vdots & \vdots \\ c_{00_N} & d_{00_N} & c_{10_N} & d_{10_N} & \cdots & c_{KK_N} & d_{KK_N} \end{pmatrix} \cdot \begin{pmatrix} A_{00} \\ B_{00} \\ A_{10} \\ B_{10} \\ \vdots \\ A_{KK} \\ B_{KK} \end{pmatrix}
$$

or more simply as

$$
\boldsymbol{F} = \boldsymbol{CD} \cdot \boldsymbol{AB} \tag{9.13}
$$

where $\boldsymbol{F}$ represents an $N \times 1$ matrix of the measured quantities, ***CD*** is an $N \times (K_{max} + 1)^2$ matrix, and AB is a $(K_{max} + 1)^2 \times 1$ matrix composed of the A_{km} and B_{km} fitting coefficients. In the case of electrostatic potential, electric field, and plasma flow considered in this Chapter, parameter f_i is taken to represent measured values of Φ_E, or measured components of $\boldsymbol{E}$ or $\mathbf{v}$, and CD is derived from Eq. (9.4) for Φ_E, and Eqs. (9.7) and (9.8) for $\boldsymbol{E}$, or Eqs. (9.9) and (9.10) for $\boldsymbol{v}$.

It is worth noting that matrix ***CD*** is entirely dependent on measurement coordinates. Applications of SCHA requiring the repeated mapping of data from fixed stations may take advantage of this fact as ***CD*** need only be calculated once. Such a consideration greatly enhances the speed of calculation.

9.2.3 Associated Legendre Functions

It is useful to define the Associated Legendre Functions and their calculation with a non-integer degree term. Haines (1985a, 1988) describe the Associated Legendre Functions of non-integer degree, $n_k(m)$, and order, m, as a power series given by

$$
P^m_{n_k(m)}(\cos\theta) = \sum_{j=0}^{\infty} A_j(m,n)\sin^{2j}\frac{\theta}{2} \tag{9.14}
$$

with a derivative with respect to θ given by

$$\frac{dP^m_{n_k(m)}(\cos\theta)}{d\theta} = \begin{cases} \frac{sin\theta}{2}\sum\limits_{j=1}^{\infty} jA_j(m,n)\sin^{2(j-1)}\frac{\theta}{2} & m=0 \\ \frac{sin\theta}{2}\sum\limits_{j=1}^{\infty} jA_j(m,n)\sin^{2(j-1)}\frac{\theta}{2} + \cos\theta\left[\frac{m}{sin\theta}P^m_{n_k(m)}(\cos\theta)\right] & m>0 \end{cases} \tag{9.15}$$

where $A_j(m,n)$ is given by

$$A_j(m,n) = \begin{cases} K_n^m \sin^m\theta & j=0 \\ A_{j-1}(m,n)\frac{(j+m-1)(j+m)-n(n+1)}{j(j+m)} & j>0 \end{cases} \tag{9.16}$$

and K_n^m is a normalization constant. Normalization is necessary to ensure manageable calculations in determining the solution of the expansion coefficients. Sample normalizations include the Schmidt or Neumann normalizations which take $K_n^m = 1$ for m = 0, and

$$K_n^m = \frac{2^{\frac{1}{2}}}{2^m m!}\left[\frac{(n+m)!}{(n-m)!}\right]^{\frac{1}{2}} \tag{9.17}$$

$$K_n^m = (-1)^m\frac{1}{2^m m!}\left[\frac{(n+m)!}{(n-m)!}\right] \tag{9.18}$$

for m ≠ 0 where Eq. (9.17) represents the Schmidt normalization and Eq. (9.18) represents the Neumann normalization with the Condon–Shortley phase term. Ultimately, the choice of normalization coefficient is not in itself crucial, as long as the same normalization is used consistently throughout the calculation.

When dealing with a non-integer degree term, the factorials in the normalizations constants given by Eqs. (9.17) and (9.18) are replaced by

$$n! = \Gamma(n+1) \tag{9.19}$$

where the gamma function (Γ) is given by

$$\Gamma(x) = \int\limits_0^{\infty} t^{x-1}e^{-t}dt. \tag{9.20}$$

In practice, the power series in Eq. (9.16) is truncated at J determined by evaluating each term during computation. Haines (1988) suggest a truncation limit of 60 terms.

9.2.4 Boundary Conditions and Basis Functions

The solution of Eq. (9.4) is, of course, subject to boundary conditions. The first boundary condition requires continuity in longitude such that

$$\begin{aligned}\Phi_E(\mathrm{r}, \theta, \varphi) &= \Phi_E(\mathrm{r}, \theta, \varphi + 2\pi) \\ \frac{\partial \Phi_E(\mathrm{r}, \theta, \varphi)}{\partial \varphi} &= \frac{\partial \Phi_E(\mathrm{r}, \theta, \varphi + 2\pi)}{\partial \varphi}\end{aligned} \tag{9.21}$$

restricting m to be both real and integer valued. The second boundary condition requires that Φ_E be independent of φ at the spherical cap pole

$$\begin{aligned}\Phi_E(\mathrm{r}, 0, \varphi) &= 0 \ \ \mathrm{m} \neq 0 \\ \tfrac{\partial \Phi_E(\mathrm{r}, 0, \varphi)}{\partial \theta} &= 0 \quad \mathrm{m} = 0\end{aligned} \tag{9.22}$$

thereby ensuring regularity of Φ_E. This condition has already been satisfied by using Associated Legendre Functions of the first kind in Eq. (9.4) and excluding those of the second kind. The final boundary condition requires Φ_E and $\frac{\partial \Phi_E}{\partial \theta}$ be arbitrary at the boundary of the spherical cap ($\theta = \theta_c$):

$$\Phi_E(\mathrm{r}, \theta_c, \varphi) = \mathrm{f}(\mathrm{r}, \varphi) \tag{9.23}$$

$$\frac{\partial \Phi_E(\mathrm{r}, \theta_c, \varphi)}{\partial \theta} = g(r, \varphi) \tag{9.24}$$

where f(r, φ) and g(r, φ) are arbitrary functions. This boundary condition is met by choosing $n_k(m)$ such that

$$P^m_{n_k(m)}(cos\theta)\Big|_{\theta=\theta_c} = 0 \quad \mathrm{k} - \mathrm{m} = \mathrm{odd} \tag{9.25}$$

$$\left.\frac{dP^m_{n_k(m)}(cos\theta)}{d\theta}\right|_{\theta=\theta_c} = 0 \quad \mathrm{k} - \mathrm{m} = \mathrm{even} \tag{9.26}$$

forming two sets of basis functions having a real degree $n_k(m)$ which is not necessarily an integer (Haines 1985a).

Although two basis functions are defined (Eqs. (9.25) and (9.26)) it is not necessary, and sometimes not advisable, to use both; there may be a physical significance behind choosing only one basis function, or it might be desirable to simplify the solution. The choice of basis functions depends on the parameter being mapped with consideration of the significance and cost of using one or two basis functions.

There are many examples in the literature where only the k–m = odd basis function (Eq. (9.25)) is used. Green et al. (2007) apply SCHA to mapping ionospheric conductance combining magnetic data from satellites and magnetometers, and electric field data from the Super Dual Auroral Radar Network (SuperDARN) and the

Defense Meteorological Satellite Programs (DMSP) satellites. SCHA techniques are applied to all data sets using only the k–m = odd basis function, which forces the mapped parameter to zero at the boundary of the spherical cap. They do this by setting the spherical cap boundary to zero equatorward of the current systems to reduce the impacts on the mapped current systems. In one instance they set $\theta_c = 50°$ MLAT, but limited the plotting region to $\theta > 40°$. Limiting the solution by calculating $n_k(m)$ using the k–m = 0 condition can be useful if it physically significant for the mapped parameter to be zeroed at the boundary of the spherical cap. Green et al. (2006) follow similar techniques in mapping the Birkeland currents based on Iridium and SuperDARN data. In another example, Weimer (2005) and Fiori et al. (2010) make use of this technique for mapping the ionospheric convection pattern using a spherical cap centred at the geomagnetic pole and extending to the equatorward boundary of the convection zone, where the plasma flow is expected to be zero. It should be noted that using only the k–m = odd basis function can lead to Gibb's phenomenon (often referred to as ringing) at the boundary of the spherical cap. Torta et al. (1992) propose using a spherical cap much larger than the region mapped to accommodate the larger wavelength to solve this problem. Solutions outside of the mapping region are ignored. Limiting the solution using the k–m = 0 basis function forces the mapped parameter to zero at the boundary of the spherical cap which is a useful tool when it is physically significant to do so.

Other situations have required the use of only the *k-m* = even basis function (Eq. (9.26)), thereby forcing the derivative of the mapped parameter to zero at the spherical cap. Examples include Weimer et al. (2010), Weimer (2013) and Pothier et al. (2015) who use the *k-m* = even basis functions to map geomagnetic perturbations from ground-based magnetometer data for cap-sizes ranging from 50° to 90°. These three papers are related and use the SCHA algorithm described here with only the *k-m* = even basis functions to reduce unnatural oscillations at the low-latitude boundary of the spherical cap where data are more sparsely distributed. A widely used application of the *k-m* = even basis function is in the Assimilative Mapping of Ionospheric Electrodynamics (AMIE) technique which maps various electrodynamic quantities in the high-latitude ionosphere. AMIE fits data over the hemisphere using SCHA functions with *k-m* = even basis functions poleward of 56° magnetic latitude followed by another tapering function from 56° magnetic latitude to the equator (Richmond and Kamide 1988).

The selection of what basis function(s) to use requires a discussion of both uniform convergence and orthogonality of the function. Haines (1990) explains that independently these basis functions are uniformly convergent and orthogonal, but their derivatives are not uniformly convergent unless of course the solution really does have a zero slope at the boundary of the spherical cap. Considering both basis functions in the solution ensures that both the parameter being mapped, and its derivative is uniformly convergent.

The consequence of the uniform convergence gained by using both sets of basis functions is sparse non-orthogonality of the system over the expansion interval. The two functions $P^m_{n_j(m)}(cos\theta)$ and $P^m_{n_k(m)}(cos\theta)$ are orthogonal when $j \neq k$ and when j-m and k-m are either both even or both odd but they are not orthogonal when one is even and one is odd. Non-orthogonality means that the expansion terms must be expressed using a specific truncation limit (i.e. not infinity) and the coefficients determined are therefore dependent on the maximum truncation level and are dependent on the choice of K_{max}. In terms of computation of the solution, non-orthogonality causes non-diagonal terms (ill-conditioning) in the least-squares matrix affecting the solution of the fitting coefficients. Korte and Holme (2003) criticize the commonly used practice of coefficient rejection discussed in Sect. 9.2.2 as having no physical justification in SCHA due to the incomplete orthogonality of the basis functions. Problems are more pronounced for high truncation levels, particularly if the field being mapped has a large amplitude, and for small spherical caps. Although this non-orthogonality is sparse, and it is possible to minimize the impacts, it should be considered when using both sets of basis functions.

Extreme care should be taken in determining whether or not one or both basis functions should be used. In general, limiting the solution with the k-m = odd basis functions severely limits the solution and should only be applied when it is physically significant to force the solution to zero at the boundary of the spherical cap. To allow the solution to take on an arbitrary value at the boundary of the spherical cap (unless of course the solution is required to be zero at the boundary), it is, therefore, recommended that if SCHA is to be applied with only one boundary condition, then the k-m = even boundary condition should be used. Although it may be desirable to promote orthogonality by using only one set of basis functions, keep in mind that the consequence is essentially throwing out half the solution, and this is a consequence that must be considered.

9.2.5 Non-integer Degree

SCHA requires the solution of a non-integer degree term $n_k(m)$ to describe the Associated Legendre Functions and their derivatives along the boundary of the spherical cap (i.e. Eqs. (9.25) and (9.26)). $P^m_{n_k(m)}(\cos\theta)\Big|_{\theta=\theta_c}$ and $\frac{dP^m_{n_k(m)}(\cos\theta)}{d\theta}\Big|_{\theta=\theta_c}$ are oscillating functions of $n_k(m)$ given m and θ_c. The degree-index term k starts at m and is incremented by one every time a root to either Eqs. (9.25) or (9.26) is found. These roots represent the $n_k(m)$ values and occur at a zero amplitude for k-m = odd (Eq. (9.25)) and at maximum amplitude for k-m = even (Eq. (9.26)). In this way, the difference k-m fluctuates between even and odd values as $P^m_{n_k(m)}(\cos\theta)\Big|_{\theta=\theta_0}$ fluctuates between maximum and zero amplitude. Figure 9.3 shows an example of $P^m_{n_k(m)}(\cos\theta)\Big|_{\theta=\theta_0}$ versus $n_k(m)$ for $m = 0$ and $\theta_0 = 10°, 20°$, and $30°$. Table 9.1 provides an example of the $n_k(m)$ values for all combinations of k and m for $\theta_0 = 30°$.

Before continuing, it is useful to take a moment to fully understand what the non-integer degree term $n_k(m)$ represents. In ordinary SHA, $n_k(m)$ is equivalent to the integer-valued degree-indexing term k and describes the sectioning of features in both latitude and longitude across the sphere; there are k-m divisions in latitude and

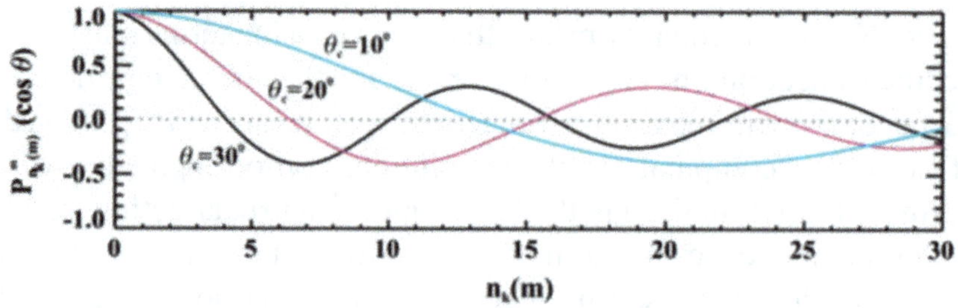

Fig. 9.3 Associated Legendre function $P^{m}_{n_k(m)}(\cos\theta)\Big|_{\theta=\theta_c}$ versus non-integer degree $n_k(m)$ for m=0 and $\theta_c = 10°, 20°$, and $30°$

2 m divisions in longitude. In SCHA, k and m also describe sectioning, but over the spherical cap opposed to the entire sphere; there are (k-m + 1)/2 divisions in latitude and 2 m divisions in longitude. Here sectioning defines the resolution of features that can be mapped using either SHA or SCHA, with a minimum resolution described for $k = K_{max}$ and $m = 0$ (for example see Fig. 9.2 of Fiori et al. 2014).

Regional mapping of ionospheric parameters using SCHA has advantages over global-scale mapping of the same parameters. Most notably, the distribution of ground-based observatories is highly dependent on the location of land and on political boundaries. It is, therefore, possible to obtain a higher resolution map over a specific region of interest (i.e. a specific country or continent) than over the globe. Both Haines (1990) and Amm and Viljanen (1999) point out that since $n_k(m) > k$ (see Table 9.1), the SCHA techniques uses higher order Associated Legendre functions than the SHA technique, and is, therefore, able to achieve a higher spectral resolution over SHA. As an example, the CGRF, which is a regional version of the IGRF with a spherical cap centred over Canada, is able to achieve a higher resolution depicting spatial variations of 833 km compared to 4000 km for the IGRF. In the case of a sparse global data set with irregularly spaced pockets of high-resolution data an SHA solution may require more basis function than data points resulting in a poor fit whereas an SCHA fit over a specific region of dense measurement offers a higher resolution fit with fewer basis functions. Practical considerations

In a theoretical setting, data are uniformly distributed about the spherical cap and the resolution of the mapped structures is a known quantity. Realistically conditions are much less ideal. This Section addresses some of the considerations necessary to apply SCHA based on a more realistic data set.

In practice, data are not uniformly distributed about the entire spherical cap as instrument locations are limited by physical parameters such as a satellite orbit or the availability of land. Mapping using a measurement-only data set has an optimal solution in the sense that the differences between the measured and calculated parameters are minimized. However, the coefficients are only well constrained over the region of measurement. Over regions not constrained by measurements, the same coefficients can cause wildly unrealistic solutions for the mapped parameter. If a smooth solution is required over the entire spherical cap, the fitting must be constrained over regions lacking measurements, using, for example, data from a statistical or empirical model.

Table 9.1 Values of $n_k(m)$ for $\theta_c = 30°$

k	m									
	0	1	2	3	4	5	6	7	8	9
0	0.00	–	–	–	–	–	–	–	–	–
1	4.08	3.12	–	–	–	–	–	–	–	–
2	6.84	6.84	5.49	–	–	–	–	–	–	–
3	10.04	9.71	9.37	7.75	–	–	–	–	–	–
4	12.91	12.91	12.37	11.81	9.96	–	–	–	–	–
5	16.02	15.82	15.62	14.92	14.18	12.13	–	–	–	–
6	18.94	18.94	18.58	18.22	17.39	16.50	14.29	–	–	–
7	22.02	21.87	21.72	21.25	20.76	19.81	18.80	16.42	–	–
8	24.95	24.95	24.69	24.42	23.84	23.24	22.19	21.07	18.55	–
9	28.01	27.90	27.78	27.42	27.05	26.38	25.69	24.55	23.31	20.66

There are numerous examples where a statistical model has been used to constrain a non-uniform data set (e.g. Richmond and Kamide 1988; Shepherd and Ruohoniemi 2000). Although the use of a model is sometimes statistically significant or necessary, care should be taken to ensure there are enough data points to overwhelm the model. Alternatively, weighting factors can be used to put more emphasis on measured data compared to the model. For an example of weighting applied to spherical harmonic mapping see Ruohoniemi and Baker (1998) who mapped ionospheric plasma flow using a modified form of spherical harmonic analysis, or Rogers and Honary (2015) who mapped absorption derived from riometer measurements.

Often it is not desirable to constrain an under sampled parameter with contributions from a statistical or empirical model, and other solutions must be explored. Walker (1989) faced a similar problem in applying SCHA techniques for mapping magnetic data based on a sparse distribution of magnetometer stations across Canada. He performed his mapping by reducing the number of longitudinal coefficients by selecting $M_{max} < K_{max}$ in the spherical cap harmonic expansion to effectively reduce the longitudinal resolution of the model. He argues that this is applicable to magnetic data as the mapped features tend to have more latitudinal than longitudinal structure and reducing the number of longitudinal coefficients enhances zonal features. Such a process also serves to prevent the number of coefficients from exceeding the number of observations and provides a first order estimate of the mapped parameter. Weimer (1995) also proposes smoothing the data by selecting $M_{max} < K_{max}$ for the purpose of mapping electrostatic potentials at high-latitudes, although it should be noted that their technique relies on a modified SHA algorithm opposed to SCHA. Weimer et al. (2010), Weimer (2013) and Pothier et al. (2015) handle the problem of sparse data at the boundary of a spherical cap by limiting the solution using only the k-m = 0 basis functions to reduce ringing at the boundary of the spherical cap. Although these sparse mapping techniques work well in the examples discussed, their use is not universal.

Another solution to handle an under sampled spherical cap is to apply a masking algorithm to map the mapped parameter only where constrained by data. Fiori et al. (2014) describe such an algorithm which they apply to limit mapping to a region of reliability surrounding measurement data points. They call the mapped region the 'region of constraint' or RoC. The method of determining the RoC is summarized here.

Essentially, a radius $\Delta\theta$ is defined to describe the region surrounding each data point around which the function is considered to be well constrained:

$$\Delta\theta = \frac{4\theta_c}{pK_{max}} \tag{9.27}$$

where θ_c and K_{max} are described as before, and p represents the number of points that must be known along a given wavelength to accurately map a given wavelength. For each data point, the number of additional data points located within $\Delta\theta$ of the data point being evaluated is counted. If there are at least p data points, the region defined by $\Delta\theta$ is considered valid. The sum of all valid regions constitutes the RoC.

For a given θ_c and K_{max}, accurate portrayal of the mapped parameter requires an appropriate definition of the RoC, which, from Eq. (9.27) is entirely dependent on the selection of p. The appropriate choice of p is somewhat ambiguous. Theoretically, a uniform distribution of data points would require $p = 2$ (i.e. the Nyquist frequency) for each spatial dimension being mapped. For a more realistic distribution of points, larger values of p are recommended. An incorrect selection of p can unnecessarily restrict (p too large) or inflate (p too small) the RoC. Restricting the RoC would potentially cause the loss of valuable information, resulting in a smoothed solution, whereas inflating the RoC risks the inclusion of poorly constrained portions of the mapped function, which are not representative of the true mapped parameter.

Figure 9.4 presents examples of the RoC applied to hypothetical coordinates for the Swarm satellite mission. For the purpose of this Figure, data are considered at a 2 Hz resolution over a 4-min interval along a possible satellite path where the lower satellite pair (Swarm A/C) is illustrated in blue and green and the upper satellite (Swarm B) is illustrated in red. Figure 9.4a, b are for the case where the spherical cap is large and covers the entire high-latitude convection zone with $\theta_c = 32°$, and the RoC is calculated for $K_{MAX} = 6$. The RoC in Fig. 9.4a is shown for $p = 6$ and is much larger than the RoC of Fig. 9.4b where a more conservative value of $p = 28$ was chosen. In Fig. 9.4c a spherical cap centred over the area of measurement is considered and the RoC is shown for $K_{max} = 2$ and $p = 6$.

Regardless of the data distribution, it is necessary to select an appropriate value for K_{max}, and therefore $n_k(m)$, to ensure an accurate representation of the mapped parameter based on the available data. Selecting too low a value of K_{max} will result in an overly smooth map whereas too large a value will cause overfitting of the data. Consider, for example, mapping some arbitrary function on a spherical cap having a surface area A where

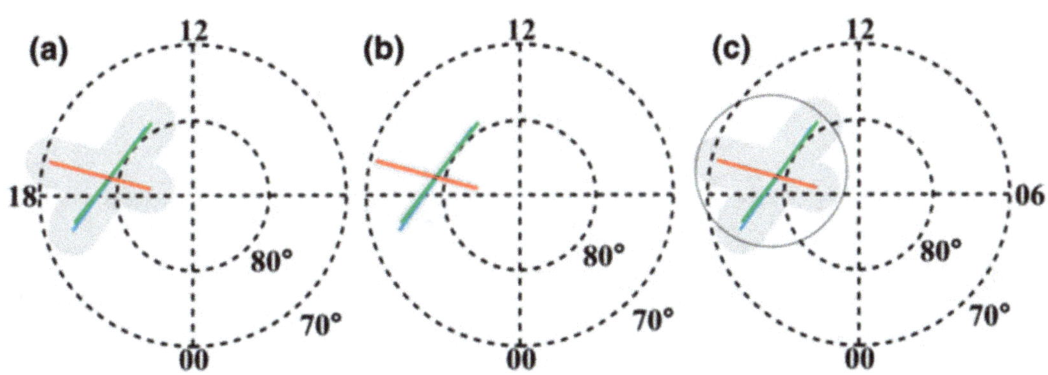

Fig. 9.4 Grey shading indicates the region of constraint (RoC) for measurements taken at a 2 Hz resolution along the hypothetical path of the Swarm A (blue), C (green), and B (red) satellites for **a** $\theta_c = 32°$, $K_{MAX} = 6$, and $p = 6$, **b** $\theta_c = 32°$, $K_{MAX} = 6$, and $p = 28$, and **c** $\theta_c = 10°$, $K_{MAX} = 2$, and $p = 6$. In **a** and **b** the spherical cap is centred on the north magnetic pole, and in **c** the spherical cap is indicated by a grey solid line. Illustration is in the magnetic latitude/MLT coordinate system

$$\begin{aligned} A &= \int_0^{2\pi} ds = \int_0^{2\pi}\int_0^{\theta_0} r^2 \sin\theta d\theta d\varphi \\ &= 2\pi r^2(1-\cos\theta_0) \\ &= 4\pi r^2 \sin^2\frac{\theta_0}{2} \\ &= 4\pi\left(\frac{180^\circ}{\pi}\right)^2 \sin^2\frac{\theta_c}{2} \end{aligned} \tag{9.28}$$

If there are n data points uniformly distributed about A, then the spacing, $\Delta\theta$, between points is given by

$$\Delta\theta = \sqrt{\frac{A}{n}}, \tag{9.29}$$

and the minimum wavelength ($\lambda_{\min}$) that can be mapped is given by (e.g. Bullard 1967)

$$\begin{aligned} \lambda_{min} &= p\Delta\theta \\ &= \frac{360^\circ}{n_k(m)} \end{aligned} \tag{9.30}$$

where p is the minimum number of points that must be present on a given wavelength. By rearranging Eq. (9.30), $n_k(m)$ can be written as

$$\begin{aligned} n_k(m) &= \frac{360^\circ}{p\Delta\theta} \\ &= \frac{360^\circ}{p}\sqrt{\frac{n}{A}} \\ &= \frac{2\pi r}{p}\sqrt{\frac{n}{4\pi r^2 \sin^2\frac{\theta_0}{2}}} \\ &= \frac{\sqrt{\pi n}}{p\sin\frac{\theta_c}{2}}. \end{aligned} \tag{9.31}$$

Equation (9.31) defines the maximum values of $n_k(m)$ appropriate for mapping a given data set assuming a uniform distribution of points.

For the case of small m, $n_k(m)$ can be approximated by (Haines 1988)

$$n_k(m) \approx \frac{k90^\circ}{\theta_c}. \tag{9.32}$$

Equation (9.31) can then be used to approximate the maximum degree-index as

$$K_{MAX} = \frac{4}{p}\sqrt{\frac{n}{\pi}}. \tag{9.33}$$

In reality, data are not distributed uniformly and it is necessary to select a lower value of K_{max}, and therefore $n_k(m)$. Too large a value results in the formation of unphysical flow structures in the mapped field in regions of the spherical cap where the $\lambda_{\min}$ defined by K_{max} is less than the $\Delta\theta$ characterizing measurement locations within that specific region of the spherical cap. Such problems are more pronounced along the edges of the spherical cap as the solution is only bounded on one 'side' (i.e. within the spherical cap). However, this problem can be improved by including data from outside the spherical cap where possible, even though the solution to Φ_E is limited to within the spherical cap. Alternatively, or additionally, a coefficient rejection procedure can be implemented to reject any insignificant coefficients which, although they may add a slight reduction to the minimization performed in the fitting procedure, do not contribute to the solution (Haines and Fiori 2013).

Green (2006) points out that the choice of K_{max} must also necessarily depend on the number of available observations within the spherical cap. As K_{max} increases (i.e. the spatial resolution decreases), the number of basis functions (i.e. the number of coefficients in the series expansion) increases. Care should be taken to ensure that the number of basis functions does not exceed the number of data points. Green (2006) provides an excellent example in their Sect. 9.3.

From Eq. (9.33) it is obvious that determination of the appropriate maximum value of $n_k(m)$ to map a given data set is dependent on an appropriate choice of p. As noted above, $p = 2$ is the absolute minimum, requiring 2 observations per wavelength. However, for a real data set with irregularly spaced observations most likely contaminated by noise, a larger value of p is more practical to ensure a good fit (Fiori et al. 2010). Bendat (1958) actually suggests that p should be chosen in the range of 10–20, but the author's experience suggest that such a dense distribution of measurements is an unreasonable assumption. In practice the choice of p is largely dependent on the typical distribution and quality of data for a given data set and is up to the discretion of the user. As an example, Fiori et al. (2010) suggest $p = 6$ for mapping two-dimensional plasma flow based on a data set smoothly distributed throughout the spherical cap. In examining satellite data confined to a narrow track transecting the spherical cap, Fiori et al. (2014) provide recommendations for selecting p based on evaluating the goodness-of-fit of the resultant map.

An additional consideration is the location and size of the spherical cap, given by θ_c. Ideally, the cap should be placed either centred over the data set having a radius which fully encompasses the data, or over a specific geographic (or geomagnetic) region of interest. Errors associated with cap-size may arise when attempting to map wavelengths that are larger than the spherical cap, for example, when mapping secular variation. Such a problem is commonly encountered when mapping magnetic field data and is solved by subtracting a known reference field, such as the IGRF, to represent the residual value, and adding the reference field back after solving for the fitting coefficients. Haines (1985a) proposes this as a solution to a poor fit at the boundary of the spherical cap due to not including data outside of the spherical cap.

9.3 Practical Example of Mapping Ionospheric Plasma Flow Using SHA and SCHA Approaches

The Super Dual Auroral Radar Network (SuperDARN) (Greenwald et al. 1995; Chisham et al. 2007) samples the high-latitude plasma flow pattern every 1-2 min in both the northern and southern hemisphere. The advantage of the large spread of measurements has been modelling of the ionospheric electric potential. Restriction of measurements to only roughly 1/3 of the sphere leads to ill-conditioning of the least-squares modelling of the potential using SHA approaches. Ruohoniemi and Baker (1998) instead take the approach of redistributing thc data in co-latitude across a larger region in a method commonly referred to as either the map potential or FIT technique. The FIT technique redistributes the data located poleward of a predefined boundary (θ_{FIT}) across the entire sphere opposed to the hemisphere. The redistribution is performed by scaling co-latitude by $\alpha = 180°/\theta_{FIT}$ to effectively 'stretch' the data across the sphere. Following redistribution, SHA is applied in the usual way by representing the electrostatic potential as a series expansion built with Associated Legendre Functions of integer degree and order. Once the electrostatic potential is modelled, the model is carefully 'un-stretched' from the entire sphere to the high-latitude region poleward of θ_{FIT}.

Redistribution of data across the entire sphere forces the boundary defined by θ_{FIT} to a single point at $\theta = 180°$ thereby forcing the electrostatic potential at θ_{FIT} to zero. Although undesirable for general mapping, the consequence to mapping the electrostatic potential does have physical significance, as the ionospheric convection pattern is expected to close at some low-latitude boundary (i.e. Hairston and Heelis 1990; Heppner and Maynard 1987; Rich and Hairson, 1994; Weimer 1995; Fiori et al. 2016). In the FIT technique, the low-latitude boundary given by θ_{FIT} is determined based on the distribution of low-velocity (<100 m/s) measured velocities. Zeroing the electrostatic potential at θ_{FIT} would be difficult to achieve using SHA without redistribution of the data and would require a very-high-order fit. In contrast, the SCHA algorithm could be used with only the k-m = odd basis functions, which has been investigated by Weimer et al. (2005) and Fiori et al. (2010, 2013).

Care should be used when applying the FIT technique in general. It is not clear what impact the changing data density caused by the redistribution of data has on the overall fit. For example, consider two pairs of observations separated by 111 km located at $\theta_1 = 29.5°$ and $\theta_2 = 10°$ for $\theta_{FIT} = 30°$. In the transformed regime, $\theta_1' = 177°$ and $\theta_2' = 60°$, and the pair of points are now separated by 12 km and >500 km, respectively. The impact of this redistribution is unclear. Mapping electrostatic potential through the FIT technique also requires the exact location of the low-latitude boundary of the ionospheric convection pattern, defined by θ_{FIT} to be known. For example, Fiori (2011) show that one consequence of a boundary located too far equatorward is the reduction of the calculated cross-polar cap potential (CPCP), which is a significant finding as SuperDARN electrostatic potential maps are shown to consistently report lower CPCP compared to values determined by other means using other instruments (i.e. Shepherd 2007).

Fiori et al. (2010) point out that constraints applied by the FIT technique to ensure a smooth fit over the high-latitude region, namely forcing a low-latitude boundary to the flow and the use of a statistical convection model (see Sect. 9.5), potentially distort the mapped convection. They apply an SCHA algorithm for mapping ionospheric convection based on SuperDARN measurements using the techniques discussed above and examine the impacts on both global-scale and regional-scale mapping. The SCHA technique is shown to perform comparably to the FIT technique over regions of good data coverage, and to provide a better solution over regions of highly variable flow, particularly near the low-latitude boundary of the convection zone near θ_{FIT}.

9.4 Application of SCHA for Mapping Ionospheric Plasma Flow Based on Swarm Satellite Ion Drift Measurements

An application of the SCHA algorithm discussed in this Chapter is the mapping of ionospheric data from multi-spacecraft missions. Satellite data from various missions have been incorporated into single or multi-instrument data sets using techniques not limited to SCHA for magnetic field mapping (e.g. An et al. 1998; AMIE; Green et al. 2007; Haines 1985b; Kotzé 2002), or for mapping statistical convection patterns based on satellite data collected over a long period of time (e.g. Heppner and Maynard 1987; Rich and Maynard 1989; Hairston and Heelis 1990; Rich and Hairston 1994; Papitashvili et al. 1999; Papitashvili and Rich 2002; Weimer 1995). However, as of the time of writing, the authors of this Chapter are not aware of the application of SCHA to the instantaneous mapping of satellite data. However, extensive work has been performed in Fiori et al. (2013, 2014) to evaluate the potential for mapping ionospheric plasma flow based on artificially generated Swarm ion drift measurements. This Section provides a summary of this work.

One objective of the Swarm mission is an investigation of electric currents flowing in the Earth's ionosphere which drive ionospheric electric fields and ion drift. Each satellite is therefore equipped with an electric field instrument (EFI) capable of measuring the three-dimensional ion drift (Knudsen et al. 2003, 2017). Ion drift measured in the along-track and cross-track directions by two orthogonal bi-dimensional Thermal Ion Imagers (TIIs) are combined to create a two-dimensional ion drift vectors describing plasma flowing in the horizontal plane. Such vectors may be processed using the SCHA mapping procedures described in this Chapter to map the two-dimensional plasma flow. The algorithm is fairly straightforward and is described as follows. (1) For the interval of interest, combine along-track and cross-track ion drift measurements to create two-dimensional vectors of the horizontal ion drift. If desirable these ion drift vectors can be mapped down from the satellite altitude to the F-region peak of the ionosphere (e.g. Walker and Sofko 2015). For example, ion drifts mapped down from 470 km to 300 km, would be reduced by approximately 4%. (2) Build the matrices described by Eq. (9.13). If there are N measurements, then matrix $\boldsymbol{F}$ represents a $2\,N \times 1$ matrix of the measured ion drift in the θ (v_θ) and φ (v_φ) directions. Note that each ion drift component is calculated from the two-dimensional ion drift and then considered separately. $\boldsymbol{CD}$ is a $2\,N \times (K_{max}+1)^2$ matrix derived from Eqs. (9.9) and (9.10) for v_θ and v_φ, and $\boldsymbol{AB}$ is a $(K_{max}+1)^2 \times 1$ matrix composed of the A_{km} and B_{km} fitting coefficients. (3) The fitting coefficients are solved as described in Sect. 9.2.2. (4) Fitting coefficients are used to calculate the two-dimensional ion drift within the spherical cap where adequately constrained by data. This SCHA algorithm has been applied to (1) examine convection mapping on a global-scale spanning the entire convection zone in the high-latitude region based on both a Swarm-only data set and a multi-instrument data set (Fiori et al. 2013), and (2) explore the possibility of convection mapping with a Swarm-only data set over a localized region centred over the satellites' orbit, and determine parameters for evaluating the appropriate size of the localized region and the goodness-of-fit that can be obtained (Fiori et al. 2014). Note that due to the unavailability of definitive Swarm data at the time of publication, Fiori et al. (2013, 2014) rely on artificially generated ion drifts emulated from a statistical model along hypothetical Swarm satellite tracks. For a full description of the emulation technique, see either paper. The SCHA techniques described here could equivalently be applied to hypothetical or real data.

Fiori et al. (2013) examined SCHA for mapping the high-latitude ionospheric convection pattern based on a Swarm data set using a global-scale spherical cap. In order to map the entire high-latitude convection patter, the spherical cap was centred at the north magnetic pole extending to ~60° MLAT ($\theta_c = 30°$). They show that even when all three satellites cross the spherical cap at the same approximate time, the Swarm data coverage is insufficient to support such large-scale convection mapping based on a Swarm-only data set; the RoC does not exceed the measurement location. However, they do show that inclusion of the Swarm data with a complementary data set has the potential to increase the region of the spherical cap populated by data thereby improving resultant convection maps.

In the case of Fiori et al. (2013), the complementary data set under consideration was the SuperDARN (Greenwald et al. 1995; Chisham et al. 2007) which samples the high-latitude plasma flow pattern continuously every 1–2 min in both the northern and southern hemisphere. Currently, SuperDARN consists of radars with a collective field of view that covers the majority of the high-latitude region (within ~30° of either magnetic pole), extending to mid-latitudes. However, despite this superior coverage, there are frequently gaps in the data due to a variety of circumstances including geomagnetic activity, solar activity, season, and magnetic local time (MLT) of observation (Ruohoniemi and Greenwald 1996; Danskin et al. 2002; Shepherd et al. 2002; Koustov et al. 2004; Shepherd 2007). It is common practice within the SuperDARN community to supplement real data with data from a statistical convection model to produce a well-constrained and smooth map (Ruohoniemi and Baker 1998; Shepherd and Ruohoniemi 2000; Ruohoniemi and Greenwald 2005; Cousins and Shepherd 2010; Pettigrew et al. 2010). Statistical models are useful in expanding convection data so that the RoC covers the entire high-latitude region during periods of stable convection but do not always accurately represent the true plasma flow during more turbulent conditions (e.g. Fiori et al. 2010). In such instances, it is more useful to include data from other instruments, such as is available from the Swarm satellites to fill in gaps in the SuperDARN data set in order to increase the overall RoC that can be mapped.

Examples presented in Fiori et al. (2013) demonstrate how effectively data from the Swarm satellite can increase the RoC for global-scale convection mapping with SuperDARN. For instance, Fiori et al. (2013) present one case (reproduced in Fig. 9.5) where despite excellent SuperDARN coverage the RoC is insufficient to determine the cross-polar cap potential (CPCP) without adding contributions from a statistical model or other instruments. Figure 9.5a illustrates hypothetical measurement locations for the SuperDARN and Swarm instruments. In Fig. 9.5b, convection is mapped based on a global-scale spherical cap ($\theta_c = 32°$) centred at the north magnetic pole for $K_{max} = 6$ and $p = 6$. Black contours indicate the electrostatic potential associate with the convection pattern, and the plus sign indicates the maximum potential. Because the minimum potential cannot be mapped, it is not possible to determine the CPCP. Addition of data from a single Swarm satellite (here data has been artificially generated based on a statistical model based on solar wind and IMF conditions), slightly enlarges the RoC. Inclusion of hypothetical data from all three Swarm satellites increases the RoC sufficiently to allow determination of the CPCP.

Fiori et al. (2013) show that the addition of Swarm ion drift data to a SuperDARN data set can cause an average relative increase in the RoC of 12.5% for periods when 1–3 Swarm satellites are located in the high-latitude region, with the greatest contributions being made when Swarm data are located at auroral latitudes.

In addition to supplementing other data sets for the purpose of global-scale convection mapping, Swarm data may also be considered independently for mapping more localized regions. For localized convection mapping data may be considered for three different satellite configurations: a single satellite track (upper satellite Swarm B), a double satellite track (lower satellite pair Swarm A/C), or for the case where all three satellites overlap. As an example, consider a regional spherical cap with $\theta_c = 10°$, which corresponds to roughly 4-min of observation processed with $K_{max} = 3$. Figure 9.6 shows two statistical convection models within such a spherical cap centred at 80° magnetic latitude and 12 MLT. Figure 9.6c indicates the location of sample satellite tracks. Artificially generated Swarm ion drift vectors generated from these statistical models along the satellite paths illustrated in Fig. 9.6c can be used to map convection on the localized spherical cap. Such maps are shown in Fig. 9.7 where the RoC is described by $p = 6$. Maps in Fig. 9.7a, b can be compared to the original statistical model in Fig. 9.6a, b (and mapped using black vectors in Fig. 9.7), respectively. The visual comparison indicates good overall agreement between the mapped vectors and the statistical model, with the greatest difference occurring at the outlying region of the RoC, suggesting the RoC could be reduced by increasing the value of p.

Although a three-satellite map offers more data for convection mapping, and is, therefore, more likely to produce a more accurate map, it is not a frequent occurrence. In the next example, consider the case where only the lower satellite pair is available for mapping. In Fig. 9.8a the IMF $B_z > 0$ convection pattern illustrated in Fig. 9.6b is

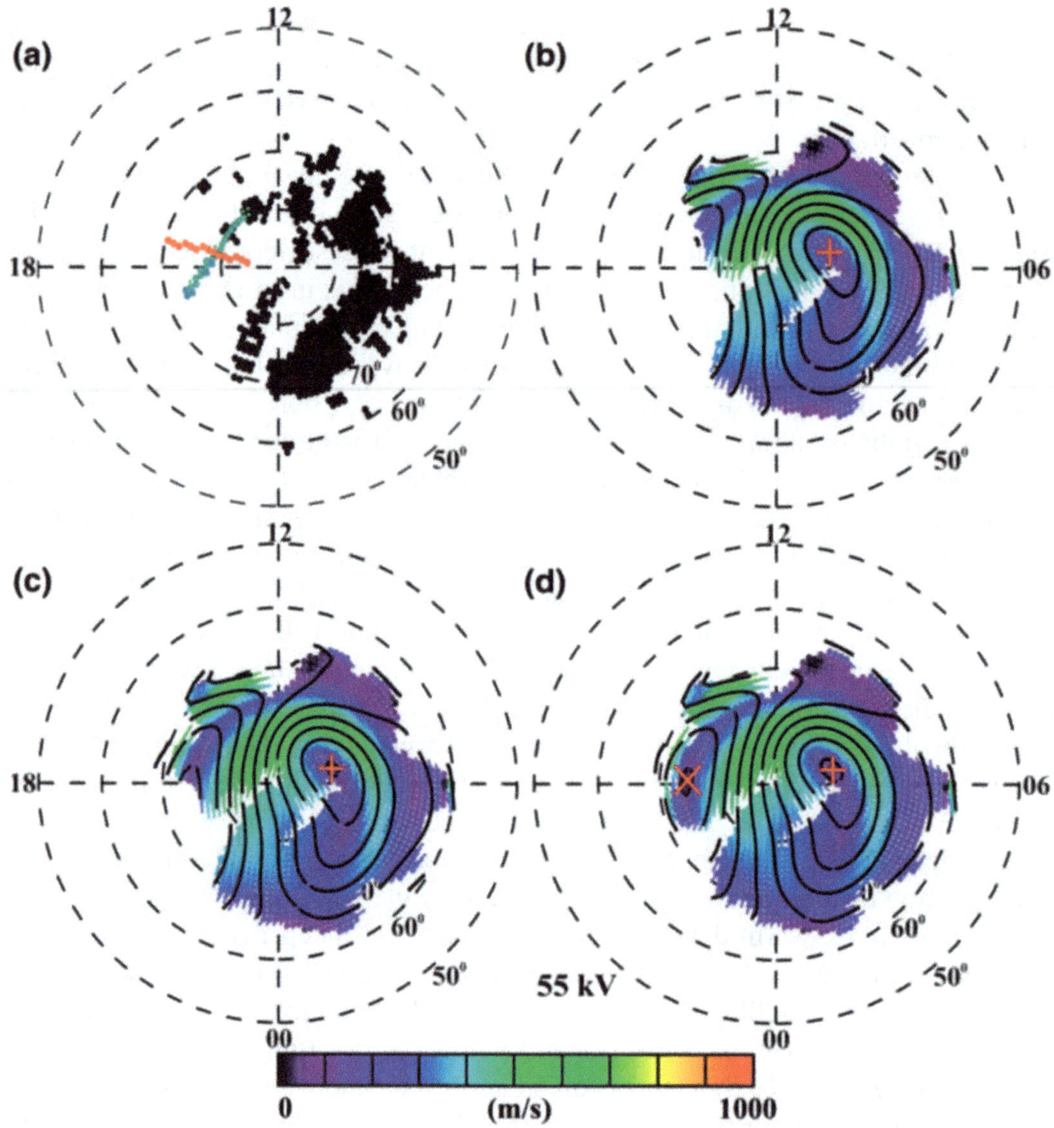

Fig. 9.5 **a** Gridded coordinate locations for the Swarm A (blue), Swarm C (green), Swarm B (red), and SuperDARN (black) data sets. SuperDARN data are based on gridded measurement coordinates for 20 January 2011 11:20–11:24 UT. **b–d** Convection calculated for $\theta_c = 32°$, $K_{max} = 6$, $p = 6$ based on artificial measurements generated from a statistical mode at possible coordinates of **b** SuperDARN, **c** SuperDARN and Swarm B, and **d** SuperDARN and Swarm A, B and C measurements shown in (**a**). Convection vectors are plotted along an evenly spaced grid of size 1° of magnetic latitude throughout the RoC. Contours are plotted with a 6 kV contour spacing. The plus and cross indicate the maximum and minimum potential and the CPCP is indicated where it can be determined. (Reproduction of Fiori et al. 2013, Fig. 9.6.)

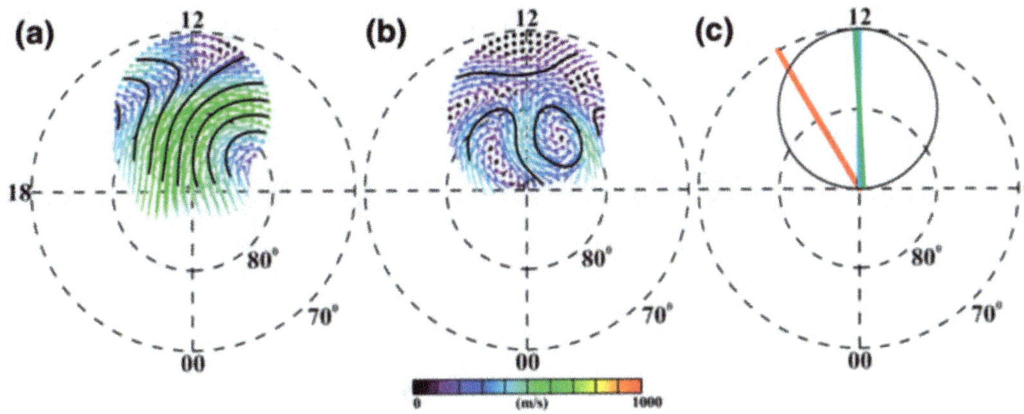

Fig. 9.6 Portion of two sample statistical model convection patterns for the case of **a** IMF $B_Z < 0$ and **b** IMF $B_Z > 0$ within a spherical cap centred at 80° MLAT and 12 MLT with $\vartheta_c = 10°$. The Cousins and Shepherd (2010) models employed in Fiori et al. (2013, 2014) are used. **c** shows a 4-min segment of the Swarm A (blue), C (green), and B (red) satellite orbits

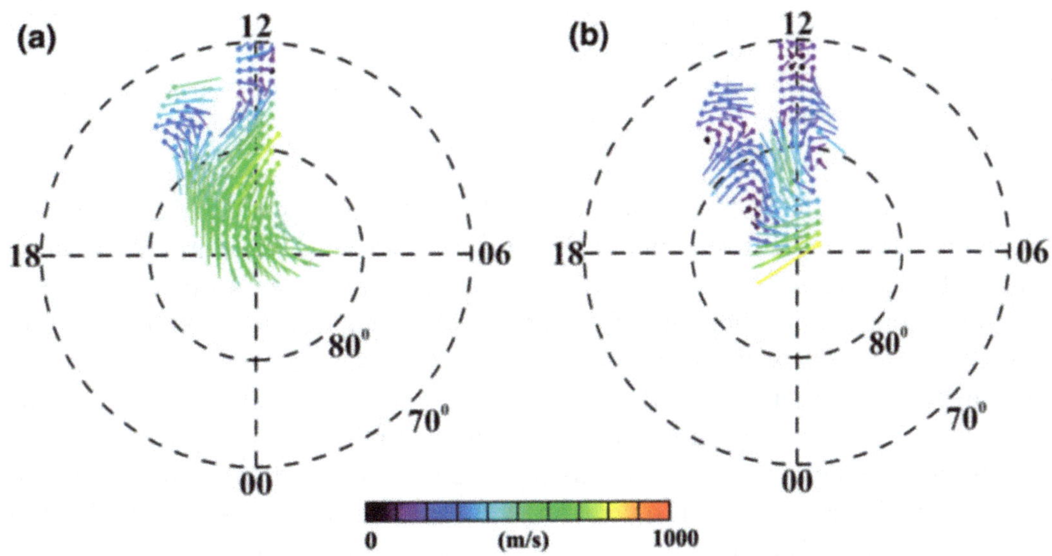

Fig. 9.7 Coloured vectors indicate convection determined based on Swarm measurements artificially generated along the satellite tracks shown in Fig. 9.6c based on the statistical models mapped in **a** Figs. 9.6a, b for a spherical cap centred at 80° MLAT and 12 MLT with $\theta_c = 10°$ and $K = 3$. The RoC is defined with $p = 6$

used to artificially generate Swarm ion drifts for the Swarm A/C satellite pair which follow a dawn/dusk orientation. Artificially generated ion drifts are processed to create the convection pattern shown where the RoC is limited by $p = 12$. In Fig. 9.8b a similar map is generated for a noon/midnight satellite orientation using the IMF $B_Z < 0$ statistical convection model shown in Fig. 9.6a. Both maps indicate good agreement between the mapped convection and the statistical convection model.

Fiori et al. (2014) examined the use of SCHA for mapping convection using a localized spherical cap centred over the lower Swarm satellite pair. Convection maps

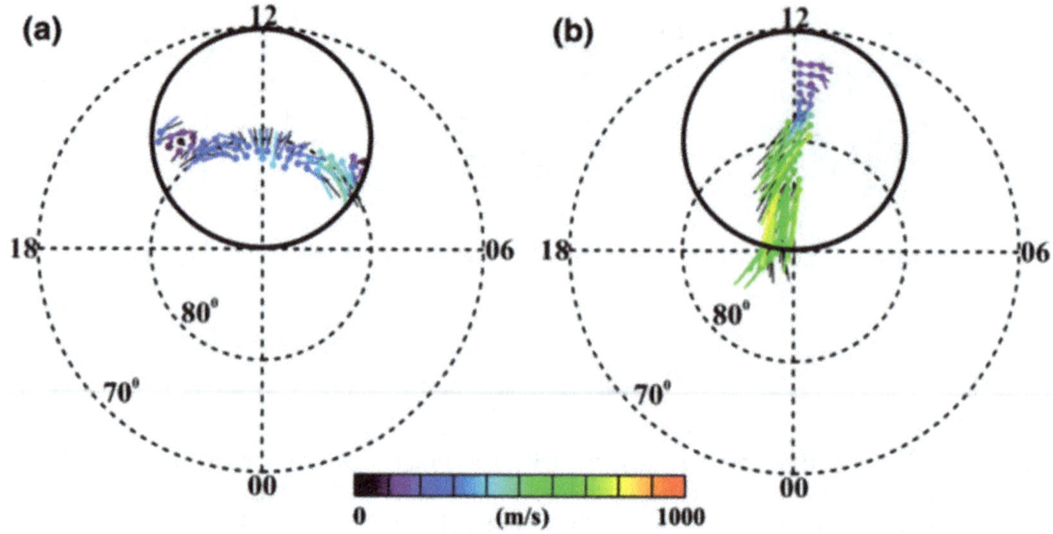

Fig. 9.8 Coloured vectors indicate convection calculated for $\theta_c = 10°$ and $K_{max} = 3$ for a spherical cap centred at 70° MLAT and 12 MLT for Swarm A and C measurements artificially generated using the **a** IMF $B_Z > 0$ convection pattern for a Swarm A satellite track offset 80° from the 12 MLT meridian and $p = 12$ and **b** IMF $B_Z < 0$ convection pattern for a Swarm A satellite track offset 170° from the 12 MLT meridian and $p = 12$. Black vectors in (**a**) and (**b**) indicate the direction of the statistical model convection

were created using the SCHA technique based on artificially generated Swarm measurements within a localized spherical cap having an angular radius, $\theta_{c,}$ ranging from 4° to 15°. These maps used a K_{MAX} that was calculated so that the SCHA mapping resolution matched that of the statistical model used to generate measurements. Within the study, the artificially generated Swarm data were used to generate 2502 maps for a variety of satellite orientations with RoC determined for varying values of p. Comparisons were made between the statistical model used to create the artificially generated measurements and the convection mapped using SCHA. These comparisons were used to categorize the maps as being well constrained or poorly constrained. Based on this analysis it was determined that the SCHA algorithm ideally maps Swarm data for $\theta_{c,} = 10°$, and was unreliable for $\theta_{c,} \geq 14°$. Fiori et al. (2014) determined that Swarm-based measurements could successfully map convection over a localized region surrounding the satellite track to examine small-scale features in the flow, and that larger scale convection mapping can be achieved through combining multiple spherical caps.

9.5 Summary

This Chapter provides detailed explanations of how to map ionospheric plasma flows in the high-latitude region based on hypothetical ion drifts from the Swarm satellites using spherical cap harmonic analysis. SCHA is a useful technique for mapping data limited in space, and is closely related to spherical harmonic analysis which is commonly applied on a larger global scale. Although originally developed for mapping the three-dimensional magnetic field on a regional scale, SCHA techniques have been used to map a variety of one-, two- and three-dimensional properties. This Chapter provides a summary of the relatively new application of SCHA to two-dimensional mapping of ion drift.

Section 9.2 provides the necessary theory for applying SCHA to mapping ionospheric plasma flow, or convection. Equations for mapping electrostatic potential, electric field, and plasma flow in the ionosphere are provided along with discussion regarding the relevant basis functions and boundary conditions considered in the solution. Practical considerations such as the appropriate cap-size to use for a given data set and how to determine where the solution is adequately constrained by measurements are discussed. Section 9.3 provides an example of SHA and SCHA mapping of ionospheric convection based on the SuperDARN data. In Sect. 9.4 the theory developed in this Chapter is applied to the mapping of ionospheric ion drift measured by the Swarm satellites. Sample convection maps are presented which illustrate SCHA mapping of Swarm ion drift data (1) on a global scale in combination with measurements from additional ground-based instruments and (2) on a local scale using satellite measurements alone.

This Chapter clearly demonstrates the application of spherical-cap harmonic analysis techniques for mapping plasma flow patterns based on ion drift measurements from the Swarm satellite.

Acknowledgements This work was supported by the Natural Resources Canada (NRCan) Lands and Minerals Sector and Public Safety Geosciences program, and the Canadian Space Agency. The author wishes to thank G. V. Haines for communications regarding spherical cap harmonic analysis and D. Knudsen and J. Burchill for communications regarding Swarm. Useful contributions and discussion from David Boteler were appreciated. The author acknowledges the use of SuperDARN data. SuperDARN is a collection of radars funded by national scientific funding agencies of Australia, Canada, China, France, Italy, Japan, Norway, South Africa, United Kingdom and the United States of America. SuperDARN data are available from the SuperDARN website hosted by Virginia Tech (http://vt.superdarn.org). We thank the International Space Science Institute (ISSI) in Bern, Switzerland for supporting the Working Group "Multi Satellite Analysis Tools—Ionosphere" from which this chapter resulted. The Editors thank David Green for his assistance in evaluating this chapter.

References

Amm, O., and A. Viljanen. 1999. Ionospheric disturbance magnetic field continuation from the ground to the ionosphere using spherical elementary current systems. *Earth Planets Space* 51: 431–440. https://doi.org/10.1186/BF03352247.

An, Z.-C. 1993. Spherical cap harmonic analysis of geomagnetic field for China. *Acta Geophysica Sinica* 36 (6): 753–764. https://doi.org/10.1002/cjg2.1602.

An, Z.-C., D.H. Tan, N.M. Rotanova, and V.P. Golovkov. 1998. Spherical cap harmonic analysis of magsat magnetic anomalies over Asia. *Acta Geophysica Sinica* 41 (2): 172–173.

Bendat, J.S. 1958. *Principles and applications of random noise theory*. New York: Wiley.

Brandt, S. 1998. *Data analysis: Statistical and computational methods for scientists and engineers*, 3rd ed. New York: Springer.

Bullard, E.C. 1967. The removal of trend from magnetic surveys. *Earth and Planetary Science Letters* 2 (4): 293–300. https://doi.org/10.1016/0012-821X(67)90145-8.

Chen, B., Z.-W. Gu, J.-T. Gao, J.-H. Yuan, and C.Z. Di. 2011. Analysis of geomagnetic field and its secular variation over China for 2005.0 epoch using spherical cap harmonic method. *Chinese Journal of Geophysics* 54 (3): 771–779. https://doi.org/10.1002/cjg2.1602.

Chisham, G., et al. 2007. A decade of the Super Dual Auroral Radar Network (SuperDARN): scientific achievements, new techniques and future directions. *Surveys In Geophysics* 28 (1): 22–109. https://doi.org/10.1007/s10712-007-9017-8.

Cousins EDP, Shepherd SG. 2010. A dynamical model of high-latitude convection derived from SuperDARN plasma drift measurements. *Journal of Geophysical Research* 115 (12). https://doi.org/10.1029/2010JA016017.

Danskin, D.W., A.V. Koustov, T. Ogawa, N. Nishitani, S. Nozawa, S.E. Milan, M. Lester, and D. Andre. 2002. On the factors controlling occurrence of F-region coherent echoes. *Annales Geophysicae* 20: 1385–1397.

De Santis, A., G. De Franceschi, B. Zolesi, S. Pau, and L.J.R. Cander. 1991. Regional mapping of the critical frequency of the F2 layer by spherical cap harmonic expansion. *Annales Geophysicae* 9: 401–406.

De Santis, A., G. De Franceschi, B. Zolesi, and L.J.R. Cander. 1992. Regional modelling and mapping of the ionospheric characteristic parameters by spherical cap harmonic expansion. *Advances in Space Research* 12 (6): 279–282. https://doi.org/10.1016/0273-1177(92)90073-7.

De Santis, A., G. De Franceschi, and D.J. Kerridge. 1994. Regional spherical cap modelling of 2D functions: The case of the critical frequency of the F2 ionospheric layer. *Computers & Geosciences* 20: 849–871. https://doi.org/10.1016/0098-3004(94)90117-1.

De Santis, A., and J.M. Torta. 1997. Spherical cap harmonic analysis: A comment on its proper use for local gravity field representation. *Journal of Geodesy* 71: 526–532. https://doi.org/10.1007/s001900050120.

De Santis, A., C. Falcone, and J.M. Torta. 1997a. SHA vs. SCHA for modelling secular variation in a small region such as Italy. *Journal of Geomagnetism and Geoelectricity* 49: 359–371. https://doi.org/10.5636/jgg.49.359.

De Santis, A., M. Chiappini, G. Dominici, and A. Meloni. 1997b. Regional geomagnetic field modelling: The contribution of the Istituto Nazionale di Geofisica. *Annali di Geofisica* XL(5). https://doi.org/10.4401/ag-3854.

Feng, L., M. Gao, and B. Chen. 2015. Influence of boundary points selection on the accuracy of spherical cap harmonic model of Mongolia magnetic field. *Acta Seismologica Sinica* 37 (4): 588–598.

Fiori, R.A.D., D.H. Boteler, A.V. Koustov, G.V. Haines, and J.M. Ruohoniemi. 2010. Spherical cap harmonic analysis of Super Dual Auroral Radar Network (SuperDARN) observations for generating maps of ionospheric convection. *Journal of Geophysical Research* 115(A07307). https://doi.org/10.1029/2009JA015055.

Fiori, R.A.D., D.H. Boteler, D. Knudsen, J. Burchill, A.V. Koustov, E.D.P. Cousins, and C. Blais. 2013. Potential impact of Swarm electric field data on global 2D convection mapping in combination with SuperDARN radar data. *Journal of Atmospheric and Solar-Terrestrial Physics* 93: 87–99. https://doi.org/10.1016/j.jastp.2012.11.013.

Fiori, R.A.D., D.H. Boteler, A.V. Koustov, D. Knudsen, and J.K. Burchill. 2014. Investigation of localized 2D convection mapping based on artificially generated Swarm ion drift data. *Journal of Atmospheric and Solar-Terrestrial Physics* 114: 30–41. https://doi.org/10.1016/j.jastp.2014.04.004.

Fiori, R.A.D., A.V. Koustov, D.H. Boteler, D.J. Knudsen, and J.K. Burchill. 2016. Calibration and assessment of Swarm ion drift measurements using a comparison with a statistical convection model. *Earth, Planets and Space* 68: 100. https://doi.org/10.1186/s40623-016-0472-7.

Garcia, A., J.M. Torta, J.J. Curto, and E. Sanclement. 1991. Geomagnetic secular variation over Spain 1970-1988 by means of spherical cap harmonic analysis. *Physics of the Earth and Planetary Interiors* 68 (1–2): 65–75. https://doi.org/10.1016/0031-9201(91)90008-6.

Gaya-Pique, L.R., J.J. Curto, J.M. Torta, and A. Chulliat. 2008. Equivalent ionospheric currents for the 5 December 2006 solar flare effect determined from spherical cap harmonic analysis. *Journal of Geophysical Research* 113(A07304). https://doi.org/10.1029/2007JA012934.

Ghoddousi-Fard, R., P. Heroux, D. Danskin, and D. Boteler. 2011. Developing a GPS TEC mapping service over Canada. *Space Weather* 9(S06D11). https://doi.org/10.1029/2010SW000621.

Green, D.L. 2006. The Mie and Helmholtz representation of vector fields in the context of magnetosphere-ionosphere coupling, Ph. D Thesis, University of Newcastle.

Green, D.L., C.L. Waters, B.J. Anderson, H. Korth, and R.J. Barnes. 2006. Comparison of large-scale Birkeland currents determined from Iridium and SuperDARN data. *Annales Geophysicae* 24: 941–959.

Green, D.L., C.L. Waters, H. Korth, B.J. Anderson, A.J. Ridley, and R.J. Barnes. 2007. Technique: Large-scale ionospheric conductance estimated from combined satellite and ground-based electromagnetic data. *Journal of Geophysical Research* 112(A05303). https://doi.org/10.1029/2006ja012069.

Greenwald, R.A., et al. 1995. DARN/SuperDARN: A global view of the dynamics of high-latitude convection. *Space Science Reviews* 71: 763–796. https://doi.org/10.1007/BF00751350.

Haines, G.V. 1985a. Spherical cap harmonic analysis. *Journal Geophysical Research* 90 (B3): 2583–2591. https://doi.org/10.1029/JB090iB03p02583.

Haines, G.V. 1985b. Magsat vertical field anomalies above 40°N from spherical cap harmonic analysis. *Journal Geophysical Research* 90 (B3): 2593–2598. https://doi.org/10.1029/JB090iB03p02593.

Haines, G.V. 1985c. Spherical cap harmonic analysis of geomagnetic secular variation over Canada 1960-1983. *Journal Geophysical Research* 90 (B14): 12563–12574. https://doi.org/10.1029/JB090iB14p12563.

Haines, G.V. 1988. Computer programs for spherical cap harmonic analysis of potential and general fields. *Computers & Geosciences* 14 (4): 413–447. https://doi.org/10.1016/0098-3004(88)90027-1.

Haines, G.V. 1990. Modelling by series expansion: A discussion. *Journal of Geomagnetism and Geoelectricity* 42: 1037–1049.

Haines, G.V. 1993. Modelling geomagnetic secular variation by main-field differences. *Geophysical Journal International* 114 (3): 490–500. https://doi.org/10.5636/jgg.42.1001.

Haines, G.V., and L.R. Newitt. 1997. The Canadian Geomangetic reference field 1995. *Journal of Geomagnetism and Geoelectricity* 49: 317–336. https://doi.org/10.5636/jgg.49.317.

Haines, G.V. 2007. *Encyclopedia of geomagnetism and paleomagnetism, chap*. Spherical Cap Harmonics, pp. 395–397, Encyclopedia of Earth Sciences, Springer.

Haines, G.V., and R.A.D. Fiori. 2013. Modeling by singular value decomposition and the elimination of statistically insignificant coefficients. *Computers & Geosciences* 58: 19–28. https://doi.org/10.1016/j.cageo.2013.04.021.

Hairston, M.R., and R.A. Heelis. 1990. Model of the high-latitude ionospheric convection pattern during southward interplanetary magnetic field using DE 2 data. *Journal Geophysical Research* 95 (3): 2333–2343. https://doi.org/10.1029/JA095iA03p02333.

Haines, G.V., and J.M. Torta. 1994. Determination of equivalent current sources from spherical cap harmonic models of geomagnetic field variations. *Geophysical Journal International* 118: 499–514. https://doi.org/10.1111/j.1365-246X.1994.tb03981.x.

Han, S.-C. 2008. Improved regional gravity fields on the moon from lunar prospector tracking data by means of localized spherical harmonic functions. *Journal of Geophysical Research* 113(E11012). https://doi.org/10.1029/2008JE003166.

Heppner, J.P., and N.C. Maynard. 1987. Empirical high-latitude electric field models. *Journal Geophysical Research* 92 (A5): 2267–4489. https://doi.org/10.1029/JA092iA05p04467.

Hwang, C., and S.-K. Chen. 1997. Fully normalized spherical cap harmonics: Application to the analysis of sea-level data from TOPEX/POSEIDON and ERS-1. *Geophysical Journal International* 129: 450–460. https://doi.org/10.1111/j.1365-246X.1997.tb01595.x.

Hwang, J.S., H.-C. Han, S.-C. Han, K.-O. Kim, J.-H. Kim, M.-H. Kang, and C.H. Kim. 2012. Gravity and geoid model in South Korea and its vicinity by spherical cap harmonic analysis. *Journal of Geodynamics* 53 (1): 27–33. https://doi.org/10.1016/j.jog.2011.08.001.

Ji, X., M. Utsugi, H. Hirai, A. Suzuki, J. He, S. Jufiwara, and Y. Fukuzaki. 2006. Modelling of spatial-temporal changes of the geomagnetic field in Japan. *Earth, Planets and Space* 58 (6): 757–763. https://doi.org/10.1186/BF03351979.

Jiancheng, L., C. Dingbo, and N. Jinsheng. 1995. Spherical cap harmonic expansion for local gravity field representation. *Manuser Geod* 20: 265–277.

Knudsen, D., J.K. Burchill, K. Berg, T. Cameron, G.A. Enno, C.G. Marcellus, E.P. King, I. Weavers, and R.A. King. 2003. A low-energy charged particle distribution imager with a compact sensor for space applications. *Review of Scientific Instruments* 74: 202–211. https://doi.org/10.1063/1.1525869.

Knudsen, D.J., J.K. Burchill, S.C. Buchert, A. Eriksson, R. Gill, J.-E. Wahlund, L. Åhlen, M. Smith, and B. Moffat. 2017. Thermal ion imagers and Langmuir probes in the Swarm electric field instruments. *Journal of Geophysical Research* 122(2) https://doi.org/10.1002/2016JA022571.

Korte, M., and V. Haak. 2000. Modeling European magnetic repeat station and survey data by SCHA in search of time-varying anomalies. *Physics of the Earth and Planetary Interiors* 122 (3–4): 205–220. https://doi.org/10.1016/S0031-9201(00)00194-1.

Korte, M., and R. Holme. 2003. Regularization of spherical cap harmonics. *Geophysical Journal International* 153: 253–262. https://doi.org/10.1046/j.1365-246X.2003.01898.x.

Kotzé, P.B. 2002. Modelling and analysis of Ørsted total field data over southern Africa. *Geophysical Research Letters* 29(15). https://doi.org/10.1029/2001GL013868.

Kotzé, P.B. 2014. Modelling and analysis of southern African geomagnetic field observations: 1840 until 1903. *South African Journal of Geology* 117 (2): 211–218. https://doi.org/10.2113/gssajg.117.2.211.

Koustov, A.V., G.J. Sofko, D. André, D.W. Danskin, and L.V. Benkevitch. 2004. Seasonal variation of HF radar F region echo occurrence in the midnight sector. *Journal of Geophysical Research* 109(A06305). https://doi.org/10.1029/2003JA010337.

Lazo, B., A. Calzadila, K. Alazo, M. Rodriguez, and J.S. Gonález. 2004. Regional mapping of F2 peak plasma frequency by spherical harmonic expansion. *Advances in Space Research* 33 (6): 880–883. https://doi.org/10.1016/j.asr.2003.03.023.

Liu, J., R. Chen, and Z. Wang. 2011. Spherical cap harmonic model for mapping and predicting regional TEC. *GPS Solut* 15: 109–119. https://doi.org/10.1007/s10291-010-0174-8.

Liu, J., R. Chen, J. An, and Z. Wang. 2014. Spherical cap harmonic analysis of the Arctic ionospheric TEC for one solar cycle. *Journal Geophysical Research* 119 (1): 601–619. https://doi.org/10.1002/2013JA019501.

Nahavo, E., P.B. Kotzé, and M.J. Alport. 2011. An investigation into the use of satellite data to develop a geomagnetic secular variation model over Southern Africa. *Data Science Journal* 10. https://doi.org/10.2481/dsj.IAGA-11.

Otsuki, S., M. Kamimura, M. Ohashi, Y. Kubo, and S. Sugimoto. 2011. Local models for ionospheric VTEC estimation based on GR models and spherical cap harmonic analysis. *Journal of Aeronautics Astronautics and Aviation* 43 (1): 1–7.

Papitashvili, V.O., F.J. Rich, M.A. Heinemann, and M.R. Hairston. 1999. Parameterization of the Defense Meteorological Satellite Program ionospheric electrostatic potentials by the interplanetary magnetic field strength and direction. *Journal Geophysical Research* 104: 177–184. https://doi.org/10.1029/1998JA900053.

Papitashvili, V.O. and F. J. Rich. 2002. High-latitude ionospheric convection models derived from Defense Meteorological Satellite Program ion drift observations and parameterized by the interplanetary magnetic field strength and direction. *Journal of Geophysical Research* 107(A8). https://doi.org/10.1029/2001JA000264.

Pavón-Carrasco, FcoJ, M.L. Osete, J.M. Torta, L.R. Gaya-Piqué, and Ph Lanos. 2008. Initial SCHA.DI.00 regional archaeomagnetic model for Europe for the last 2000 years. *Physics and Chemistry of the Earth* 33 (6–7): 596–608. https://doi.org/10.1016/j.pce.2008.02.024.

Pavón-Carrasco, F.J., J.M. Torta, M. Catalán, T. Talarn, and T. Ishihara. 2013. Improving total field geomagnetic secular variation modeling from a new set of cross-over marine data. *Physics of the Earth and Planetary Interiors* 216: 21–31. https://doi.org/10.1016/j.pepi.2013.01.002.

Pettigrew, E.D., S.G. Shepherd, J.M. Ruohoniemi. 2010. Climatological patterns of high-latitude convection in the northern and southern hemispheres: Dipole tilt dependencies and interhemispheric comparisons. *Journal of Geophysical Research* 115(A07305). https://doi.org/10.1029/2009JA014956.

Pothier, N.M., D.R. Weimer, and W.B. Moore. 2015. Quantitative maps of geomagnetic perturbation vectors during substorm onset and recover. *Journal Geophysical Research* 120: 1197–1214. https://doi.org/10.1002/2014JA020602.

Press, W.H., S.A. Teukolsky, W.T. Vetterling, and B.P. Flannery. 1992. *Numerical recipes in C: The art of scientific computing*, 2 ed., Cambridge University Press.

Rich, F.J., and M. Hairston. 1994. Large-scale convection patterns observed by DMSP. *Journal Geophysical Research* 99 (A3): 3827–3844. https://doi.org/10.1029/93JA03296.

Rich, F.J., and N.C. Maynard. 1989. Consequences of using simple analytical functions for the high-latitude convection electric field. *Journal Geophysical Research* 94 (A4): 3687–3701. https://doi.org/10.1029/JA094iA04p03687.

Richmond, A.D., and Y. Kamide. 1988. Mapping electrodynamic features of the high-latitude ionosphere from localized observations: technique. *Journal Geophysical Research* 93 (A6): 5741–5759. https://doi.org/10.1029/JA093iA06p05741.

Rogers, N.C., and F. Honary. 2015. Assimilation of real-time riometer measurements into models of 20 MHz polar cap absorption. *Journal Space weather Space Clim* 5: A8. https://doi.org/10.1051/swsc/2015009.

Ruohoniemi, J.M., and K.B. Baker. 1998. Large-scale imaging of high-latitude convection with Super Dual Auroral Radar Network HF radar observations. *Journal Geophysical Research* 103: 20797–20811. https://doi.org/10.1029/98JA01288.

Ruohoniemi, J.M., and R.A. Greenwald. 1996. Statistical patterns of high-latitude convection obtained from Goose Bay HF radar observations. *Journal Geophysical Research* 101 (A10): 21743–21763. https://doi.org/10.1029/96JA01584.

Ruohoniemi, J.M., and R.A. Greenwald. 2005. Dependencies of high-latitude plasma convection: Consideration of interplanetary magnetic field, seasonal, and universal time factors in statistical patter. *Journal of Geophysical Research* 110(A09204). https://doi.org/10.1029/2004JA010815.

Shepherd, S.G., and J.M. Ruohoniemi. 2000. Electrostatic potential patterns in the high-latitude ionosphere constrained by SuperDARN measurements. *Journal Geophysical Research* 105 (A10): 23005–23014. https://doi.org/10.1029/2000JA000171.

Shepherd, S.G., R.A. Greenwald, and J.M. Ruohoniemi. 2002. Cross polar cap potentials measured with Super Dual Auroral Radar Network during quasi-steady solar wind and interplanetary magnetic field conditions. *Journal of Geophysical Research*. https://doi.org/10.1029/2001JA000152.

Shepherd, S.G. 2007. Polar cap potential saturation: Observations, theory, and modeling. *Journal of Atmospheric and Solar-Terrestrial Physics* 69: 234–248. https://doi.org/10.1016/j.jastp.2006.07.022.

Stening, R.J., T. Reztsova, D. Ivers, J. Turner, and D.E. Winch. 2008. Spherical cap harmonic analysis of magnetic variations data from mainland Australia. *Earth, Planets and Space* 60 (12): 1177–1186. https://doi.org/10.1186/BF03352875.

Torta, J.M., A. Garcia, J.J. Curto, and A. De Santis. 1992. New representation of the geomagnetic secular variation over restricted regions by means of spherical cap harmonic analysis: application to the case of Spain. *Physics of the Earth and Planetary Interiors* 74 (3–4): 209–217. https://doi.org/10.1016/0031-9201(92)90011-J.

Tozzi, R., A. De Santis, and L.R. Gaya-Piqué. 2013. Antarctic geomagnetic reference model updated to 2010 and provisionally to 2012. *Tectonophysics* 585: 13–25. https://doi.org/10.1016/j.tecto.2012.06.034.

Walker, J.K. 1989. Spherical cap harmonic modelling of high latitude magnetic activity and equivalent sources with sparse observations. *Journal of Atmospheric and Terrestrial Physics* 51 (2): 67–80. https://doi.org/10.1016/0021-9169(89)90106-2.

Walker, A.D.M., and G.J. Sofko. 2015. Mapping steady state electric fields and convective drifts in geomagnetic fields—1. Elementary models. *Annales Geophysicae* 34: 55–65. https://doi.org/10.5194/angeo-34-55-2016.

Waters, C.L., J.W. Gjerloev, M. Dupont, and R.J. Barnes. 2015. Global maps of ground magnetometer data. *Journal of Geophysical Research* 120. https://doi.org/10.1002/2015JA021596.

Weimer, D.R. 1995. Models of high-latitude electric potentials derived with a least error fit of spherical harmonic coefficients. *Journal of Geophysical Research* 100(A10): 19,595–19,607. https://doi.org/10.1029/95JA01755.

Weimer, D.R. 2005. Predicting surface geomagnetic variations using ionospheric electrodynamic models. *Journal of Geophysical Research* 110(A12307). https://doi.org/10.1029/2005JA011270.

Weimer, D.R., C.R. Clauer, M.J. Engebretson, T.L. Hansen, H. Gleisner, I. Mann, K. Yumoto. 2010. Statistical maps of geomagnetic perturbations as a function of the interplanetary magnetic field. *Journal of Geophysical Research* 115(A10320). https://doi.org/10.1029/2010JA015540.

Weimer, D.R. 2013. An empirical model of ground-level geomagnetic perturbations. *Space Weather* 11(3). https://doi.org/10.1002/swe.20030.

Chapter 10
Recent Progress on Inverse and Data Assimilation Procedure for High-Latitude Ionospheric Electrodynamics

Tomoko Matsuo

Abstract Polar ionospheric electrodynamics plays an important role in the Sun–Earth connection chain, acting as one of the major driving forces of the upper atmosphere and providing us with a means to probe physical processes in the distant magnetosphere. Accurate specification of the constantly changing conditions of high-latitude ionospheric electrodynamics has long been of paramount interest to the geospace science community. The Assimilative Mapping of Ionospheric Electrodynamics procedure, developed with an emphasis on inverting ground-based magnetometer observations for historical reasons, has long been used in the geospace science community as a way to obtain complete maps of high-latitude ionospheric electrodynamics by overcoming the limitations of a given geospace monitoring system. This Chapter presents recent technical progress on inverse and data assimilation procedures motivated primarily by availability of regular monitoring of high-latitude electrodynamics by space-borne instruments. The method overview describes how electrodynamic state variables are represented with polar-cap spherical harmonics and how coefficients are estimated from the point of view of the Bayesian inferential framework. Some examples of the recent applications to analysis of SuperDARN plasma drift, Iridium, and DMSP magnetic fields, as well as DMSP auroral particle precipitation data are included to demonstrate the method.

10.1 Introduction

The most dynamic electromagnetic energy and momentum exchange processes between the upper atmosphere and the magnetosphere take place in the polar ionosphere. Physical processes producing aurora involve ionization and excitation of atmospheric constituents due to energetic charged particles precipitating into the upper atmosphere from the magnetosphere along the geomagnetic field lines, which

T. Matsuo (✉)
Ann and H.J. Smead Department of Aerospace Engineering Sciences, University of Colorado, Boulder, CO 80303-0429, USA
e-mail: tomoko.matsuo@colorado.edu

M. W. Dunlop and H. Lühr (eds.), *Ionospheric Multi-Spacecraft Analysis Tools*, ISSI Scientific Report Series 17,
https://doi.org/10.1007/978-3-030-26732-2_10

in turn modulates the ionosphere's ability to conduct electric currents. Polar ionospheric electrodynamics plays an important role in the Sun–Earth connection chain, acting as one of the major driving forces of the upper atmosphere and providing us with a means to probe physical processes in the distant magnetosphere. Accurate specification of the constantly changing conditions of high-latitude ionospheric electrodynamics has long been of paramount interest to the geospace science community.

Global monitoring of high-latitude geospace has dramatically improved thanks to a recent expansion of ground-based and space-based observing capability. International consortiums of ground-based instrumentation such as the Super Dual Auroral Radar Network (SuperDARN) (e.g., Greenwald et al. 1995), International Real-Time Magnetic Observatory Network (e.g., Love 2013) and SuperMAG (e.g., Gjerloev 2009) have made a large volume of quality-controlled, standardized data accessible to the public. Acquisition, processing, and distribution of engineering-grade magnetometer data from the Iridium satellite constellation for scientific purposes by the Active Magnetosphere and Polar Electrodynamics Response Experiment (AMPERE) program (Anderson et al. 2000) have been instrumental in making continuous, global monitoring of geomagnetic-field-aligned currents (FAC) possible. Defense Meteorological Satellite Program (DMSP) space environment instruments have long been providing valuable measurements of precipitating electron and ion particles, magnetic fields, and ultraviolet spectrographic images (e.g., Rich 1984; Hardy et al. 1984; Paxton et al. 2002). And the Swarm multi-satellite mission (Friis-Christensen et al. 2006) provides high precision measurements of magnetic fields that complement theses existing geospace observing systems.

Data assimilation techniques such as the Assimilative Mapping of Ionospheric Electrodynamics (AMIE) procedure of Richmond and Kamide (1988) have long been used in the geospace science community as a way to obtain complete maps of high-latitude ionospheric electrodynamics by overcoming the limitations of a given geospace monitoring system. The procedure combines a number of different types of space-based and ground-based observations with an empirical model of ionospheric electrodynamics to infer distributions of ionospheric electric fields and currents, FAC, associated geomagnetic perturbation fields at both ground and low-Earth-orbit altitudes, Hall and Pedersen conductance, and Joule heating. AMIE maps have yielded a number of important insights into the coupling of the magnetosphere, ionosphere, and thermosphere that takes place at high latitudes. Lu (2017) provides a comprehensive overview of AMIE applications.

This paper presents an overview of the recent technical developments of the inverse and data assimilation procedure for high-latitude electrodynamics. Some of these developments are a consequence of a reformulation of the best linear unbiased estimation problem presented in Richmond and Kamide (1988) as a Bayesian estimation problem (Matsuo et al. 2005). Under the assumption that electrodynamic variables are Gaussian distributed, these two estimation problems are equivalent. A Bayesian perspective has helped to clarify the role of the prior model (background) error covariance as a key component in the modeling of Gaussian processes, and thus guided modeling and estimation of prior covariance functions from a large volume of SuperDARN data (Cousins et al. 2013a), DMSP particle precipitation data

(McGranaghan et al. 2015, 2016), and Iridium magnetic perturbation data (Cousins et al. 2015b; Shi et al. 2019). Even though ionospheric conductivity serves as a critical linkage in electromagnetic energy and momentum exchange processes, direct monitoring of this conductivity is almost nonexistent. Another notable development led by McGranaghan et al. (2016) is an assimilative mapping of the conductance using the auroral ionization derived from DMSP electron energy flux spectra with help of the GLobal airglOW (GLOW) model (Solomon et al. 1988) without the assumption of Maxwellian distribution. Since the AMIE has been developed with an emphasis on inverting ground-based magnetometer observations for historical reasons (Kamide et al. 1981; Richmond and Kamide 1988), it is not tailored to analyses of space-based magnetometer data from DMSP, Iridium, and Swarm. In order to solve the optimization problem in terms of electrostatic potential, the space-based magnetometer data first need to be converted to electrostatic potential through the application of Ohm's law and current continuity. To minimize the impact of conductance on the inversion of space-based magnetometer data for FAC, the optimization problem is now being solved in terms of both magnetic potential and electrostatic potential (Matsuo et al. 2015; Cousins et al. 2015a).

10.2 Method Overview

10.2.1 Representation of Electrodynamic State Variables Using Scalar and Vector Polar-Cap Spherical Harmonic Basis Functions

The ionosphere is treated as a thin conductive slab centered at a reference height $h_r = 110\,\text{km}$, and the current above the ionosphere is assumed to be strictly radial. The effect of the neutral wind dynamo is not considered. Electrodynamic variables analyzed here include the electrostatic potential Φ, electric fields $\mathbf{E}$, Pedersen and Hall conductance (height-integrated conductivity) Σ_p, Σ_h, height-integrated horizontal ionospheric current density $\mathbf{J}_\perp$, toroidal magnetic potential Ψ associated with field-aligned current density $\mathbf{J}_\parallel$, and equivalent current potential Ω associated with ground-based magnetic fields. These variables are presumed to be related to each other as follows.

$$\mathbf{E} = -\nabla\Phi \tag{10.1}$$

$$\mathbf{J}_\perp = \boldsymbol{\Sigma} \cdot \mathbf{E} \tag{10.2}$$

$$\mathbf{J}_\parallel = -\nabla \cdot \mathbf{J}_\perp \tag{10.3}$$

$$\nabla \times \mathbf{J}_\perp = -\nabla^2_{\text{hor}}\Omega \tag{10.4}$$

$$\mathbf{J}_\parallel = \frac{1}{\mu_o}\nabla^2_{\text{hor}}\Psi \tag{10.5}$$

where $\mathbf{\Sigma} = \begin{pmatrix} \Sigma_p & -\Sigma_h \\ \Sigma_h & \Sigma_p \end{pmatrix}$ is the conductance tensor, ∇^2_{hor} is the horizontal Laplacian, and μ_o is permeability of free space. Equation (10.4) results from the assumption of strictly vertical $\mathbf{J}_{\parallel}$ that allows equating the curls of $\mathbf{J}_{\perp}$ and the equivalent current (i.e., Fukushima Theorem). If the Pedersen and Hall conductances are given, the relationship among all electrodynamic variables (10.1)–(10.5) becomes linear.

In the procedure, electrodynamic variables are expressed in terms of the polar-cap spherical harmonic basis functions developed by Richmond and Kamide (1988). Suppose that $\boldsymbol{\Psi}$ represents a matrix of the polar-cap spherical harmonic basis functions evaluated at discrete grid locations specified by the Modified Magnetic Apex longitude ϕ_m and latitude λ_m at the altitude of h_r (Richmond 1995) and that $\mathbf{x}$ denotes a vector of the coefficients. $\boldsymbol{\Psi}$ is furthermore given by a set of 244 polar-cap spherical harmonic basis functions up to order $m_\Psi = 12$, with non-integer degrees n_Ψ up to a maximum of $n_\Psi = 72.6$ for $m_\Psi = 0$, with a polar-cap co-latitude for the functions of 40°. Therefore, $\mathbf{x}$ is a column vector of 244 elements and Ψ is an $n \times 244$ matrix, where n is the number of grid points. Using the Nyquist sampling rate, the effective resolution is 15° longitude and 2.5° latitude. Let's suppose that the electrostatic potential Φ at ϕ_m and λ_m is given by

$$\Phi(\phi_m, \lambda_m) = \boldsymbol{\Psi}\mathbf{x}_{\text{E}} + \boldsymbol{\epsilon}_t \ , \tag{10.6}$$

where $\boldsymbol{\epsilon}_t$ is the truncation error, and the electric fields $\mathbf{E}$ by

$$\mathbf{E}(\phi_m, \lambda_m) = -\boldsymbol{\Psi}'\mathbf{x}_{\text{E}} + \boldsymbol{\epsilon}_t \ , \tag{10.7}$$

where a $(n \times 244)$ matrix $\boldsymbol{\Psi}'$ contains the gradients of the polar-cap spherical harmonic basis functions, which discretizes (10.1). The toroidal potential Ψ at ϕ_m and λ_m is then given by

$$\Psi(\phi_m, \lambda_m) = \boldsymbol{\Psi}\mathbf{x}_{\text{M}} + \boldsymbol{\epsilon}_t \ , \tag{10.8}$$

where $\boldsymbol{\epsilon}_t$ is the truncation error, and the FAC magnitude J by

$$J(\phi_m, \lambda_m) = \boldsymbol{\Psi}''\mathbf{x}_{\text{M}} + \boldsymbol{\epsilon}_t \ , \tag{10.9}$$

where a $(n \times 244)$ matrix $\boldsymbol{\Psi}''$ contains a simplified evaluation of (10.5) using the analytical expression of the horizontal Laplacian of polar-cap spherical harmonic basis functions applicable to spherical coordinates, rather than the full expression applicable to M(110) coordinates. As explained in (Matsuo et al. 2015), this computational simplification introduces errors on the order of 10%. For a given Σ_p and Σ_h, $\mathbf{x}_{\text{E}}$ and $\mathbf{x}_{\text{M}}$ are related linearly through the current continuity and Ohm's law (10.2)–(10.3).

10.2.2 Bayesian State Estimation for Gaussian Processes

Suppose that $\mathbf{y}$ represents a vector of j observations that may consist of electric field, ground-based magnetic field, and/or space-based magnetic field measurements at discrete observation locations. By evaluating the polar-cap spherical harmonics and their derivatives at observation locations, $\mathbf{y}$ can be expressed as

$$\mathbf{y} = \mathbf{H}\mathbf{x} + \boldsymbol{\epsilon}_r \ , \tag{10.10}$$

where $\mathbf{H}$ is a $(j \times 244)$ matrix that contains the polar-cap spherical harmonic basis functions and their spatial derivatives with corresponding vector calculus operations as specified in (10.1)–(10.5), $\mathbf{x}$ denotes a vector of the 244 coefficients, and $\boldsymbol{\epsilon}_r$ is the sum of observational and truncation errors. The objective of the Bayesian state estimation is to infer the polar-cap spherical harmonics coefficients $\mathbf{x}$ given observations $\mathbf{y}$ according to Bayes rule: $[\mathbf{x}|\mathbf{y}] \propto [\mathbf{y}|\mathbf{x}][\mathbf{x}]$ The vectors $\mathbf{x}$ and $\mathbf{y}$ are herein assumed to be distributed according to the multivariate normal distribution denoted by $\mathcal{MN}$ as

$$\mathbf{x} \sim \mathcal{MN}[\mathbf{x}_b, \mathbf{C}_b] \ , \tag{10.11}$$

$$\mathbf{y} \sim \mathcal{MN}[\mathbf{H}\mathbf{x}, \mathbf{C}_r] \ , \tag{10.12}$$

where $\mathbf{x}_b$ is the prior mean, $\mathbf{C}_b$ is the prior (background) model error covariance $< (\mathbf{x}_b - \mathbf{x})(\mathbf{x}_b - \mathbf{x})^T >$, and $\mathbf{C}_r$ is the observational error covariance $< \boldsymbol{\epsilon}_r \boldsymbol{\epsilon}_r^T >$. $\mathbf{x}_b$ is specified by using an empirical model. $\mathbf{C}_b$ is described in the following section. The errors $\boldsymbol{\epsilon}_r$ are assumed to be uncorrelated, so $\mathbf{C}_r$ is given by a diagonal matrix of the variance of observational error. The posterior distribution or the conditional distribution of $\mathbf{x}$ given observations $\mathbf{y}$ is given by the multivariate normal distribution as

$$[\mathbf{x}|\mathbf{y}] \sim \mathcal{MN}[\mathbf{x}_a, \mathbf{C}_a] \ , \tag{10.13}$$

where $\mathbf{x}_a$ is the posterior mean or the data assimilation analysis and $\mathbf{C}_a$ is the analysis error covariance $< (\mathbf{x}_a - \mathbf{x})(\mathbf{x}_a - \mathbf{x})^T >$. In the case of normally distributed $\mathbf{x}$ and $\mathbf{y}$ and linear $\mathbf{H}$, there are closed formulae for $\mathbf{x}_a$ and $\mathbf{C}_a$ (e.g., Jazwinski 1970; Lorenc 1986):

$$\mathbf{x}_a = \mathbf{x}_b + \mathbf{C}_b\mathbf{H}^T(\mathbf{H}\mathbf{C}_b\mathbf{H}^T + \mathbf{C}_r)^{-1}(\mathbf{y} - \mathbf{H}\mathbf{x}_b) \ , \tag{10.14}$$

$$\mathbf{C}_a = [\mathbf{I} - \mathbf{C}_b\mathbf{H}^T(\mathbf{H}\mathbf{C}_b\mathbf{H}^T + \mathbf{C}_r)^{-1}\mathbf{H}]\mathbf{C}_b \ . \tag{10.15}$$

By specifying $\mathbf{C}_b$, $\mathbf{C}_r$, $\mathbf{H}$, and $\mathbf{x}_b$, the analysis $\mathbf{x}_a$ and error covariance $\mathbf{C}_a$ can be computed for given observations $\mathbf{y}$. The prior model error covariance $\mathbf{C}_b$ plays an important role here, not only balancing the weighting between observations and the prior model but also spreading the observation-model discrepancy information spatially according to the correlation represented in the covariance.

10.2.3 Nonstationary Covariance Modeling

Following the approach adopted in Matsuo et al. (2005) as a way to incorporate anisotropic and inhomogeneous characteristics of the prior (background) model errors into the analysis (10.14) in a computationally tractable manner, $\mathbf{C}_b$ is modeled using the empirical orthogonal functions (EOFs, i.e., principal components). EOFs and their coefficients are estimated in advance of the data assimilation, for instance, from 50 million total SuperDARN plasma drift data points over January 2011 through August 2012 for electrostatic potential (Cousins et al. 2013a), from over 60 million DMSP electron energy flux spectra during the solar cycles 22 and 24 for conductance (McGranaghan et al. 2015), and from over 300 days of Iridium magnetic perturbation data from 2010 to 2015 for field-aligned currents (Shi et al. 2019).

Since observation sampling is often irregular and incomplete, a straightforward eigenvalue decomposition of sample covariance cannot be applied to the dataset. Instead, the nonlinear regression analysis of Matsuo et al. (2002) is used, wherein p principal components are expressed by a linear combination of the polar-cap spherical harmonic basis functions of Richmond and Kamide (1988), and each component is estimated sequentially by a back-fitting technique along with orthonormalization of the regression coefficients for each component. Each EOF can be expressed as $\mathbf{\Psi}\boldsymbol{\beta}$, where $\boldsymbol{\beta}$ is a $244 \times p$ matrix. Then $\mathbf{C}_b$ is given as

$$\mathbf{C}_b \approx \boldsymbol{\beta}\mathbf{C}_\gamma\boldsymbol{\beta}^T , \tag{10.16}$$

where $\mathbf{C}_\gamma$ is the covariance $< \boldsymbol{\gamma}\boldsymbol{\gamma}^T >$ of the EOF coefficients $\boldsymbol{\gamma}$, where $\boldsymbol{\gamma}$ is a $p \times 1$ column vector. EOFs estimated by the method of Matsuo et al. (2002) are equivalent to the eigenfunctions of a covariance matrix computed from observational data.

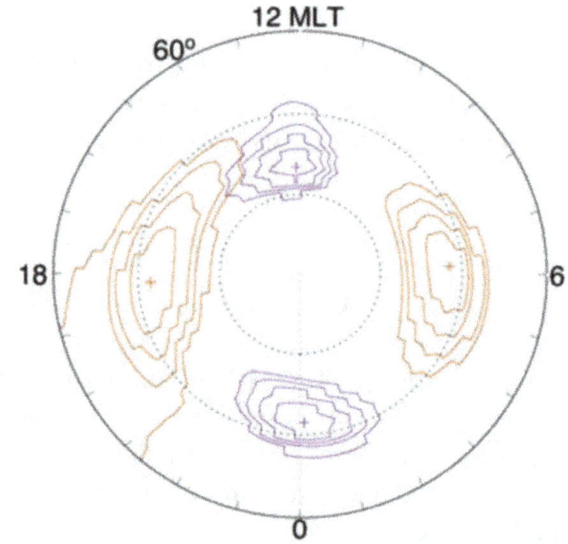

Fig. 10.1 Two-dimensional correlation functions of electrostatic potential with respect to the 4 points indicated by crosses. The contour lines represent the correlation level of 0.9, 0.8, 0.7, and 0.6 (Fig. 7 of Cousins et al. 2013a)

As with other principal component analysis methods, a certain replication of data samples is required to estimate $\mathbf{C}_\gamma$ and $\boldsymbol{\beta}$ from the observations.

Figure 10.1 shows 2-dimensional correlation maps for electrostatic potential computed from the EOF-based covariance derived from SuperDARN data (Cousins et al. 2013a) where p is set to 30. It is evident that the correlation structures are highly anisotropic with a larger correlation length scales in the zonal direction in comparison to the meridional direction, and correlations vary depending on reference point locations. These are features of strong nonstationary correlation, which will enable the data assimilation procedure to spatially distribute the impact of observations with consideration of realistic location-specific correlation structures of SuperDARN plasma drifts or electric fields.

10.3 Analysis of Electrostatic Potential and Electric Fields

Cousins et al. (2013b) presents an inverse and data assimilation procedure designed to specifically estimate $\mathbf{x}_\mathrm{E}$ as defined in (10.6) and (10.7) from SuperDARN data. A comprehensive cross-validation study (Cousins et al. 2013b) wherein observations are systematically set aside for validation and compared to predictions by data assimilation outperforms the standard SuperDARN mapping procedure (Ruohoniemi and Baker 1998; Shepherd and Ruohoniemi 2000). The inverse and data assimilation procedure is found to reduce median prediction errors by up to 43% as compared to the standard SuperDARN mapping procedure. The procedure is built using the prior covariance modeled with EOFs obtained by Cousins et al. (2013a) and the prior mean specified by the empirical plasma convection model of Cousins and Shepherd (2010). Figure 10.2 compares the maps of electrostatic potentials obtained by the standard SuperDARN mapping procedure (Ruohoniemi and Baker 1998; Shepherd and Ruohoniemi 2000 to the ones by Cousins et al. (2013b) along with maps of the uncertainty associated with assimilative mapping as given by the diagonal elements of $\mathbf{C}_a$ (10.15). The uncertainty reflects the observation distributions with higher uncertainty found in the area of the SuperDARN data gap. The comparison also highlights the role of the nonstationary covariance in the inverse and data assimilation procedure that help regularize assimilative mapping analysis.

10.4 Analysis of Toroidal Magnetic Potential and Field-Aligned Currents

Matsuo et al. (2015) presents an inverse and data assimilation analysis of space-based magnetometer data that directly solves for $\mathbf{x}_\mathrm{M}$ as defined in (10.8) and (10.9) to circumvent the need to use conductance in analysis of space-based magnetometer data for FAC as has been originally done in Richmond and Kamide (1988).

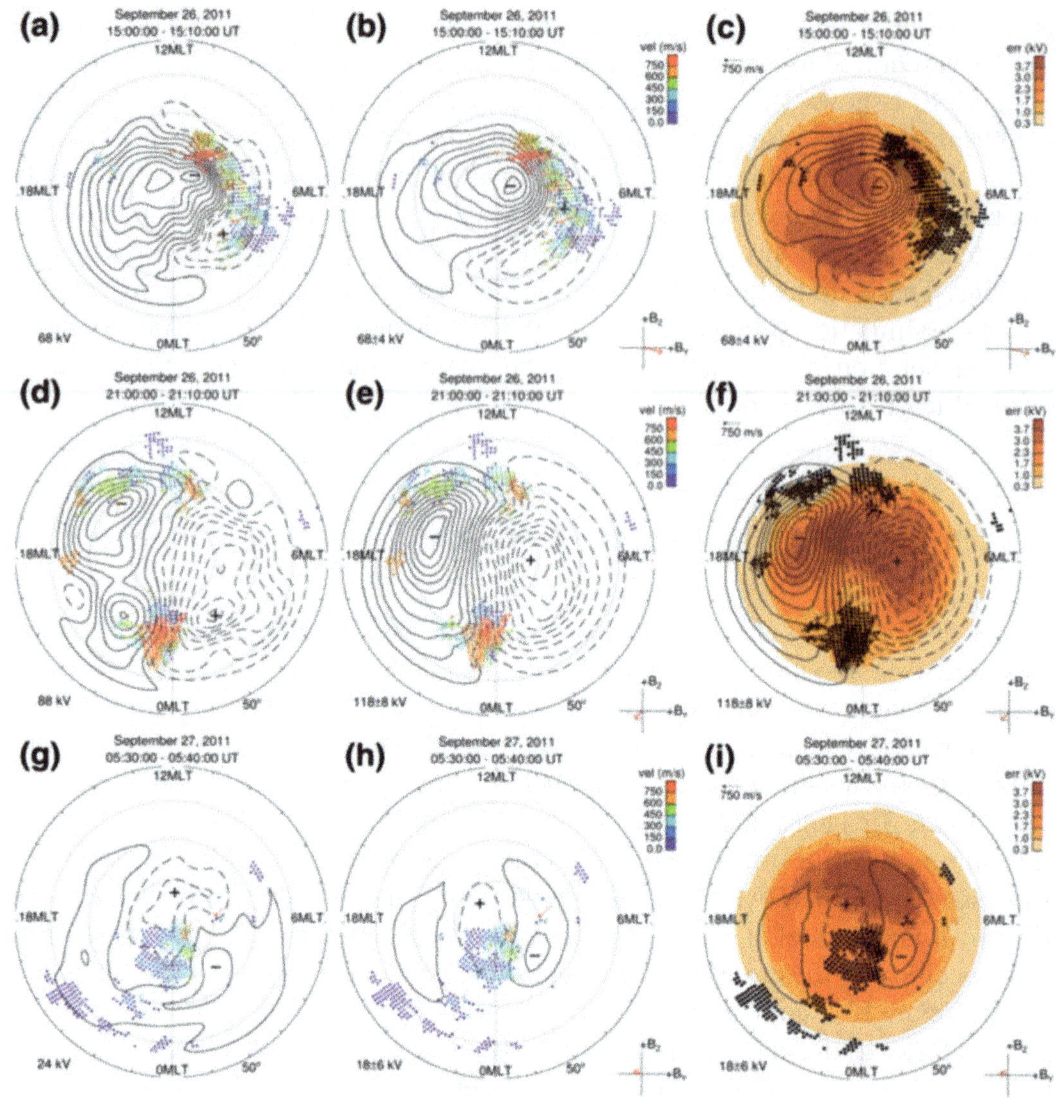

Fig. 10.2 Potential distributions for three selected times. **a**, **d**, and **g** Results from the standard SuperDARN mapping procedure. **b**, **c**, **e**, **f**, **h**, and **i** Results from the inverse and data assimilation procedure, with associated uncertainty shown as background coloring on the right side (Fig. 5 of Cousins et al. 2013b)

Figure 10.3 displays maps of the toroidal magnetic potential and FAC estimated from both AMPERE and DMSP data under four distinctive interplanetary magnetic field conditions during a magnetic cloud event on May 29, 2010, and demonstrates the Interplanetary Magnetic Field (IMF) control of high-latitude electrodynamics. Note that the uncertainty associated with the magnetic potential analysis is shown in the black-and-white contour in the background, with darker shades indicating greater errors. For comparison, the bottom row shows maps of the FAC provided by the AMPERE program. The AMPERE data product obtained from the spherical harmonic fit has an effective resolution of 3° latitude and 36° longitude (Anderson et al. 2014). As discussed in Matsuo et al. (2015), the overall distribution of FAC

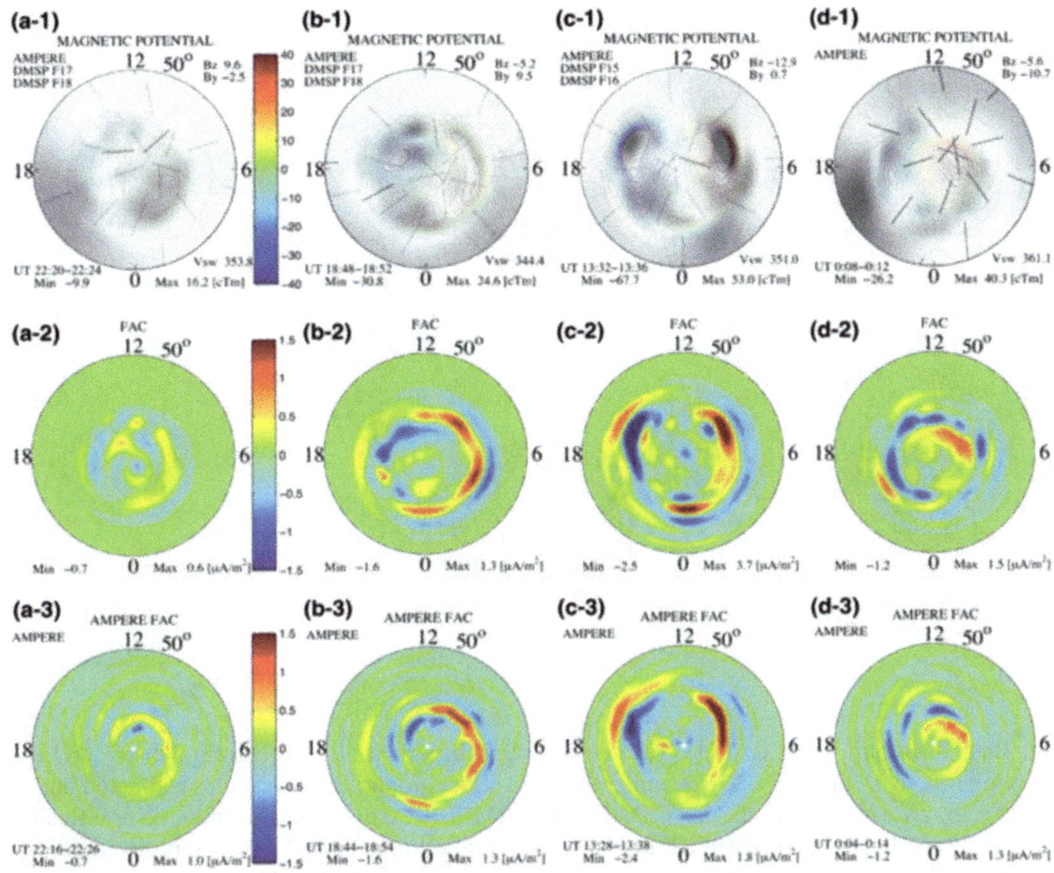

Fig. 10.3 The plots in the top and middle rows are maps of toroidal magnetic potential and FAC on May 29, 2010, estimated from the AMPERE and DMSP data over a 4 min interval: **a-1**, **a-2** IMF Bz positive, **b-1**, **b-2** IMF By positive, **c-1**, **c-2** IMF Bz negative, and **d-1**, **d-2** IMF By negative. The plots in the bottom row are maps of the FAC provided by the AMPERE program, estimated from the AMPERE data over a 10 min interval using Altitude Adjusted Corrected Geomagnetic Coordinates (Fig. 5 of Matsuo et al. 2015)

is similar to the one obtained by the current procedure, except for a few notable differences in the detail, such as the absence of high-frequency features and more longitudinally continuous FAC spatial structures are seen in the present analysis. Thanks to the regularization through the use of the prior model error covariance in solving the inverse problem, there is no need to fill the data gap with synthetic data to make a regression analysis stable, as is required in the AMPERE inversion.

10.5 Dual Optimization Approach

The framework for the inverse and data assimilation procedure described in Sect. 10.2 has thus far been applied to assimilative analysis of individual electromagnetic variables. In this section, the same framework is applied to the analysis of multiple

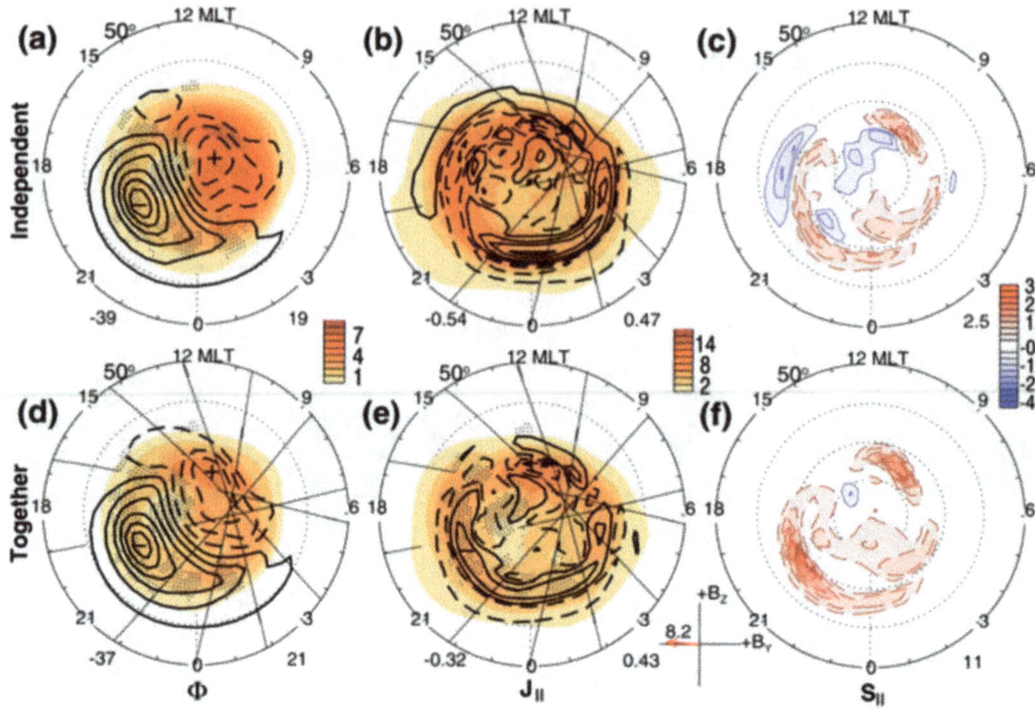

Fig. 10.4 Distributions of **a**, **d** electrostatic potential, **b**, **e** field-aligned current density, and **c**, **f** Poynting flux for 07400750 UT, November 29, 2011. Background color indicates estimated uncertainty. **a**, **b**, **c** Results using the SuperDARN and AMPERE data, independently. **d**, **e**, **f** Results using the SuperDARN and AMPERE data together. While SuperDARN observation locations are indicated by black dots, Iridium satellite tracks (where there are AMPERE data) are indicated by black lines (Fig. 5 of Cousins et al. 2015a)

variables. The relationship among electrodynamic variables given in (10.1)–(10.5) is nonlinear, requiring a nonlinear optimization approach. As an intermediate step toward implementing a fully nonlinear solver, Cousins et al. (2015a) presents a dual optimization approach by combining the two linear optimization approaches presented in Sects. 10.3 and 10.4 but using both SuperDARN and Iridium magnetic perturbation data. For a given conductance Σ_p and Σ_h, optimal values for $\mathbf{x}_\mathrm{E}$ and $\mathbf{x}_\mathrm{M}$ are estimated independently. Specifically, the optimal interpolation (or Kalman filter update) Eqs. (10.14) and (10.15) are applied to $\mathbf{x}_\mathrm{E}$ with the prior error covariance for electrostatic potential estimated from the SuperDARN data (Cousins et al. 2013a) and with $\mathbf{y}$ being composed of SuperDARN plasma drifts and Iridium magnetic perturbation fields. For estimation of $\mathbf{x}_\mathrm{M}$, (10.14) and (10.15) are applied with the prior error covariance for toroidal magnetic potential estimated from the Iridium magnetic perturbation data (Cousins et al. 2015b).

Figure 10.4 demonstrates the benefit of incorporating both SuperDARN and Iridium magnetic perturbation observations into the estimation of both electrostatic and magnetic potential (Cousins et al. 2015a). For example, as shown in the orange-shaded background contour in Fig. 10.4a and d, the uncertainty for electrostatic potential distributions estimated from SuperDARN data alone is higher in comparison

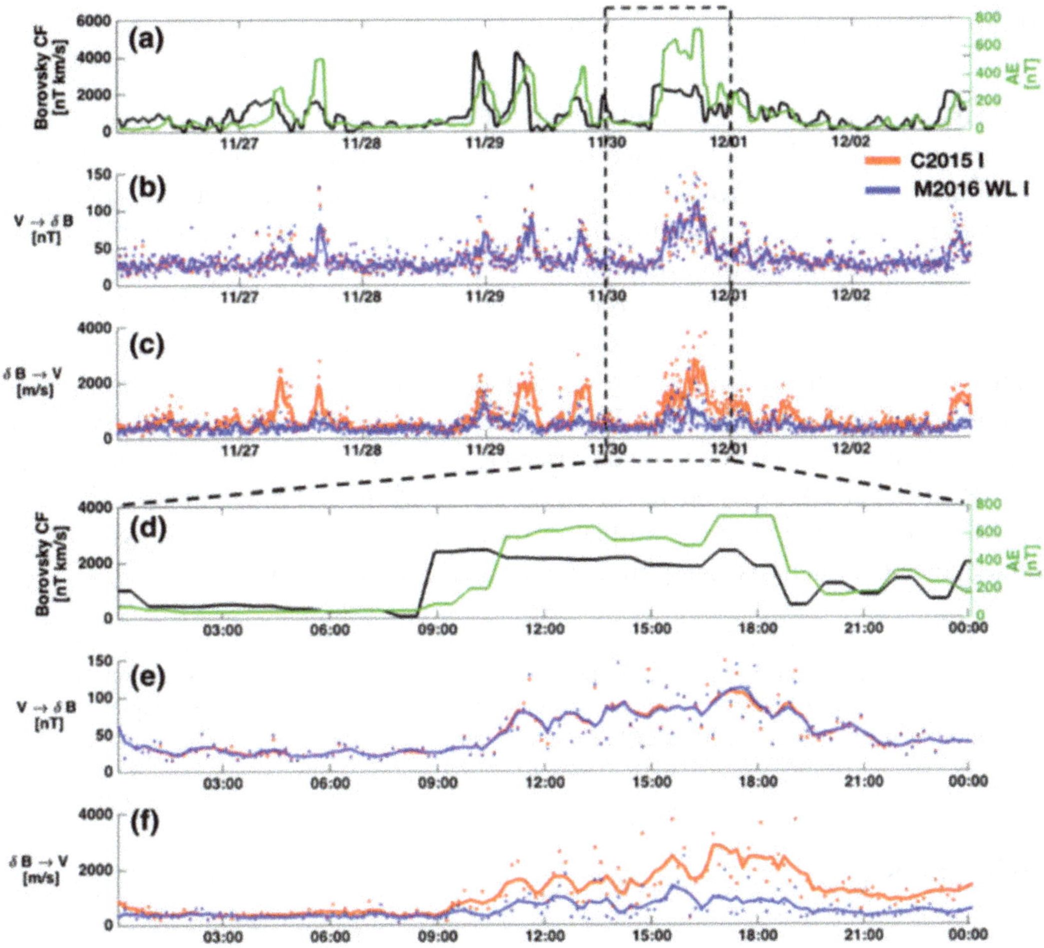

Fig. 10.5 Median Absolute Differences (MADs) between the prediction by data assimilation and the validation observation over the November 26–December 2, 2011 analysis time period. MADs are computed over the polar region. **a**, **d** Borovsky coupling function (black trace, left y-axis), **b**, **e** SuperDARN plasma drifts are used to predict Iridium magnetic fields, and **c**, **f** Iridium magnetic fields are used to predict SuperDARN plasma drifts (Fig. 8 of McGranaghan et al. 2016)

to the uncertainty when both data are assimilated. This is particularly evident in the dawn cell where there is no SuperDARN data but there is Iridium data. McGranaghan et al. (2016) have examined the effects of using different conductances in this dual optimization approach to assimilative mapping. When Σ_p and Σ_h are estimated by assimilation of the DMSP electron precipitation data (blue in Fig. 10.5) rather than specified by a climatological model (red in Fig. 10.5), the prediction of SuperDARN plasma drifts by assimilative analysis of Iridium magnetic perturbation data becomes more consistent with SuperDARN plasma drifts observations, as shown in Fig. 10.5c and f. Note that SuperDARN data are not used here for prediction of Iridium magnetic perturbation data, and vice versa.

10.6 Summary

This paper demonstrates that simultaneous analysis of multiple types of space-based and ground-based global geospace observations enabled by the inverse and data assimilation procedure provides a global perspective of high-latitude ionospheric electrodynamics. The paper summarizes important technical developments that have been made in response to the expansion of high-latitude geospace observing systems. The primary areas of the methodological extension to the AMIE (Richmond and Kamide 1988) are (a) the optimization in terms of both magnetic and electrostatic potential to minimize the impact of conductance on the inversion of space-based Iridium and DMSP magnetometer data for FAC mapping (Matsuo et al. 2015; Cousins et al. 2015a); (b) the use of realistic prior error covariance estimated from a large data set of SuperDARN (Cousins et al. 2013a), DMSP (McGranaghan et al. 2015) and Iridium magnetic perturbation data (Shi et al. 2019); (c) improved assimilative conductance/conductivity mapping (McGranaghan et al. 2016).

Acknowledgements This study is supported by the NSF grants PLR-1443703 and ICER-1541010. We thank the International Space Science Institute (ISSI) in Bern, Switzerland for supporting the Working Group "Multi Satellite Analysis Tools—Ionosphere" from which this chapter resulted. The Editors thank Gang Lu for her assistance in evaluating this chapter.

References

Anderson, B.J., K. Takahashi, and B.A. Toth. 2000. Sensing global birkeland currents with Iridium engineering magnetometer data. *Geophysical Research Letters* 27 (24): 4045–4048. https://doi.org/10.1029/2000gl000094.

Anderson, B.J., H. Korth, C.L. Waters, D.L. Green, V.G. Merkin, R.J. Barnes, and L.P. Dyrud. 2014. Development of large-scale birkeland currents determined from the active magnetosphere and planetary electrodynamics response experiment. *Geophysical Research Letters* 41: 30173025. https://doi.org/10.1002/2014GL059941.

Cousins, E.D.P., and S.G. Shepherd. 2010. A dynamical model of high-latitude convection derived from SuperDARN plasma drift measurements. *Journal of Geophysical Research* 115: A12329. https://doi.org/10.1029/2010JA016017.

Cousins, E.D.P., T. Matsuo, and A.D. Richmond. 2013a. Mesoscale and large-scale variability in high-latitude ionospheric convection: Dominant modes and spatial/temporal coherence. *Journal of Geophysical Research* 118: 78957904. https://doi.org/10.1002/2013JA019319.

Cousins, E.D.P., T. Matsuo, and A.D. Richmond. 2013b. SuperDARN assimilative mapping. *Journal of Geophysical Research* 118: 1–9. https://doi.org/10.1002/2013JA019321.

Cousins, E.D.P., T. Matsuo, and A.D. Richmond. 2015a. Mapping high-latitude ionospheric electrodynamics with SuperDARN and AMPERE. *Journal of Geophysical Research* 120: 5854–5870. https://doi.org/10.1002/2014JA020463.

Cousins, E.D.P., T. Matsuo, A.D. Richmond, and B.J. Anderson. 2015b. Dominant modes of variability in large-scale birkeland currents. *Journal of Geophysical Research* 120: 6722–6735. https://doi.org/10.1002/2014JA020462.

Friis-Christensen, E., H. Lühr, and G. Hulot. 2006. Swarm: A constellation to study the Earth's magnetic field. *Earth Planets Space* 58: 351–358.

Gjerloev, J. 2009. A global ground-based magnetometer initiative. *EOS* 90 (27):

Greenwald, R.A., K.B. Baker, J.R. Dudeney, M. Pinnock, T.B. Jones, and E.C. Thomas. 1995. DARN/SuperDARN. *Space Science Reviews* 71: 761796. https://doi.org/10.1007/BF00751350.

Hardy, D.A., L.K. Schmidt, M.S. Gussenhoven, F.J. Marshall, H.C. Yeh, T.L. Shumaker, A. Huber, and J. Pantazis. 1984. Precipitating electronand ion detectors (SSJ/4) for block 5D/flights 4-10 DMSP satellites: Calibration and data presentation. Technical Report AFGL-TR-84-0317, Air Force Research Laboratory, Hanscom Air Force Base, Mass

Jazwinski, A.H. 1970. *Stochastic processes and filtering theory*. Academic Press.

Kamide, Y., A.D. Richmond, and S. Matsushita. 1981. Estimation of ionospheric electric fields, ionospheric currents, and field-aligned currents from ground magnetic records. *Journal of Geophysical Research* 86 (A2): 801–813. https://doi.org/10.1029/JA086iA02p00801.

Lorenc, A. 1986. Analysis methods for numerical weather prediction. *Quarterly Journal of the Royal Meteorological Society* 112: 205–240.

Love, J. 2013. An international network of magnetic observations. *EOS* 94 (42):

Lu, G. 2017. Large scale high-latitude ionospheric electrodynamic fields and currents. *Space Science Reviews* 206: 431–450. https://doi.org/10.1007/s11214-016-0269-9.

Matsuo, T., A.D. Richmond, and D.W. Nychka. 2002. Modes of high-latitude electric field variability derived from DE-2 measurements: Empirical Orthogonal Function (EOF) analysis. *Geophysical Research Letters* 29 (7): 1107. https://doi.org/10.1029/2001GL014077.

Matsuo, T., A.D. Richmond, and G. Lu. 2005. Optimal interpolation analysis of high-latitude ionospheric electrodynamics using empirical orthogonal functions: Estimation of dominant modes of variability and temporal scales of large-scale electric fields. *Journal of Geophysical Research* 110: A06301. https://doi.org/10.1029/2004JA010531.

Matsuo, T., D.J. Knipp, A.D. Richmond, L. Kilcommons, and B.J. Anderson. 2015. Inverse procedure for high-latitude ionospheric electrodynamics: Analysis of satellite-borne magnetometer data. *Journal of Geophysical Research* 5241–5251: https://doi.org/10.1002/2014JA020565.

McGranaghan, R., D.J. Knipp, T. Matsuo, H. Godinez, R.J. Redmon, S.C. Solomon, and S.K. Morley. 2015. Modes of high-latitude auroral conductance variability derived from DMSP energetic electron precipitation observations: Empirical orthogonal function analysis. *Journal of Geophysical Research* 120: 11013–11031. https://doi.org/10.1002/2015JA021828.

McGranaghan, R., D.J. Knipp, T. Matsuo, and E. Cousins. 2016. Optimal interpolation analysis of high-latitude ionospheric Hall and Pedersen conductivities: Application to assimilative ionospheric electrodynamics reconstruction. *Journal of Geophysical Research* 121: 4898–4923. https://doi.org/10.1002/2016JA022486.

Paxton, L.J., D. Morrison, Y. Zhang, H. Kil, B. Wolven, B.S. Ogorzalek, D.C. Humm, C.I. Meng. 2002. Validation of remote sensing products produced by the Special Sensor Ultraviolet Scanning Imager (SSUSI): A far UV-imaging spectrograph on DMSP F-16. In: *Proceedings of SPIE Optical Spectroscopic Techniques, Remote Sensing, and Instrumentation for Atmospheric and Space Research IV*, 30 January 2002, vol. 4485. https://doi.org/10.1117/12.454268

Rich, F.J. 1984. Fluxgate magnetometer (SSM) for the Defense Meteorological Satellite Program (DMSP) block 5D-2, Flight 7. Technical Report AFGL-TR-84-0225, Air Force Research Laboratory, Hanscom Air Force Base, Mass.

Richmond, A.D. 1995. Ionospheric electrodynamics using magnetic Apex coordinates. *Journal of Geomagnetism and Geoelectricity* 47: 191–212.

Richmond, A.D., and Y. Kamide. 1988. Mapping electrodynamic features of the high-latitude ionosphere from localized observations: Technique. *Journal of Geophysical Research* 93 (A6): 5741–5759.

Ritter, P., and H. Lühr. 2006. Curl-B technique applied to Swarm constellation for determining field-aligned currents. *Earth Planets Space* 58: 463–476.

Ruohoniemi, J.M., and K.B. Baker. 1998. Large-scale imaging of high-latitude convection with super dual Auroral Radar network HF radar observations. *Journal of Geophysical Research* 103 (20): 797.

Shepherd, S.G., and J.M. Ruohoniemi. 2000. Electrostatic potential patterns in the high latitude ionosphere constrained by SuperDARN measurements. *Journal of Geophysical Research* 105 (23): 005.

Shi, Y., D.J. Knipp, T. Matsuo, L. Kilcommons, B.J. Anderson. 2019. Modes of field-aligned currents variability and their hemispheric asymmetry revealed by inverse and assimilative analysis of Iridium magnetometer data. *Journal of Geophysical Research* Submitted.

Solomon, S.C., P.B. Hays, and V.J. Abreu. 1988. The auroral 6300 Å emission: Observations and modeling. *Journal of Geophysical Research* 93 (A9): 9867–9882. https://doi.org/10.1029/JA093iA09p09867.

Chapter 11
Estimating Currents and Electric Fields at Low Latitudes from Satellite Magnetic Measurements

Patrick Alken

Abstract Low-latitude ionospheric electric currents produce prominent signatures in the magnetic field measurements made by low Earth-orbiting satellites. Analyzing these magnetic signatures not only provides insight into the currents themselves, but also many other important and interesting phenomena in the low-latitude ionosphere and thermosphere. The low-latitude currents are modulated by thermospheric winds, so attaining a global knowledge of the spatial structure of the currents can give insight into the neutral tidal harmonics present at ionospheric altitudes. Furthermore, the equatorial electrojet (EEJ) current is driven by an equatorial electric field which in turn is generated by a dynamo process. This electric field is additionally responsible for the vertical plasma fountain and equatorial ionization anomaly at low-latitudes. Magnetic measurements of the EEJ, therefore, allows the study of low-latitude plasma motion in the E and F regions of the ionosphere. This chapter will present techniques developed for processing magnetic measurements of the EEJ to extract information about the low-latitude currents and their driving electric fields. This chapter will present a line current approach to recover the EEJ current strengths, with an emphasis on cleaning the satellite data and minimizing magnetic fields from other internal and external sources. The electric fields will be determined using a combination of physical modeling and fitting the EEJ current strengths from the satellite measurements.

11.1 Introduction

This chapter will be concerned with the calculation of ionospheric current flow and electric fields at low-latitudes, using magnetic field measurements from low Earth orbiting (LEO) satellite missions, such as Swarm (Friis-Christensen et al. 2006). In the ionosphere, neutral particles are ionized by solar extreme ultraviolet (EUV) radiation. The resulting charged plasma then interacts with the Earth's electromagnetic

P. Alken (✉)
University of Colorado at Boulder, Boulder, CO, USA
e-mail: alken@colorado.edu

M. W. Dunlop and H. Lühr (eds.), *Ionospheric Multi-Spacecraft Analysis Tools*, ISSI Scientific Report Series 17,
https://doi.org/10.1007/978-3-030-26732-2_11

field, neutral wind field, gravitational forces, and pressure-gradient forces. Each of these forces drives electric current flow, which has complex spatial and temporal structure. In the ionospheric E-region, which extends from about 90 to 120 km altitude, ion-neutral collisions are significant, while the electrons are mostly frozen to magnetic field lines. The frictional forces between the ions and the neutral wind field drive current which in general is not divergence free. Therefore, polarization electric fields build up globally to ensure divergence-free current flow. At low and mid-latitudes, the large-scale current system resulting from this ion-neutral coupling is called Solar-quiet (Sq). "Solar" refers to the current system's dependence on solar local time, since ionospheric conductivity at mid-latitudes peaks during daytime hours and diminishes during the night. The term "quiet" indicates geomagnetically quiet conditions, since during strong geomagnetic storms, the Sq system can experience large perturbations. At the magnetic equator, the horizontal geometry of the field lines leads to an enhanced zonal current called the equatorial electrojet (EEJ). An eastward component of the electric field at the equator, when coupled with the northward geomagnetic field, will drive vertical drift of electrons. Above about 120 km altitude, the Hall conductivity decreases substantially, since the reduced density of neutral particles result in less ion-neutral collisions, and the ions are essentially free to move with the electrons. This effect causes a nonconducting layer at the top of the E-region, and so the charge will accumulate at about 120 km altitude near the magnetic equator. This charge accumulation will cause a strong vertical polarization electric field which will drive zonal electron drift. Because the vertical polarization electric field is typically about 10 times stronger than the eastward component, the zonal $\mathbf{E} \times \mathbf{B}$ drift results in a strong current system, the EEJ. The EEJ has prominent magnetic signatures in both LEO satellite and ground observatory data, offering a convenient means of studying equatorial E-region dynamics. Since the EEJ is driven by the eastward component of the equatorial electric field (EEF), and is modulated by the neutral wind field, studying its magnetic signature reveals a lot of important information about the equatorial electrodynamics and also tidal features in the neutral winds.

This chapter will discuss recent methods of fitting an equivalent current model, based on line current geometry, to Swarm satellite magnetic measurements in order to recover the EEJ current strength flowing in the E-region. Line current methods have long been applied to studies of both the equatorial electrojet (Lühr et al. 2004; Alken et al. 2013a, 2014) and polar electrojets (Olsen 1996; Ritter et al. 2004; Vennerstrom and Moretto 2013; Aakjær et al. 2016). Since the electrojets typically follow lines of constant magnetic latitude, they have relatively simple spatial flow patterns, and line current models are particularly suitable for their study. For current systems which have more spatial complexity, such as Sq, pressure-gradient currents, inter-hemispheric field-aligned currents (IHFAC), and high-latitude field-aligned currents (FAC), it is necessary to utilize other methods to determine equivalent current flow, such as spectral methods (Fiori and Boteler 2018) or spherical elementary current systems (SECS) (Vanhamäki et al. 2018). This chapter will follow the work of Alken et al. (2013a, 2014) in order to define line currents which follow lines of constant quasi-dipole (QD) latitude in the equatorial region to estimate realistic EEJ equivalent current flow. These current estimates will then be inverted using a physics-based modeling procedure in order to recover the eastward component of the equatorial electric field.

11.2 Satellite Data Preprocessing

Any scientific data collection system, such as LEO satellite magnetometers, can contain erroneous measurements not related to the physical system under study. Therefore in order to recover accurate low-latitude currents and electric fields from satellite measurements, the magnetic data must be carefully preprocessed to isolate the signal of interest and remove other contamination as much as possible. Satellite magnetic data can be preprocessed in many different ways. The approach used in this chapter closely follows previous work in studying magnetic perturbations from low-latitude ionospheric current systems (Alken and Maus 2010a, b; Alken et al. 2013b, 2015). We will focus only on the scalar field measurements in this chapter, since they contain enough information at low-latitudes to determine equivalent current flow in the E-region. First, the satellite data are separated into half-orbital tracks, extending from the south to north pole or vice versa. This is convenient since the low-latitude currents are analyzed on an orbit-by-orbit basis. In order to identify potential problems with the measurements, a convenient quantity to examine is the along-track root-mean-square (rms) difference between the scalar magnetic field measurements and a recent main field model, typically from 60°S–60°N QD latitude. The along-track rms is computed only at mid and low latitudes, since high-latitude scalar field data are influenced heavily by polar electrojets which are difficult to model and would add large contributions to the rms differences. Tracks with a large rms difference in the above latitude range are discarded from analysis. A typical threshold for rms scalar differences is 150 nT. This threshold is large enough to keep good geophysical data that may be perturbed significantly during geomagnetic storms, but small enough to discard completely erroneous measurements that may occur due to instrument problems, satellite maneuvers, etc. An example of such events is shown in Fig. 11.1. Panel (a) shows scalar field residuals from Swarm A on a single day, 30 January 2014, when orbit maneuvers perturbed the absolute scalar magnetometer (ASM) measurements. Some tracks exhibit up to 1000 nT difference with the main field model at mid-latitudes, which is highly unlikely to be due to real geophysical signal. These data are not flagged in any way in the Level-1b Swarm dataset, and so this type of analysis is required in order to detect such events. Panel (b) shows the remaining scalar field residuals after discarding all data with an along-track rms greater than 150 nT. Note the reduced scale on the vertical axis. The remaining tracks differ with the main field model only up to about 20 nT at low and mid-latitudes, which is normal for ionospheric signals. Panel (c) shows a time series of Swarm A scalar residuals from November 2013 until December 2014, plotting residuals below 60° QD latitude. There are a small number of localized perturbations in the residuals, including the event on 30 January 2014. This shows that these events are uncommon, but must

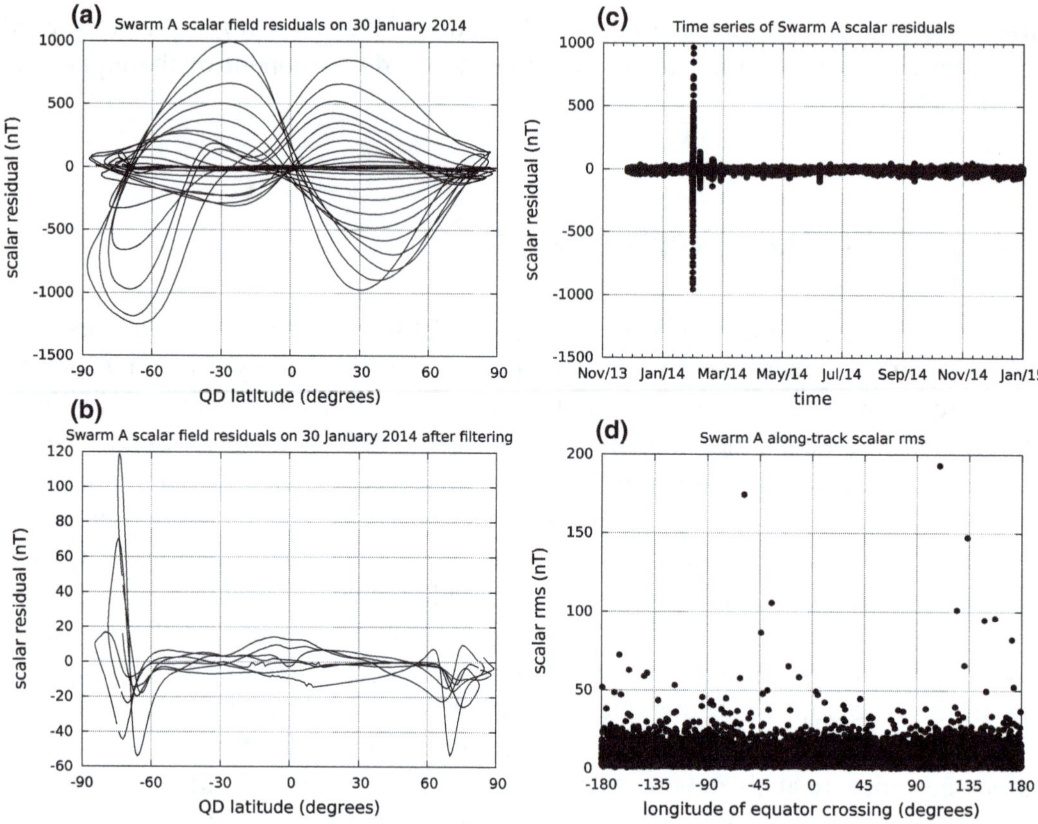

Fig. 11.1 **a** Scalar residuals for all tracks for Swarm A on 30 January 2014. **b** Swarm A scalar field residuals after removing tracks with large along-track rms differences with main field model. **c** Time series of Swarm A scalar residuals from the beginning of the mission to the end of 2014. **d** Along-track scalar rms values plotted versus the longitude of the geographic equator crossing for same time period as (**c**)

nevertheless be detected and removed from the data. A useful way to determine an appropriate rms threshold is shown in panel (d), where the along-track rms value is plotted versus the longitude of the satellite's geographic equator crossing, for the same time period as panel (c). Most of the along-track scalar rms values are well below 50 nT, and so outliers are fairly straightforward to identify in such a plot. Experience has shown that the along-track rms can easily exceed 100 nT during strong geomagnetic storms, which leads to the choice of 150 nT as a threshold value.

After discarding erroneous measurements as discussed above, the next step is to isolate the ionospheric field signal from other sources in the data. The primary non-ionospheric sources at satellite altitudes are the main field, lithospheric field, and magnetospheric field with its induced counterpart. The main field and its secular variation can be removed using high-quality models such as CHAOS (Finlay et al. 2016) or POMME (Maus et al. 2010). The lithospheric field can similarly be removed using a high-resolution model such as MF7 (Maus et al. 2008). The magnetospheric field can be removed using models of the ring current and tail currents, based on the techniques of Maus and Lühr (2005), Lühr and Maus (2010). Denoting the main, crustal, and magnetospheric models as $\mathbf{B}_{core}$, $\mathbf{B}_{crust}$, and $\mathbf{B}_{ext}$ respectively, an appropriate model of the non-ionospheric sources is given by

$$\mathbf{B}_{model} = \mathbf{B}_{core} + \mathbf{B}_{crust} + \mathbf{B}_{ext} \tag{11.1}$$

with the scalar residual given by

$$F^{(1)} = F_{sat} - |\mathbf{B}_{model}| \tag{11.2}$$

Here, F_{sat} is the scalar field measurement made by a LEO satellite. The core and crustal field sources at satellite altitude are generally well represented by current field models, but the magnetospheric sources can vary with complex temporal behavior which is still not completely understood. Therefore while $\mathbf{B}_{ext}$ will remove most of the external magnetospheric field, some residual could remain. This can be mitigated to some extent by fitting low-degree spherical harmonic models in solar magnetic (SM) and geocentric solar magnetospheric (GSM) coordinates on a track-by-track basis to attempt to further remove any remaining ring and tail current fields (see recent review by Lühr et al. 2016). After the $F^{(1)}$ residual is computed, it will primarily represent the ionospheric field plus its induced counterpart. This is then the starting point to isolate the EEJ contribution to the ionospheric field, so that it can be modeled with line currents, and then used to extract information about the low-latitude electric field.

11.3 Removing the Sq Field

The $F^{(1)}$ residual defined in Sect. 11.2 will contain ionospheric signals from Sq, EEJ, inter-hemispheric field-aligned current (IHFAC), polar electrojets (PEJ), high-latitude field-aligned currents (FAC), induced fields originating inside the solid Earth, as well as unmodeled field contributions from the magnetosphere. Since our goal is to recover the EEJ current at low-latitudes, the main concern is removing the Sq signal from the data, as well as unmodeled magnetospheric ring current fields. To do this, assume that the Sq field and its induced counterpart can be well represented by a magnetic scalar potential for sources internal to the satellite orbit, defined in a spherical coordinate system (r, θ, ϕ) as

$$V_{int}(r, \theta, \phi; g_n^m) = a \sum_{n=1}^{N_I} \sum_{m=-1}^{1} g_n^m \left(\frac{a}{r}\right)^{n+1} Y_n^m(\theta, \phi) \tag{11.3}$$

where a is taken to be an Earth radius of 6371.2 km, g_n^m are model coefficients to be determined, N_I is the maximum spherical harmonic degree needed to model the Sq field, and $Y_n^m(\theta, \phi)$ are defined as

$$Y_n^m(\theta,\phi) = \begin{cases} S_n^m(\cos\theta)\cos(m\phi) & m \ge 0 \\ S_n^{|m|}(\cos\theta)\sin(|m|\phi) & m < 0 \end{cases} \tag{11.4}$$

where $S_n^m(\cos\theta)$ are the Schmidt-normalized associated Legendre functions. The Sq field is normally large-scale without sharp localized features, and so $N_I = 12$ is a typical choice for the cutoff. Since V_{int} is fitted to a single satellite orbit, it can be expanded only to spherical harmonic order 1, since a polar orbiting satellite does not provide enough longitudinal coverage to use higher orders. In order to account for unmodeled magnetospheric fields, we use the external scalar potential

$$V_{ext}(r,\theta,\phi; q_n^m) = a \sum_{n=1}^{N_E} \sum_{m=-1}^{1} q_n^m \left(\frac{r}{a}\right)^n Y_n^m(\theta,\phi) \tag{11.5}$$

where q_n^m are coefficients to be determined, and N_E is the maximum spherical harmonic degree needed for the external fields. Since the magnetospheric sources are multiple Earth radii away from the satellite measurements, their fields can usually be well represented by a low-order spherical harmonic degree, such as 1 or 2. The magnetic fields corresponding to the internal and external scalar potentials are

$$\mathbf{M}(r,\theta,\phi; g_n^m) = -\nabla V_{int}(r,\theta,\phi; g_n^m) = \sum_{nm} \left(\frac{a}{r}\right)^{n+2} g_n^m \begin{bmatrix} (n+1)Y_n^m(\theta,\phi) \\ -\partial_\theta Y_n^m(\theta,\phi) \\ -\frac{m}{\sin\theta}\partial_\phi Y_n^m(\theta,\phi) \end{bmatrix} \tag{11.6}$$

$$\mathbf{K}(r,\theta,\phi; q_n^m) = -\nabla V_{ext}(r,\theta,\phi; q_n^m) = -\sum_{nm} \left(\frac{r}{a}\right)^{n-1} q_n^m \begin{bmatrix} nY_n^m(\theta,\phi) \\ \partial_\theta Y_n^m(\theta,\phi) \\ \frac{m}{\sin\theta}\partial_\phi Y_n^m(\theta,\phi) \end{bmatrix} \tag{11.7}$$

where the vector components are defined in geocentric spherical coordinates, ordered with respect to r, θ, ϕ. The magnetospheric ring current field is a primary contributor to **K**, and since it is more efficiently parameterized in the solar magnetic (SM) coordinate system (Lühr et al. 2016), it is advantageous to expand the scalar potential V_{ext} using SM coordinates, and then rotate the resulting **K** to geographic coordinates for fitting the satellite data. This is straightforward since the SM coordinate system is an Earth-centered Earth-fixed Euclidean system, which just involves a rotation from standard geocentric NEC coordinates. Similarly, the Sq field could be efficiently decomposed in QD coordinates, as it is aligned with the geomagnetic main field geometry at low and mid-latitudes, but great care must be taken when working with vector quantities in QD coordinates since they are a nonorthogonal system (Richmond 1995; Laundal and Richmond 2016) and that discussion goes beyond the scope of this chapter. Therefore, we will define the combined Sq and external field model to be fitted to the $F^{(1)}$ satellite residuals as

$$\mathbf{T}(r,\theta,\phi; g_n^m, q_n^m) = \mathbf{M}(r,\theta,\phi; g_n^m) + \mathbf{K}(r,\theta^{SM},\phi^{SM}; q_n^m) \tag{11.8}$$

where θ^{SM} and ϕ^{SM} are the colatitude and longitude of the point (r, θ, ϕ) in SM coordinates. Equations for converting from geocentric spherical to SM coordinates may be found in Laundal and Richmond (2016). The Sq and external field model $\mathbf{T}$ can be fitted to the scalar residuals $F^{(1)}$ by minimizing the cost function

$$\chi^2 = \sum_i \left[F_i^{(1)} - \hat{\mathbf{b}}_i \cdot \mathbf{T}(\mathbf{r}_i; \mathbf{x}) \right]^2 + \lambda_{Sq}^2 ||L_{Sq}\mathbf{x}||^2 \tag{11.9}$$

where $\hat{\mathbf{b}}_i$ is a unit vector in the main field direction at the measurement point $\mathbf{r}_i$, $\mathbf{x}$ is a vector containing the model coefficients g_n^m and q_n^m, λ_{Sq} is a regularization parameter, L_{Sq} is a regularization matrix, and i is summed over all scalar residuals to be used in the Sq fitting for a particular orbit. The model fit is typically restricted to use data below about 60° QD latitude to exclude effects from the high-latitude currents. Additionally one should exclude data in the EEJ region (between $\pm 12°$ QD latitude), since the goal is to fit the large-scale Sq and external fields and preserve as much EEJ signal as possible. The model $\mathbf{T}$ is projected onto the main field direction in order to compare with the computed scalar field residuals $F_i^{(1)}$. Any standard main field model, such as IGRF (Thébault et al. 2015) or CHAOS (Finlay et al. 2016) could be used to compute $\hat{\mathbf{b}}_i$. When inverting scalar data over a single satellite orbit, it can be challenging to separate the internal Sq signal from the external magnetospheric signal, and so it is often useful to include the regularization term in the least-squares minimization. A typical choice for the regularization matrix is $L_{Sq} = I$ to prevent the solution norm $||\mathbf{x}||$ from growing too large, resulting in nonphysical fields. However other choices could also work well if additional constraints are imposed on the internal and external fields (for example, minimizing latitudinal gradients of current flow). The regularization parameter λ_{Sq} represents a tradeoff between minimizing the model residuals and the solution norm. Choosing the right value of λ_{Sq} in an automated fashion can be a challenging problem, as the low and mid-latitude ionospheric and magnetospheric fields can change drastically during different local times, seasons, and geomagnetic activity levels. Selecting λ_{Sq} using L-curve analysis (Hansen and O'Leary 1993) often produces good results, although sometimes it is necessary to visually look at the data and fitted model to ensure a physically realistic fit is achieved. Once the model coefficients g_n^m, q_n^m are determined for a particular satellite track, the Sq and external field contribution are removed from the residuals to isolate only the EEJ contribution to the ionospheric field. This is done by defining a new scalar residual

$$F_i^{(2)} = F_i^{(1)} - \hat{\mathbf{b}}_i \cdot \mathbf{T}(\mathbf{r}_i; g_n^m, q_n^m) \tag{11.10}$$

The residuals $F^{(2)}$ will be used to invert for the EEJ current flow, discussed in the next section. Figure 11.2 (top panel) shows a single latitude profile recorded by Swarm B on 22 June 2015, when the satellite was in a 12:57 local time. The $F^{(1)}$ residual, computed by removing the core, crustal and magnetospheric field models from the scalar measurements is shown in purple as a function of QD latitude. At the magnetic equator, we see the characteristic sharp trough of the equatorial electrojet

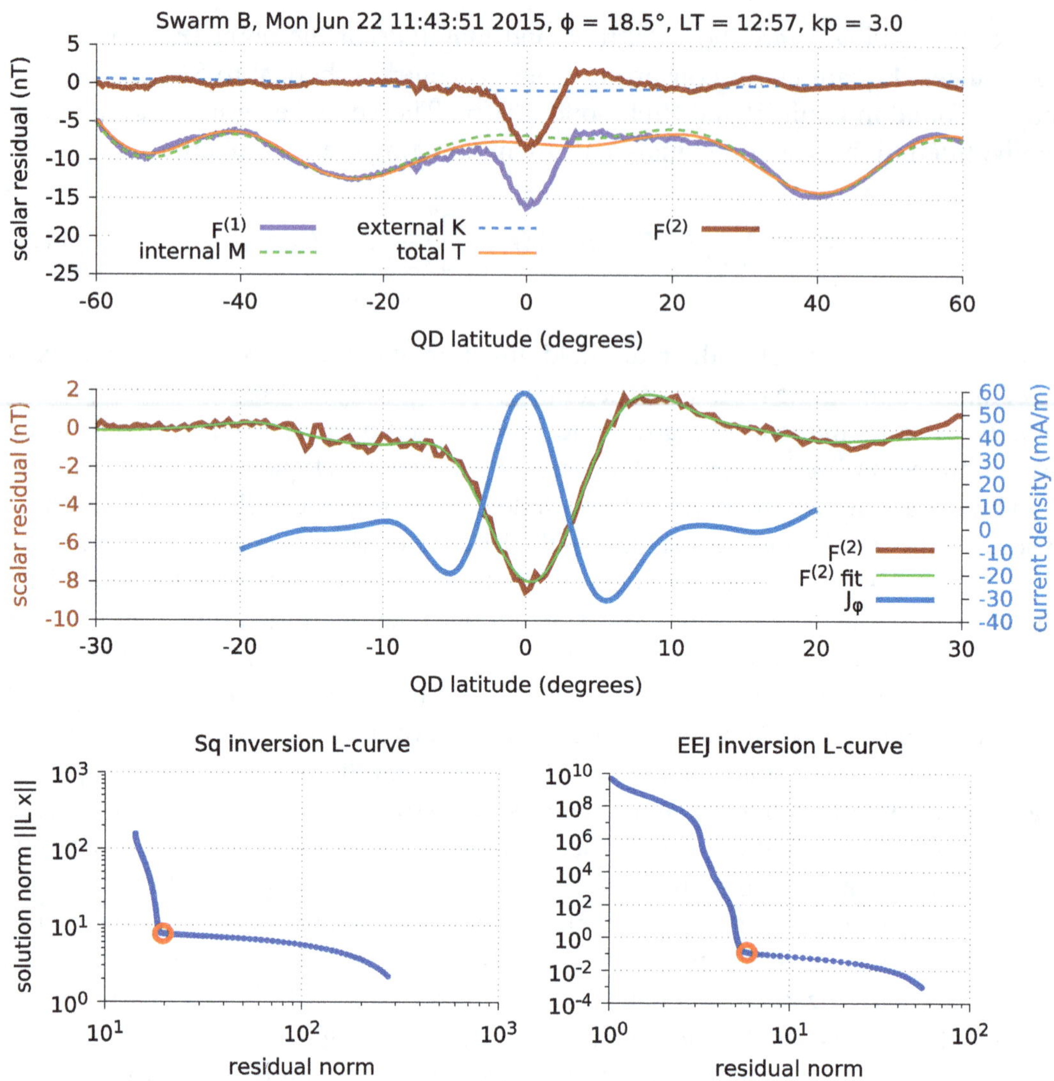

Fig. 11.2 Top: scalar residual $F^{(1)}$ (purple) with fitted model components $\mathbf{M}$ (dashed green), $\mathbf{K}$ (dashed blue), and total model $\mathbf{T} = \mathbf{M} + \mathbf{K}$ (orange) projected onto main field direction. Final residual $F^{(2)}$ plotted in red. Middle: Zoomed in view of $F^{(2)}$ residual (red) with fitted EEJ line current model (green) and equivalent E-region zonal current flow (blue). Bottom: L-curves from the fitted Sq/external model (left) and EEJ line current model (right). Computed L-curve corners, used to estimate regularization parameters, circled in red. Data recorded by Swarm B on 22 June 2015

signature. This data was recorded shortly before a strong geomagnetic storm and so the variations seen above 20° QD latitude likely contain magnetospheric fields as well as the Sq signature. The result from fitting the internal and external source models $\mathbf{M}$ and $\mathbf{K}$ are shown as dashed green and blue lines, respectively. The internal $\mathbf{M}$ model closely tracks the higher latitude variations due to its degree 12 expansion in spherical harmonics. The external model $\mathbf{K}$ is expanded only to degree 1 in this calculation and so it exhibits only a large-scale slow variation with latitude. The combined model $\mathbf{T}$ is shown in orange and the final residual $F^{(2)}$ after removing the

Sq and external models are shown in red. We see that the $F^{(2)}$ residual has close to zero mean at non-equatorial latitudes except for short-scale variations. This is a result of removing the background Sq field and leaving only the EEJ peak at the equator. The bottom left panel of Fig. 11.2 shows the L-curve, which is a log–log plot of the residual norm versus the solution norm of the least-squares function. The corner of the L-curve, circled in red, often represents a good tradeoff between minimizing the residuals and minimizing the solution norm (Hansen and O'Leary 1993). The remaining panels of the figure will be discussed in the next section.

11.4 Estimating EEJ Flow with Line Currents

The peak equatorial electrojet current flow follows the magnetic equator, since the horizontal field geometry enhances the zonal conductivity. Observationally, this was confirmed from CHAMP measurements (Lühr et al. 2004). Due to the magnetic eastward (or westward) flow of the EEJ, the spatial geometry of the current flow can be well represented by line currents. However, defining straight line currents tangent to the Earth at the longitude of the satellite's equator crossing would not take into account the Earth's curvature, nor would it account for the magnetic equator geometry. Therefore, we use instead small straight line segments, which can be placed along lines of constant magnetic latitude. Also, the endpoints of each segment can be fixed to a given E-region altitude to account for the spherical Earth geometry. Therefore, we define a set of N_C "segmented" line currents following lines of constant quasi-dipole latitude at an altitude of 110 km, and covering the low-latitude EEJ current flow region. The segmented line currents, hereafter referred to as simply line currents, are spaced apart at equal intervals in QD latitude. These line currents will represent equivalent height-integrated EEJ current flow, since the EEJ current system also has vertical structure which cannot be resolved by satellite measurements far above the current region. Figure 11.3 depicts the line current geometry along the magnetic equator. Each of the N_C line currents is divided into 360 segments, with each segment spanning 1° of longitude. The unit current vector for longitude segment $k \in [1, ..., 360]$ of line current $j \in [1, ..., N_C]$ is then given by

$$I^r_{jk} \approx 0 \tag{11.11}$$

$$I^\theta_{jk} = -\cos\alpha_{jk} \tag{11.12}$$

$$I^\phi_{jk} = \sin\alpha_{jk} \tag{11.13}$$

where α_{jk} is the angle between geographic north and the linear current segment k of current j. The unit current vector $\mathbf{I}_{jk}$ defined above represents simply the spherical coordinate components of the direction of current flow for that particular line segment. Next, the unit magnetic field perturbation due to this single line segment is given by the Biot-Savart law:

$$d\mathbf{B}_{ijk} = \frac{\mu_0}{4\pi}\delta_{jk}\frac{\mathbf{I}_{jk} \times (\mathbf{r}_i - \mathbf{r}_{jk})}{|\mathbf{r}_i - \mathbf{r}_{jk}|^3} \tag{11.14}$$

where δ_{jk} is the distance in meters of segment k of line current j, $\mathbf{r}_i$ is the position vector of satellite observation i, and $\mathbf{r}_{jk}$ is the position vector pointing to the midpoint of line segment k of line current j. The total magnetic field contribution from line current j is simply the sum over all longitudinal segments k:

$$\mathbf{B}_{ij} = \sum_k d\mathbf{B}_{ijk} \tag{11.15}$$

The sum is taken only over a subset of all 360 longitudinal segments, because current flowing far away from the satellite observation has less influence on the measured magnetic field. Experience has shown that only including contributions from segments within $\pm 30°$ longitude of the satellite's equator crossing provide adequate estimates of the current strength. Next, we project $\mathbf{B}_{ij}$ onto the internal field direction using a main field model, since only the scalar field satellite measurements are used to fit the EEJ currents. The result is

$$F_{ij} = \mathbf{B}_{ij} \cdot \hat{\mathbf{b}}_i \tag{11.16}$$

Finally, assume the same current flows along each segment of one line current. Then there will be j unknown line current strengths s_j to determine from the magnetic measurements. These can be computed by minimizing the objective function

$$\chi^2 = ||\mathbf{b} - F\mathbf{s}||^2 + \lambda_{EEJ}^2 ||L_{EEJ}\mathbf{s}||^2 \tag{11.17}$$

where $\mathbf{b}$ is a vector of scalar field residuals $F_i^{(2)}$, $\mathbf{s}$ is a vector of length N_C containing the line current strengths s_j, λ_{EEJ} is a regularization parameter, L_{EEJ} is a regularization matrix, and F is the least-squares matrix relating the line current strengths to the scalar field observations, whose elements are the F_{ij} defined in Eq. 11.16. The regularization term is added to prevent nonphysical solutions during the least-squares inversion. The regularization matrix L_{EEJ} is typically set to a second order finite difference operator, to ensure a smooth variation along the latitudinal current profile and prevent neighboring currents from exhibiting large oscillations. The regularization parameter λ_{EEJ} provides a tradeoff between minimizing the residual norm $||\mathbf{b} - F\mathbf{s}||$ and the solution norm $||L_{EEJ}\mathbf{s}||$. Similarly to the Sq field inversion, L-curve methods for determining λ_{EEJ} work well in practice. Figure 11.2 (middle panel) shows a zoomed-in view of the $F^{(2)}$ residual (red) computed by removing the Sq field, discussed in Sect. 11.3. The magnetic field of the line current model fitted to this residual is shown in green, while the line current profile is shown in blue. The line current profile was transformed into the geographic frame to show J_ϕ eastward flow. The L-curve for the EEJ line current inversion is shown in the bottom right panel, with the computed corner circled in red, which determined the regularization parameter for this fit.

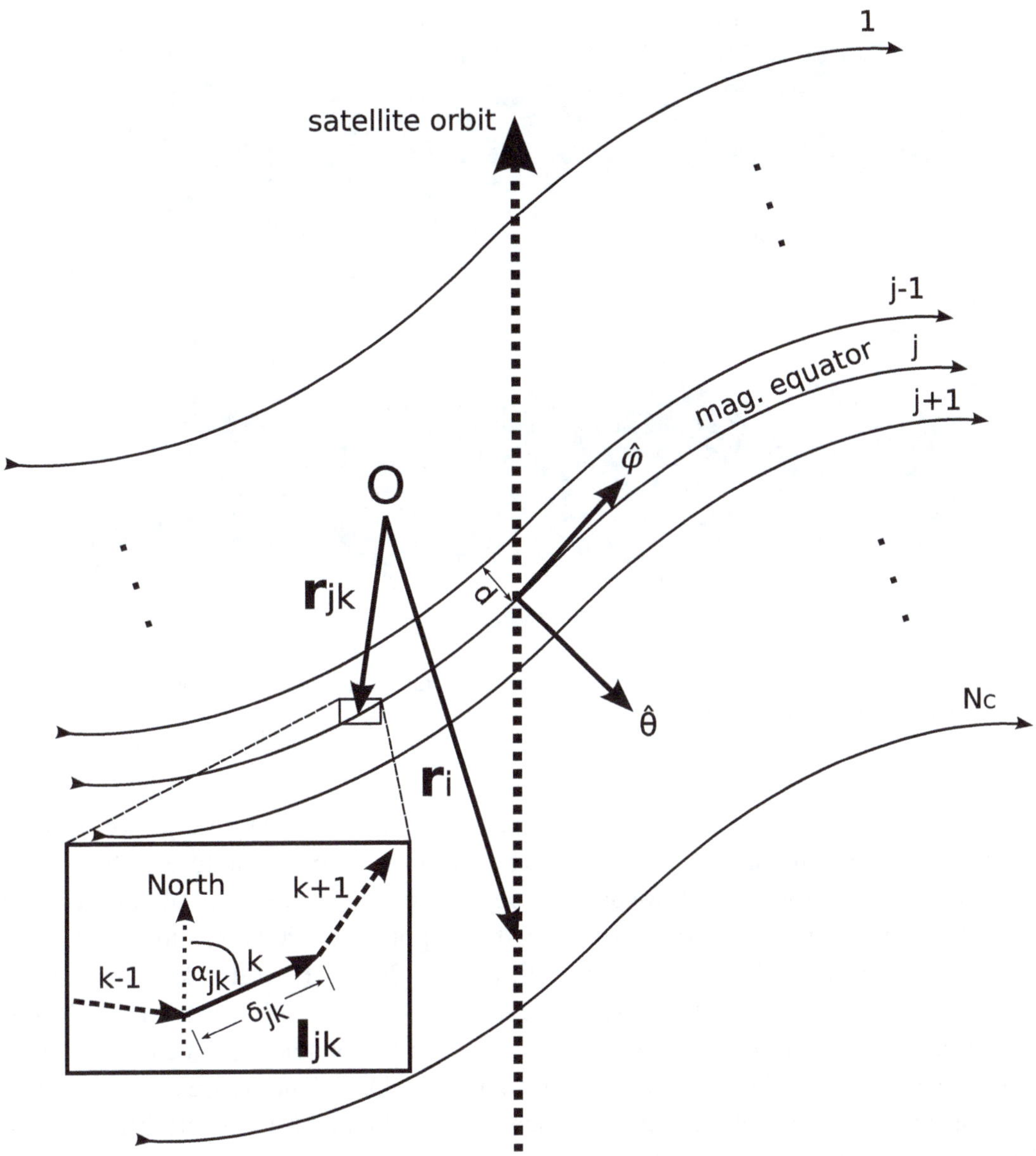

Fig. 11.3 Current model used for inversion. Currents $1 \ldots N_C$ shown following lines of constant quasi-dipole latitude, with the satellite crossing the magnetic equator (shown as current j). Origin O represents the Earth's center with vector $\mathbf{r}_{jk}$ pointing to linear current segment k of arc current j, and $\mathbf{r}_i$ pointing to satellite observation point i. Unit current vector $\mathbf{I}_{jk}$ shown in enlarged region with relevant parameters (see text). Basis vectors $\hat{\phi}$ and $\hat{\theta}$ are shown for use during the modeling step). Reproduced from Alken et al. (2013b)

The equatorial electrojet current strength is known to be modulated in longitude by atmospheric tides originating from latent heat release in deep convective tropical clouds. A number of different tidal components have been found to contribute to ionospheric longitudinal variability at low-latitudes, including the prominent eastward propagating diurnal tide with zonal wavenumber 3 (DE3) (Forbes et al. 2008; Lühr et al. 2008; Zhou et al. 2016). Figure 11.4 shows Swarm-derived height-integrated

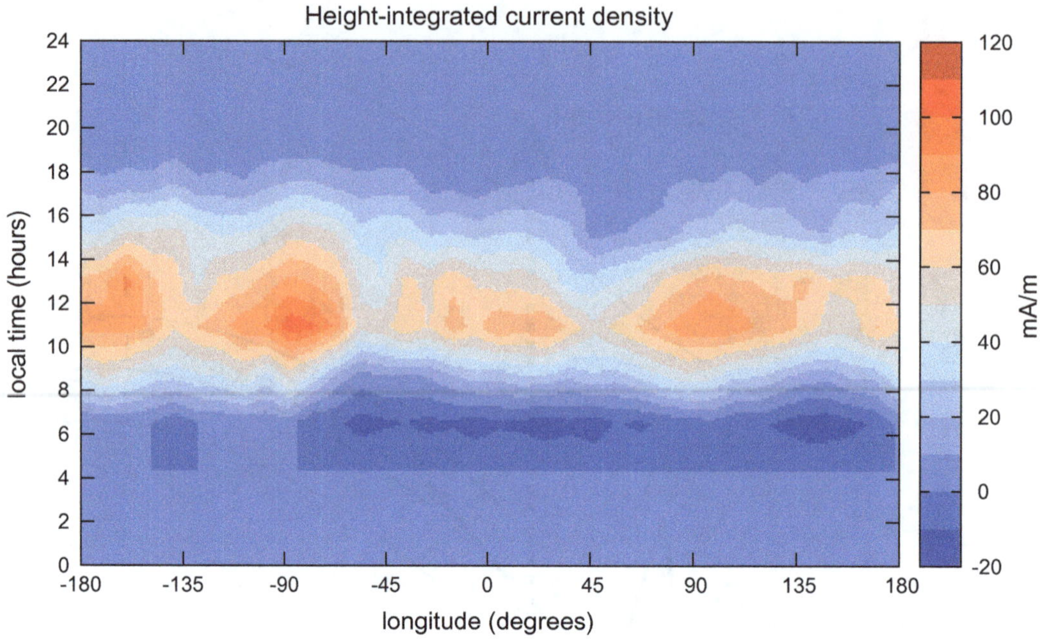

Fig. 11.4 Swarm derived height-integrated peak equatorial electrojet current for March equinox plotted as longitude versus local time

EEJ current estimates on the magnetic equator around the March equinox season and binned in longitude and local time. The evident wave-4 structure in longitude is characteristic of the DE3 atmospheric tide. There are also other tidal components present in the EEJ dataset, which have been thoroughly analyzed by Lühr et al. (2008, 2012), Zhou et al. (2016), Yamazaki et al. (2017, 2018). It is also known that the tidal components exhibit a strong seasonal dependence. Notably for the EEJ, during the months of December and January the eastward propagating diurnal tide with zonal wavenumber 2 (DE2) becomes dominant over DE3. The figure also exhibits a westward flowing current around 0600 local time, indicated by the dark blue ribbon. This is known as counter electrojet (CEJ) and is often found in the early morning hours near the dawn terminator.

11.5 Estimating Low-Latitude Electric Fields

The current profile representing EEJ flow in the E-region discussed in the previous section contains a wealth of information about low-latitude ionospheric electrodynamics. Since the EEJ current system is affected by the neutral wind field and solar wind activity, the EEJ current derived from satellite observations can reveal a great deal about the underlying driving mechanisms, their seasonal and local time structure, and how they change during active and quiet times. Satellite-derived estimates of EEJ flow have been used to study atmospheric tides (Lühr and Manoj 2013; Zhou et al. 2016) and ionospheric response during geomagnetic storms (Astafyeva et al. 2016).

We can go a step further, toward a more fundamental parameter than the currents themselves, which is the low-latitude electric field. The electric field is ultimately responsible for driving the currents, so knowledge of the electric field will enable a deeper understanding of the current system, as well as other low-latitude phenomena like the vertical plasma fountain (Anderson 1981; Stening 1992; Kelley 2009). In order to infer electric field information from satellite-derived currents, we need to apply some ionospheric modeling.

11.5.1 Ionospheric Electrostatic Modeling

We will ignore ionospheric dynamics which occur on time scales of one minute or less, which will allow us to assume a steady-state system, and thus consider only electrostatic fields. In this case, the equations governing the electric fields and currents are given by

$$\nabla \times \mathbf{E} = 0 \tag{11.18}$$

$$\mathbf{J} = \sigma \left(\mathbf{E} + \mathbf{u} \times \mathbf{B} \right) \tag{11.19}$$

where $\mathbf{E}$ is the electrostatic field, $\mathbf{J}$ is the current density, σ is the anisotropic conductivity tensor (Forbes 1981, Eq. 10), $\mathbf{u}$ is the neutral wind velocity field, and $\mathbf{B}$ is the ambient geomagnetic field. Equation (11.18) comes from Faraday's law assuming a steady-state magnetic field and Eq. (11.19) is Ohm's law describing the current density driven by the neutral winds and electric field. The main idea is to solve Eqs. (11.18), (11.19) for the electric field $\mathbf{E}$ in a low-latitude region by

1. making simplifying assumptions about the longitudinal structure of the current and electric fields in order to reduce from a 3D problem to 2D
2. using global climatological models to specify the conductivity σ, neutral wind field $\mathbf{u}$, and geomagnetic field $\mathbf{B}$
3. using the satellite-derived current profile obtained by the methods of Sect. 11.4 to specify J_ϕ, the eastward component of $\mathbf{J}$

First, we will ignore longitudinal gradients of the electric field and current density ($\partial \mathbf{E}/\partial \phi = \partial \mathbf{J}/\partial \phi = 0$). This assumption is known to be incorrect on large scales, as there are many reports in the literature of 3 and 4-cell patterns at low-latitudes in many ionospheric parameters, such as vertical plasma drift velocities, EEJ currents, and plasma density (Immel et al. 2006; England et al. 2006; Lühr et al. 2007, 2008, 2012) Gradients in $\mathbf{E} \times \mathbf{B}$ drift velocities have been reported up to 3 m/s/deg (Araujo-Pradere et al. 2011). To account for the full and complex longitude structure of the ionosphere, we would need to solve the electrostatic equations in three dimensions. However, by ignoring longitudinal gradients on local scales, the problem drastically simplifies, and previous studies calculating electric fields with this assumption have demonstrated remarkable agreement with independent radar measurements at Jicamarca (Alken and Maus 2010a; Alken et al. 2013a, 2015). This assumption, coupled with the divergence-free current condition, $\nabla \cdot \mathbf{J} = 0$, allows the J_r and J_θ current components to be derived from a single current stream function ψ (Untiedt 1967; Sugiura and Poros 1969).

$$J_r = \left(\frac{R}{r}\right)^2 \frac{1}{\sin\theta} \frac{\partial \psi}{\partial \theta} \tag{11.20}$$

$$J_\theta = -\left(\frac{R}{r}\right) \frac{R}{\sin\theta} \frac{\partial \psi}{\partial r} \tag{11.21}$$

Equation (11.18) then becomes

$$\partial_r (r E_\theta) - \partial_\theta (E_r) = 0 \tag{11.22}$$

$$\left.\begin{array}{r} \partial_\theta (\sin\theta E_\phi) = 0 \\ \partial_r (r E_\phi) = 0 \end{array}\right\} \Rightarrow E_\phi = \frac{R E_{\phi_0}}{r \sin\theta} \tag{11.23}$$

where R is a constant of integration and can be taken as a reference radius, and E_{ϕ_0} is the eastward electric field at the equator at the radius R. Equation (11.23) shows that for a given value of the equatorial eastward electric field $E_{\phi_0} = E_\phi(r = R, \theta = \pi/2)$, $E_\phi(r, \theta)$ is determined everywhere in the (r, θ) plane. The unknowns to be determined are therefore E_r, E_θ, and ψ.

Next, we use a priori climatological models to specify the conductivity σ, neutral wind field $\mathbf{u}$, and geomagnetic field $\mathbf{B}$. The conductivity requires knowledge of the global densities and temperatures of the electrons, ions and neutrals. For these we use the IRI-2012 (Bilitza et al. 2011) and NRLMSISE-00 (Picone et al. 2002) models. The equations for the direct, Pedersen and Hall conductivities are given in (Kelley 1989, Appendix B) and reproduced below:

$$\sigma_0 = e^2 \left(\frac{n_e}{m_e \nu_e} + \sum_i \frac{n_i}{m_i \nu_i} \right) \tag{11.24}$$

$$\sigma_p = e^2 \left(\frac{n_e \nu_e}{m_e(\nu_e^2 + \Omega_e^2)} + \sum_i \frac{n_i \nu_i}{m_i(\nu_i^2 + \Omega_i^2)} \right) \tag{11.25}$$

$$\sigma_h = e^2 \left(\frac{n_e \Omega_e}{m_e(\nu_e^2 + \Omega_e^2)} - \sum_i \frac{n_i \Omega_i}{m_i(\nu_i^2 + \Omega_i^2)} \right) \tag{11.26}$$

Here, the i sums overall ion species in the ionosphere. e is the electron charge, n_e is the electron density, n_i is the ion density of species i, m_e and m_i are the electron and ion masses, ν_e and ν_i are the electron and ion collision frequencies, and Ω_e and Ω_i are the electron and ion gyro-frequencies around the magnetic field lines. Expressions for the collision frequencies ν_e and ν_i are given in (Kelley 1989, Appendix B). The ionospheric densities and temperatures are taken from IRI-2012. The neutral density needed to compute the collision frequencies is taken from the NRLMSISE-00 model. Previous efforts to model the equatorial electrojet have found it necessary to increase the electron collision frequency ν_e by an empirical factor of 4 during typical daytime eastward electric field conditions, to account for unmodeled nonlinear instabilities in the electrojet stream (Gagnepain et al. 1977; Ronchi et al. 1990, 1991; Fang et al. 2008; Alken and Maus 2010a, b). We adopt this same convention when modeling satellite-derived EEJ profiles. The neutral wind field **u** is supplied by the Horizontal Wind Model (HWM14) (Drob et al. 2015). HWM14 does not provide vertical wind velocities, and so they are ignored during this modeling. Any standard geomagnetic field model, such as IGRF (Thébault et al. 2015) can be used to specify **B**.

Eliminating E_r and E_θ from Eqs. (11.19)–(11.22) yields a second order partial differential equation (PDE) for the current stream function ψ:

$$f_1 \frac{\partial^2 \psi}{\partial r^2} + 2 f_2 \frac{\partial^2 \psi}{\partial r \partial \theta} + f_3 \frac{\partial^2 \psi}{\partial \theta^2} + f_4 \frac{\partial \psi}{\partial r} + f_5 \frac{\partial \psi}{\partial \theta} = g \tag{11.27}$$

where the coefficients f_1, ..., f_5 and right hand side g are functions of r, θ given by

$$\begin{aligned}
f_1 &= r^2 \sigma_{rr} \\
f_2 &= \frac{1}{2} r \left(\sigma_{r\theta} + \sigma_{\theta r} \right) \\
f_3 &= \sigma_{\theta\theta} \\
f_4 &= r\alpha \left[r \frac{\partial}{\partial r} \left(\frac{\sigma_{rr}}{\alpha} \right) + \sin\theta \frac{\partial}{\partial \theta} \left(\frac{1}{\sin\theta} \frac{\sigma_{r\theta}}{\alpha} \right) \right] \\
f_5 &= \alpha \left[r^2 \frac{\partial}{\partial r} \left(\frac{1}{r} \frac{\sigma_{\theta r}}{\alpha} \right) + \sin\theta \frac{\partial}{\partial \theta} \left(\frac{1}{\sin\theta} \frac{\sigma_{\theta\theta}}{\alpha} \right) \right] \\
g &= \frac{r}{R} \alpha \left[r \frac{\partial}{\partial r} \left(\frac{\gamma}{\alpha} \right) + \sin\theta \frac{\partial}{\partial \theta} \left(\frac{1}{\sin\theta} \frac{\beta}{\alpha} \right) \right] E_{\phi_0} + \\
&\quad \left(\frac{r}{R} \right)^2 \alpha \sin\theta \left\{ \frac{\partial}{\partial r} \left[\frac{r}{\alpha} \left(\sigma_{\theta r} W_r - \sigma_{rr} W_\theta \right) \right] + \frac{\partial}{\partial \theta} \left[\frac{1}{\alpha} \left(\sigma_{\theta\theta} W_r - \sigma_{r\theta} W_\theta \right) \right] \right\}
\end{aligned} \tag{11.28}$$

with

$$\alpha = \sigma_{rr}\sigma_{\theta\theta} - \sigma_{r\theta}\sigma_{\theta r} \tag{11.29}$$

$$\beta = \sigma_{\theta\theta}\sigma_{r\phi} - \sigma_{r\theta}\sigma_{\theta\phi} \tag{11.30}$$

$$\gamma = \sigma_{r\phi}\sigma_{\theta r} - \sigma_{rr}\sigma_{\theta\phi} \tag{11.31}$$

and $\mathbf{W} = \sigma \left(\mathbf{u} \times \mathbf{B} \right)$. The conductivity tensor σ is represented in a basis of spherical coordinates. The components of the conductivity tensor may be related to the direct, Pedersen, and Hall conductivities using (Richmond 1995, Eq. 2.1), or alternatively in coordinate-free matrix notation:

$$\sigma = \sigma_p I + \left(\sigma_0 - \sigma_p\right) \mathbf{b}\mathbf{b}^T + \sigma_h \left[\mathbf{b}\right]_\times \tag{11.32}$$

where $\mathbf{b}$ is a unit vector in the geomagnetic field direction $\mathbf{B}$ and $[\mathbf{b}]_\times$ is the skew-symmetric matrix defined by the cross product with $\mathbf{b}$. In spherical coordinates, $\mathbf{b} = \left(b_r, b_\theta, b_\phi\right)^T$ and

$$[\mathbf{b}]_\times = \begin{pmatrix} 0 & -b_\phi & b_\theta \\ b_\phi & 0 & -b_r \\ -b_\theta & b_r & 0 \end{pmatrix} \tag{11.33}$$

We solve Eq. (11.27) in spherical geocentric coordinates, however the coordinates are rotated so that the azimuthal direction $\hat{\phi}$ is tangent to the magnetic equator at the location of the satellite crossing (see Fig. 11.3). This is a first-order correction in order to allow the modeling of the currents flowing along lines of constant quasi-dipole latitude, as we calculated during the satellite data inversion step. To perform a strict comparison between our modeled current and satellite-derived current would require solving the electrostatic equations in quasi-dipole coordinates, but this first-order correction enables us to use the simplicity of spherical coordinates and captures most of the difference between magnetic and geographic east. The PDE is solved on a 2D grid in the (r, θ) plane, holding the longitude ϕ fixed where the satellite crosses the magnetic equator. The grid should be large enough to encompass the main current flow at low-latitudes, including the meridional current system, but not so large to include effects of current flow from mid and high-latitudes. Typically, a grid ranging from 65 to 500 km altitude, and $-25°$ to $25°$ latitude is sufficient for capturing the main EEJ current of interest. The boundary conditions on the PDE are that the current should vanish at the lower and upper boundaries ($\psi = 0$ at $r = r_{min}$ and $r = r_{max}$), and there is no radial current flow at the northern and southern boundaries ($\partial_\theta \psi = 0$ at $\theta = \theta_{min}$ and $\theta = \theta_{max}$). The PDE can be solved with standard finite difference methods.

11.5.2 Estimating the Electric Field

The solution for the current stream function ψ depends on the conductivities, neutral wind field, and geomagnetic field components in the solution region, but it also requires a value for the eastward electric field component on the magnetic equator, E_{ϕ_0}. The former fields are specified by climatological models, but the eastward electric field is what we wish to obtain as a final output of this procedure. There is one additional piece of information we have not yet used, which is the satellite-derived zonal current profile. The solution for the electric field now becomes an optimization problem: which value of E_{ϕ_0} produces a modeled current which best agrees with the satellite-derived current? One approach would be to solve Eq. (11.27) multiple times with different values of $E_{\phi 0}$ until we obtain the best agreement between the modeled and observed current. However, in this case we are fortunate that the current density $\mathbf{J}$ in Eq. (11.19) is a linear function of the electric field $\mathbf{E}$ when $\mathbf{u} = 0$. Therefore we only need to solve the PDE twice, once with the full wind field $\mathbf{u}$, and a second time with $\mathbf{u} = 0$. Then the unknown component $E_{\phi 0}$ can be determined from a least-squares inversion of

$$\begin{aligned} J_\phi^{SAT}(\theta) &= s J_{PDE}(\theta; E_{\phi 0} = 1 \text{ mV/m}, \mathbf{u} = 0) \\ &+ J_{PDE}(\theta; E_{\phi 0} = 0, \mathbf{u}) \\ &- J_{DC} \end{aligned} \tag{11.34}$$

where s is a scaling factor (discussed below), $J_\phi^{SAT}(\theta)$ is the latitude current profile derived from the satellite observations, $J_{PDE}(\theta)$ is the height-integrated zonal current profile calculated from the PDE solution:

$$J_{PDE}(\theta) = \sum_i J_\phi(r_i, \theta)\delta r \tag{11.35}$$

where δr is the radial grid spacing and $J_\phi(r, \theta)$ is the PDE-derived zonal current on the (r, θ) grid, and finally J_{DC} is a constant offset to allow for a difference in zero-levels between the modeled and observed current. As can be seen in Eq. (11.34), the PDE is solved twice, once with a "unit" value of $E_{\phi 0} = 1$ mV/m with the wind field turned off, and a second time with $E_{\phi 0} = 0$ with the wind field turned on. The parameters s and J_{DC} are determined by least-squares inversion of Eq. (11.34) with the additional constraint that the left and right hand sides of that equation must agree on the magnetic equator ($\theta = \pi/2$). This constraint has been found to yield more accurate electric fields (Alken and Maus 2010a; Alken et al. 2013b, 2015), since the EEF is primarily responsible for current flow near the magnetic equator, while the neutral winds affect current flow at higher latitudes (Fambitakoye and Mayaud 1976).

Figure 11.5 shows the height-integrated current density profile derived from Swarm B near 12 UT on 1 January 2017 (red) along with the modeled current density profile (green). For this orbit, the satellite was in local time of 13:06. The two current profiles agree well at the magnetic equator, which was a condition imposed during the least-squares inversion. We can see that the main peak is modeled well, while the side-lobes differ considerably. This is fairly typical in our modeling approach, since the side-lobes are mainly determined by the neutral winds at higher altitudes (Fambitakoye and Mayaud 1976), and the climatological wind model (HWM14) cannot capture the highly variable winds. However, as discussed by Fambitakoye and Mayaud (1976), Reddy and Devasia (1981), the height and width of the main peak is primarily determined by the EEF, and so an accurate model of this peak will produce reliable estimates of the eastward equatorial electric field component.

Figure 11.6 shows the eastward equatorial electric field data derived from four years of Swarm data (2014–2017), selected for March equinox season and binned

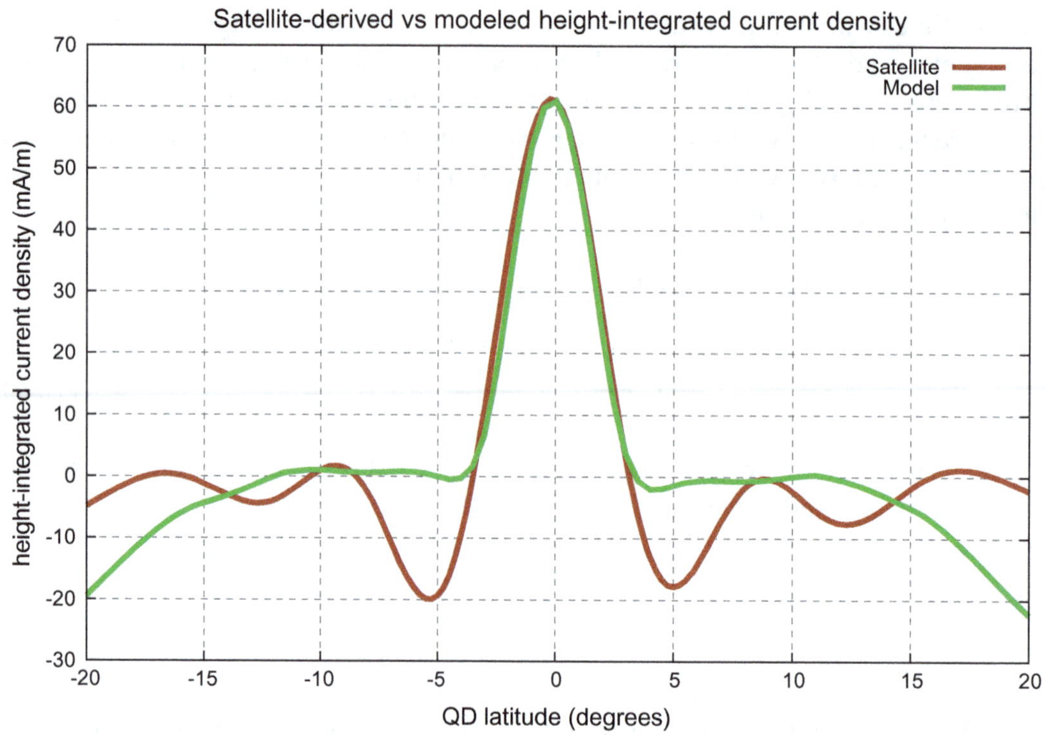

Fig. 11.5 Height-integrated EEJ current density derived from Swarm B measurements near 12 UT on 1 January 2017 in red. Corresponding modeled current density is shown in green

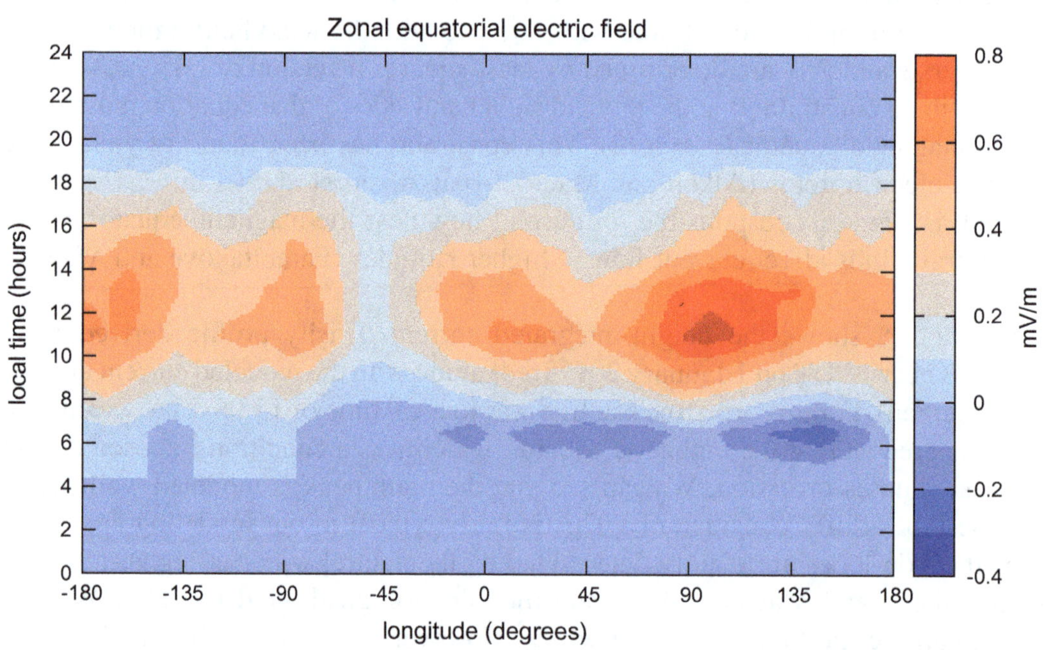

Fig. 11.6 Swarm derived zonal equatorial electric field for March equinox, plotted as longitude versus local time

as a function of longitude and local time. Similar to Fig. 11.4 we see the wave-4 longitude structure due to the DE3 tide, as well as the westward electric field around 0600 local time indicating a counter electrojet.

11.6 Conclusion

We have presented a methodology of inverting scalar magnetic measurements from LEO near-polar orbiting satellites for EEJ equivalent current flow at low-latitudes. This method is based on modeling the magnetically eastward current flow with straight line segments placed along lines of constant quasi-dipole latitude. The line currents are spaced equidistantly in QD latitude, and they have simple expressions for their magnetic field perturbations which can be fit to the satellite measurements using regularized least-squares methods. The result of this least-squares inversion is a latitude profile of zonal current flow at low-latitudes which corresponds to height-integrated equivalent EEJ current in the E-region. These latitude profiles can further be used to recover the eastward component of the equatorial electric field by using a physics-based modeling procedure which relies on input from global climatological models to specify the state of the ionosphere and neutral atmosphere in the vicinity of the satellite measurements. Both the equivalent current profiles and the EEF estimates can be used to study seasonal and local time variations of the low-latitude ionosphere, as well as tidal characteristics of the low-latitude neutral atmosphere.

Acknowledgements The European Space Agency (ESA) is gratefully acknowledged for providing Swarm data and for financially supporting the Swarm Level-2 equatorial electric field (EEF) product. We thank the International Space Science Institute (ISSI) in Bern, Switzerland for supporting the working group "Ionospheric multi-satellite analysis tools" from which this chapter resulted. The editors thank an anonymous reviewer for his assistance in evaluating this chapter.

References

Aakjær, C.D., N. Olsen, and C.C. Finlay. 2016. Determining polar ionospheric electrojet currents from Swarm satellite constellation magnetic data. *Earth, Planets and Space* 68 (1): 140. https://doi.org/10.1186/s40623-016-0509-y.

Alken, P., and S. Maus. 2010a. Electric fields in the equatorial ionosphere derived from CHAMP satellite magnetic field measurements. *Journal of Atmospheric and Solar-Terrestrial Physics* 72: 319–326. https://doi.org/10.1016/j.jastp.2009.02.006.

Alken, P., and S. Maus. 2010b. Relationship between the ionospheric eastward electric field and the equatorial electrojet. *Geophysical Research Letters* 37: L04104. https://doi.org/10.1029/2009GL041989.

Alken, P., A. Chulliat, and S. Maus. 2013a. Longitudinal and seasonal structure of the ionospheric equatorial electric field. *Journal of Geophysical Research* 118. https://doi.org/10.1029/2012JA018314.

Alken, P., S. Maus, P. Vigneron, O. Sirol, and G. Hulot. 2013b. Swarm SCARF equatorial electric field inversion chain. *Earth, Planets and Space* 65 (11): 1309–1317. https://doi.org/10.5047/eps.2013.09.008.

Alken, P., S. Maus, H. Lühr, R.J. Redmon, F. Rich, B. Bowman, and S.M. O'Malley. 2014. Geomagnetic main field modeling with DMSP. *Journal of Geophysical Research* 119. https://doi.org/10.1002/2013JA019754.

Alken, P., S. Maus, A. Chulliat, P. Vigneron, O. Sirol, and G. Hulot. 2015. Swarm equatorial electric field chain: first results. *Geophysical Research Letters* 42 (3): 673–680. https://doi.org/10.1002/2014GL062658,2014GL062658.

Anderson, D.N. 1981. Modeling the ambient, low latitude F-region ionosphere—A review. *Journal of Atmospheric and Solar-Terrestrial Physics* 43 (8): 753–762.

Araujo-Pradere, E.A., D.N. Anderson, and M. Fedrizzi. 2011. Communications/navigation outage forecasting system observational support for the equatorial E x B drift velocities associated with the four-cell tidal structures. *Radio Science* 46:RS0D09. https://doi.org/10.1029/2010RS004557.

Astafyeva, E., I. Zakharenkova, and P. Alken. 2016. Prompt penetration electric fields and the extreme topside ionospheric response to the 22-23 June 2015 geomagnetic storm as seen by the Swarm constellation. *Earth, Planets and Space* 68(152). https://doi.org/10.1186/s40623-016-0526-x.

Bilitza, D., L.A. McKinnell, B. Reinisch, and T. Fuller-Rowell. 2011. The International Reference Ionosphere (IRI) today and in the future. *Journal of Geodesy* 85: 909–920. https://doi.org/10.1007/s00190-010-0427-x.

Drob, D.P., J.T. Emmert, J.W. Meriwether, J.J. Makela, E. Doornbos, M. Conde, G. Hernandez, J. Noto, K.A. Zawdie, S.E. McDonald, J.D. Huba, and J.H. Klenzing. 2015. An update to the Horizontal Wind Model (HWM): The quiet time thermosphere. *Earth and Space Science* 2 (7): 301–319. https://doi.org/10.1002/2014EA000089,2014EA000089.

England, S.L., S. Maus, T.J. Immel, S.B. Mende. 2006. Longitudinal variation of the E-region electric fields caused by atmospheric tides. *Geophysical Research Letters* 33.

Fambitakoye, O., and P.N. Mayaud. 1976. The equatorial electrojet and regular daily variation S_R:-I. A determination of the equatorial electrojet parameters. *Journal of Atmospheric and Terrestrial Physics* 38: 1–17.

Fang, T.W., A.D. Richmond, J.Y. Liu, A. Maute, C.H. Lin, C.H. Chen, and B. Harper. 2008. Model simulation of the equatorial electrojet in the Peruvian and Philippine sectors. *Journal of Atmospheric and Solar-Terrestrial Physics* 70: 2203–2211.

Finlay, C.C., N. Olsen, S. Kotsiaros, N. Gillet, and L. Tøffner-Clausen. 2016. Recent geomagnetic secular variation from Swarm and ground observatories as estimated in the CHAOS-6 geomagnetic field model. *Earth, Planets and Space* 68 (1): 112. https://doi.org/10.1186/s40623-016-0486-1.

Fiori, R.A.D., and D.H. Boteler. 2018. Spherical cap Harmonic analysis techniques for mapping high-latitude ionospheric plasma flow—application to the Swarm mission. Submitted.

Forbes, J.M. 1981. The equatorial electrojet. *Reviews of Geophysics and Space Physics* 19 (3): 469–504.

Forbes, J.M., X. Zhang, S. Palo, J. Russell, C.J. Mertens, and M. Mlynczak. 2008. Tidal variability in the ionospheric dynamo region. *Journal of Geophysical Research* 113: A02310. https://doi.org/10.1029/2007JA012737.

Friis-Christensen, E., H. Lühr, and G. Hulot. 2006. Swarm: A constellation to study the Earth's magnetic field. *Earth, Planets and Space* 58: 351–358.

Gagnepain, J., M. Crochet, and A.D. Richmond. 1977. Comparison of equatorial electrojet models. *Journal of Atmospheric and Terrestrial Physics* 39: 1119–1124.

Hansen, P.C., and D.P. O'Leary. 1993. The use of the L-Curve in the regularization of discrete Ill-posed problems. *SIAM Journal on Scientific Computing* 14(6):1487–1503. https://doi.org/10.1137/0914086.

Immel, T.J., E. Sagawa, S.L. England, S.B. Henderson, M.E. Hagan, S.B. Mende, H.U. Frey, C.M. Swenson, and L.J. Paxton. 2006. Control of equatorial ionospheric morphology by atmospheric tides. *Geophysical Research Letters* 33(L15108). https://doi.org/10.1029/2006GL026161.

Kelley, M.C. 1989. *The Earth's ionosphere: Plasma physics and electrodynamics.*, International Geophysics Series San Diego: Academic Press Inc.

Kelley, M.C. 2009. *The Earth's ionosphere: Plasma physics and electrodynamics.*, International Geophysics Series San Diego: Academic Press Inc.

Laundal, K.M., and A.D. Richmond. 2016. Magnetic coordinate systems. *Space Science Reviews* 1–33.

Lühr, H., and C. Manoj. 2013. The complete spectrum of the equatorial electrojet related to solar tides: CHAMP observations. *Annales de Geophysicae* 31: 1315–1331.

Lühr, H., and S. Maus. 2010. Solar cycle dependence of quiet-time magnetospheric currents and a model of their near-Earth magnetic fields. *Earth, Planets and Space* 62: 843–848.

Lühr, H., S. Maus, and M. Rother. 2004. Noon-time equatorial electrojet: Its spatial features as determined by the CHAMP satellite. *Journal of Geophysical Research* 109: A01306. https://doi.org/10.1029/2002JA009656.

Lühr, H., K. Häusler, and C. Stolle. 2007. Longitudinal variation of F region electron density and thermospheric zonal wind caused by atmospheric tides. *Geophysical Research Letters* 34(16):L16102. https://doi.org/10.1029/2007GL030639,

Lühr, H., M. Rother, K. Häusler, P. Alken, and S. Maus. 2008. The influence of non-migrating tides on the longitudinal variation of the equatorial electrojet. *Journal of Geophysical Research* 113: A08313. https://doi.org/10.1029/2008JA013064.

Lühr, H., M. Rother, K. Häusler, B. Fejer, and P. Alken. 2012. Direct comparison of nonmigrating tidal signatures in the electrojet, vertical plasma drift and equatorial ionization anomaly. *Journal of Atmospheric and Solar-Terrestrial Physics* 75–76: 31–43. https://doi.org/10.1016/j.jastp.2011.07.009.

Lühr, H., C. Xiong, N. Olsen, G. Le. 2016. Near-Earth magnetic field effects of large-scale magnetospheric currents. *Space Science Reviews* 1–25. https://doi.org/10.1007/s11214-016-0267-y.

Maus, S., and H. Lühr. 2005. Signature of the quiet-time magnetospheric magnetic field and its electromagnetic induction in the rotating Earth. *Geophysical Journal International* 162: 755–763. https://doi.org/10.1111/j.1365-246X.2005.02691.x.

Maus, S., F. Yin, H. Lühr, C. Manoj, M. Rother, J. Rauberg, I. Michaelis, C. Stolle, and R.D. Müller. 2008. Resolution of direction of oceanic magnetic lineations by the sixth-generation lithospheric magnetic field model from CHAMP satellite magnetic measurements. *Geochemistry, Geophysics, Geosystems* 9(7):Q07021. https://doi.org/10.1029/2008GC001949.

Maus, S., C. Manoj, J. Rauberg, I. Michaelis, and H. Lühr. 2010. NOAA/NGDC candidate models for the 11th generation international geomagnetic reference field and the concurrent release of the 6th generation Pomme magnetic model. *Earth, Planets and Space* 62: 729–735.

Olsen, N. 1996. A new tool for determining ionospheric currents from magnetic satellite data. *Geophysical Research Letters* 23 (24): 3635–3638. https://doi.org/10.1029/96GL02896.

Picone, J.M., A.E. Hedin, D.P. Drob, and A.C. Aikin. 2002. NRLMSISE-00 empirical model of the atmosphere: statistical comparisons and scientific issues. *Journal of Geophysical Research* 107. https://doi.org/10.1029/2002JA009430.

Reddy, C.A., and C.V. Devasia. 1981. Height and latitude structure of electric fields and currents due to local eastwest winds in the equatorial electrojet. *Journal of Geophysical Research: Space Physics* 86 (A7): 5751–5767. https://doi.org/10.1029/JA086iA07p05751.

Richmond, A.D. 1995. Ionospheric electrodynamics using magnetic apex coordinates. *Journal of Geomagnetism and Geoelectricity* 47: 191–212.

Ritter, P., H. Lühr, A. Viljanen, O. Amm, A. Pulkkinen, and I. Sillanpää. 2004. Ionospheric currents estimated simultaneously from CHAMP satelliteand IMAGE ground-based magnetic field measurements: a statisticalstudy at auroral latitudes. *Annales Geophysicae* 22 (2): 417–430.

Ronchi, C., R.N. Sudan, and P.L. Similon. 1990. Effect of short-scale turbulence on kilometer wavelength irregularities in the equatorial electrojet. *Journal of Geophysical Research* 95 (A1): 189–200.

Ronchi, C., R.N. Sudan, D.T. Farley. 1991. Numerical simulations of large-scale plasma turbulence in the daytime equatorial electrojet. *Journal of Geophysical Research* 96(A12):21,263–21,279.

Stening, R. 1992. Modelling the low latitude F region. *Journal of Atmospheric and Terrestrial Physics* 54(1112):1387–1412. https://doi.org/10.1016/0021-9169(92)90147-D, url http://www.sciencedirect.com/science/article/pii/002191699290147D.

Sugiura, M., and D.J. Poros. 1969. An improved model equatorial electrojet with a meridional current system. *Journal of Geophysical Research* 74: 4025–34.

Thébault, E., C.C. Finlay, C.D. Beggan, P. Alken, J. Aubert, O. Barrois, F. Bertrand, T. Bondar, A. Boness, L. Brocco, E. Canet, A. Chambodut, A. Chulliat, P. Coïsson, F. Civet, A. Du, A. Fournier, I. Fratter, N. Gillet, B. Hamilton, M. Hamoudi, G. Hulot, T. Jager, M. Korte, W. Kuang, X. Lalanne, B. Langlais, J.M. Léger, V. Lesur, F.J. Lowes, S. Macmillan, M. Mandea, C. Manoj, S. Maus, N. Olsen, V. Petrov, V. Ridley, M. Rother, T.J. Sabaka, D. Saturnino, R. Schachtschneider, O. Sirol, A. Tangborn, A. Thomson, L. Tøffner-Clausen, P. Vigneron, I. Wardinski, and T. Zvereva. 2015. International geomagnetic reference field: The 12th generation. *Earth, Planets and Space* 67(79). https://doi.org/10.1186/s40623-015-0228-9.

Untiedt, J. 1967. A model of the equatorial electrojet involving meridional currents. *Journal of Geophysical Research* 72(23).

Vanhamäki, H., L. Juusola, K. Kauristie, A. Workayehu, and S. Käki. 2018. Spherical elementary current systems applied to Swarm data. Submitted.

Vennerstrom, S., and T. Moretto. 2013. Monitoring auroral electrojets with satellite data. *Space Weather* 11 (9): 509–519. https://doi.org/10.1002/swe.20090.

Yamazaki, Y., C. Stolle, J. Matzka, T. Siddiqui, H. Lühr, and P. Alken. 2017. Longitudinal variation of the lunar tide in the equatorial electrojet. *Journal of Geophysical Research* 122(12). https://doi.org/10.1002/2017JA024601.

Yamazaki, Y., C. Stolle, J. Matzka, and P. Alken. 2018. Quasi-6 day wave modulation of the equatorial electrojet. *Journal of Geophysical Research* 123. https://doi.org/10.1029/2018JA025365.

Zhou, Y., H. Lühr, P. Alken, and C. Xiong. 2016. New perspectives on equatorial electrojet tidal characteristics derived from the Swarm constellation. *Journal of Geophysical Research: Space Physics* 121. https://doi.org/10.1002/2016JA022713.

Chapter 12
Models of the Main Geomagnetic Field Based on Multi-satellite Magnetic Data and Gradients—Techniques and Latest Results from the Swarm Mission

Christopher C. Finlay

Abstract Magnetic field observations from low-Earth-orbiting satellites provide a unique means of studying ionospheric current systems on a global scale. Such studies require that estimates of other sources of the Earth's magnetic field, in particular, the dominant main field generated primarily in Earth's core but also due to the magnetized lithosphere and large-scale magnetospheric currents, are first removed. Since 1999 multiple low-Earth-orbit satellites including Ørsted, CHAMP, SAC-C, and most recently the *Swarm* trio have surveyed the near-Earth magnetic field in increasing detail. This chapter reviews how models of the main magnetic field are today constructed from multiple satellites, in particular discussing how to take advantage of estimated field gradients, both along-track and across-track. A summary of recent results from the *Swarm* mission regarding the core and lithospheric field components is given, with the aim of informing users interested in ionospheric applications of the options available for high accuracy data reduction. Limitations of the present generation of main field models are also discussed, and it is pointed out that further progress requires improved treatment of ionospheric sources, in particular at polar latitudes.

12.1 Introduction

Ionospheric current systems produce magnetic fields that are measured by magnetometers on low-Earth-orbit satellites, together with the magnetic fields produced by a wide range of other natural sources. The largest of these sources is the so-called 'main' magnetic field, generated in Earth's liquid metal outer core through motional induction in a process known as the geodynamo (e.g. Roberts and King 2013). For those interested in precise studies of ionospheric currents it is important to remove a high-resolution estimate of the internally-generated field, capturing as far as possible its small-scale structure and secular time dependence (see e.g. Stolle et al. 2016).

C. C. Finlay (✉)
DTU Space, Technical University of Denmark, Lyngby, Denmark
e-mail: cfinlay@space.dtu.dk

M. W. Dunlop and H. Lühr (eds.), *Ionospheric Multi-Spacecraft Analysis Tools*, ISSI Scientific Report Series 17,
https://doi.org/10.1007/978-3-030-26732-2_12

In the context of this book, models of the main magnetic field can, therefore, be considered as important tools needed for studying ionospheric physics. Moreover, some of the data processing and modelling techniques used in main field studies are themselves of interest to ionospheric physicists, since they can easily be adapted to the study of ionospheric processes.

In this chapter, I begin by reviewing how models of the main geomagnetic field are constructed, focusing on recent developments that take advantage of magnetic field data collected by the low-Earth-orbit *Swarm* satellite constellation. The aim is to provide an easily accessible account of the construction of advanced main field models, so that users can make a well-informed decision about which models may be most suitable for their specific data processing and reduction tasks. Following this, a survey is given of the latest results regarding the structure and time-dependence of the internal geomagnetic field, as derived from data collected by the *Swarm* mission. The CHAOS series of field models (Olsen et al. 2006, 2009, 2010, 2014; Finlay et al. 2015, 2016) is a regularly updated, high resolution, main field model that covers the past one and half solar cycles. It will serve here as an illustrative example of an advanced field model that may be of interest for ionospheric studies.

The development of high-resolution geomagnetic field models is a community effort, in particular, facilitated by comparisons carried out within the framework of the International Geomagnetic Reference Field (IGRF) (Thébault et al. 2015a). Aside from the CHAOS model, that is the focus of this chapter, high-resolution field models are also available from a number of other groups, for example, the GRIMM series of models (e.g. Lesur et al. 2008, 2010, 2015b), the POMME series of models (e.g. Maus et al. 2005, 2006b) and the Comprehensive Model/Inversion series of models (e.g. Sabaka et al. 2004, 2015, 2018). Interested readers should consult these references for further details on these models. Limitations of all existing main field models, and opportunities to improve them using our expanding knowledge of ionospheric processes are discussed at the end of this chapter.

12.2 Fundamentals of Main Field Modelling

12.2.1 Calibration of Vector Magnetic Field Measurements

Modern main field geomagnetic reference models are derived primarily from magnetic field observations collected by low-Earth-orbit satellites. In particular, data from the *Swarm* satellite constellation, supplemented by measurements made on ground at geomagnetic observatories, are now crucial. For studies of the main field it is essential that the measurements, from both ground and satellite, have absolute accuracy—this is in contrast to the study of ionospheric processes, where often only rapid field variations are of interest. For satellite measurements this involves careful magnetometer design, strict magnetic cleanliness procedures when constructing the spacecrafts, pre-flight characterization of stray fields (Jørgensen et al. 2008) and in-flight calibration based on comparisons between fluxgate vector magnetometers and absolute scalar magnetometers that independently measure the field intensity (e.g. Olsen et al. 2003; Yin and Lühr 2011). Accurately orientated vector field measurements are also essential, since using scalar field intensity data alone there is a fundamental ambiguity arising from lack of knowledge perpendicular to the field (Backus 1970; Lowes 1975), particularly at the magnetic equator. In magnetic mapping missions, attitude information is today provided by high precision, non-magnetic, star trackers (Jørgensen et al. 2003). For *Swarm*, after application of models describing thermal fluctuations, attitude information is available at the arc-second level (Herceg et al. 2017).

Inflight calibration of the vector field data, for example in the case of the *Swarm* satellites, is carried out by minimizing the difference between the scalar magnetic field F_{ASM} measured by an absolute scalar magnetometer and the magnitude of the vector magnetic field $|\mathbf{B}|$ measured by the vector fluxgate magnetometer (VFM) frame. Free parameters that can be adjusted during inflight calibration arise when relating $\mathbf{B}$ to the vector field $\mathbf{B}_{\text{pre-flight}}$ determined using pre-flight determined fluxgate magnetometer calibration parameters, and after correction for the pre-flight determined stray magnetic fields (e.g. Olsen and Kotsiaros 2011), via the relation

$$\mathbf{B} = \underline{\underline{P}}^{-1}\underline{\underline{S}}^{-1}\mathbf{B}_{\text{pre-flight}} + \mathbf{B}_{\text{offset}} \tag{12.1}$$

where $\underline{\underline{S}}$ is a 3×3 diagonal scaling matrix, whose elements can be time dependent. $\underline{\underline{P}}$ is the non-orthogonality matrix that makes small adjustments to the pre-flight estimated non-orthogonalities of the VFM sensor and takes the form (cf. Olsen et al. 2003)

$$\underline{\underline{P}} = \begin{pmatrix} 1 & 0 & 0 \\ -\sin u_1 & \cos u_1 & 0 \\ \sin u_2 & \sin u_3 & \sqrt{1 - \sin^2 u_2 - \sin^2 u_3} \end{pmatrix}$$

where u_i are parameters describing the rotation to the non-orthogonal magnetic sensor axes coordinate system. $\mathbf{B}_{\text{offset}}$ represents additional small vector offsets/biases. In the case of the *Swarm* satellites, a small solar-driven magnetic disturbance, thought to be due to currents flowing in the satellite body as a result of thermo-electric effects, was detected post-launch (Tøffner-Clausen et al. 2016). This has been successfully described using an empirical model that depends on the sun position relative to the satellite and it is applied as an additional offset factor during the Level 1b magnetic data calibration procedure (for more details see Tøffner-Clausen et al. 2016). Physics-based models of this disturbance have also been developed, and improvements in Swarm's on-board calibration are ongoing.

12.2.2 *Selection of Magnetic Field Data for Main Field Modelling*

When constructing models of the main field, typically only data from geomagnetically quiet times are used, in an effort to reduce as far as possible the contaminating signatures arising from magnetospheric and ionospheric current systems. Of course this is the opposite mode of operation to that of the space physicist, who is often more interested in data collected during strongly disturbed conditions. Typical quiet-time data selection criteria are that the K_p index is less than $2o$, that the rate of change of D_{st} or similar ring current indices is less than 2 nT/h, and that for data in the polar region that the merging electric field, as determined from solar wind and IMF conditions measured at the L1 point, is less than 0.8 mV/m [Olsen et al. (2006), Olsen et al. (2014)]; for a more detailed discussion of data selection in internal field modelling interested readers should consult the recent reviews by Finlay et al. (2017) and Kauristie et al. (2017).

12.2.3 *Potential Field Modelling*

The majority of the main field models presently in operational use assume that the region of interest (where magnetic measurements are collected and where an estimate of the magnetic field is to be made) is current-free, i.e. the current density $\mathbf{J} = 0$. Under these conditions, the curl of the vector magnetic field $\mathbf{B}$ is zero

$$\mu_0 \mathbf{J} = \nabla \times \mathbf{B} = 0. \tag{12.2}$$

Note that strictly $\mathbf{B}$ should be referred to as the magnetic induction, but in geomagnetism it is for simplicity called the magnetic field. In such source-free regions, it follows that the magnetic field vector $\mathbf{B}$ can be represented as the gradient of a scalar magnetic potential V

$$\mathbf{B} = -\nabla V. \tag{12.3}$$

where the $-$ sign is included as a matter of convention. Since the magnetic field is also divergence-free

$$\nabla \cdot \mathbf{B} = 0 \tag{12.4}$$

the scalar potential V must be a solution of Laplace's equation

$$\nabla \cdot (\nabla V) = \nabla^2 V = 0 \tag{12.5}$$

In spherical geometry that is relevant for describing the Earth, Laplace's equation takes the form

$$\nabla^2 V = \frac{1}{r^2}\frac{\partial}{\partial r}\left(r^2\frac{\partial V}{\partial r}\right) + \frac{1}{r^2 \sin\vartheta}\frac{\partial}{\partial \vartheta}\left(\sin\vartheta\frac{\partial V}{\partial \vartheta}\right) + \frac{1}{r^2 \sin^2\vartheta}\frac{\partial^2 V}{\partial \lambda^2} = 0 \quad (12.6)$$

where r is the distance from the centre of the Earth, ϑ is the co-latitude measured from the north pole, and λ is the longitude, all defined in an Earth-centred, Earth-fixed, geographic reference frame.

Laplace's equation in spherical geometry (12.6) may be solved by parts (see e.g. Riley et al. 2006); two solutions with different radial dependence are possible, V^{int}, describing internal sources (e.g. due to currents originating in the core or due to the magnetized lithosphere), and V^{ext}, describing external sources (e.g. due to magnetospheric currents)

$$V = V^{\text{int}} + V^{\text{ext}} \quad (12.7)$$

12.2.4 Representation of the Field Due to Internal Sources

The solution for the potential due to internal sources takes the form

$$V^{\text{int}}(r, \vartheta, \lambda, t) = a \sum_{n=1}^{N_{\text{int}}} \sum_{m=0}^{n} \left[g_n^m(t)\cos m\lambda + h_n^m(t)\sin m\lambda\right]\left(\frac{a}{r}\right)^{n+1} P_n^m(\cos\vartheta) \quad (12.8)$$

where $a = 6371.2$ km is a reference radius, taken to be Earth's mean spherical radius, (r, ϑ, λ) are the geographic coordinates, P_n^m are Schmidt semi-normalized associated Legendre functions (e.g. Langel 1987; Winch et al. 2005) of degree n and order m, and $\{g_n^m, h_n^m\}$ are the Gauss coefficients describing the amplitude of the internal sources. In principle, the maximum degree N_{int} would be infinity if one wished to represent all possible details of the field structure, but for practical reasons the expansion is truncated at some finite maximum degree, beyond which the smallest wavelengths cannot be reliably retrieved. Illustrative examples of spherical harmonic functions $Y_n^m(\vartheta, \lambda)$, i.e. $\cos m\lambda\, P_n^m(\cos\vartheta)$ or $\sin m\lambda\, P_n^m(\cos\vartheta)$ are presented in Fig. 12.1. These are fundamental building blocks that may be combined to represent global functions on a spherical surface.

In addition to spatial dependence, accurate models of the main field must take into account the slow temporal or secular variation of the internal field. In standard models such as the International Geomagnetic Reference Field (IGRF), this is accounted for by linear interpolation between Gauss coefficients g_n^m defined at reference epochs.

In more advanced models, the time dependence is often represented using B-spline basis functions of order K (de Boor 1978; Bloxham and Jackson 1992) such that

$$g_n^m(t) = \sum_p g_n^{m,p} \mathscr{B}_{K,p}(t) \quad (12.9)$$

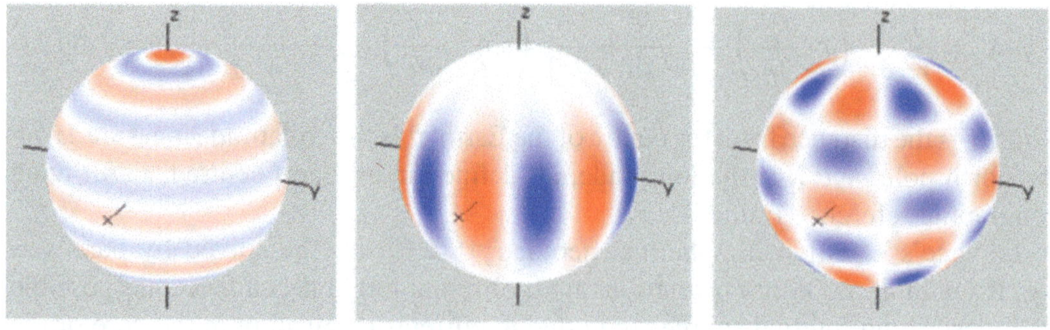

Fig. 12.1 Examples of spherical harmonic basis functions used to parameterize spatial structure in main field models. (Left), a zonal harmonic $n = 11$, $m = 0$ (centre) a sectorial harmonic $n = 6$, $n = 6$, (right), a general tesseral harmonic $n = 8$, $m = 4$

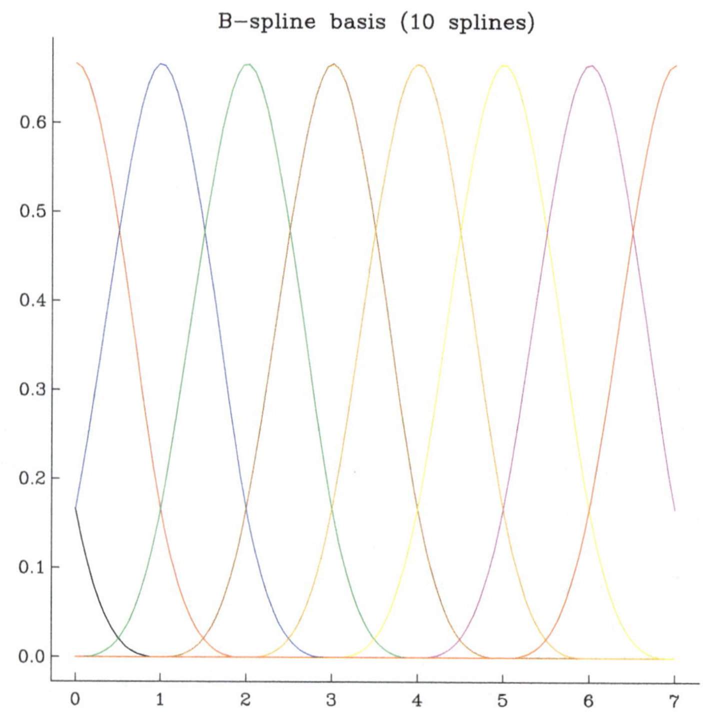

Fig. 12.2 Example of 10 cubic B-spline local basis functions that when combined with different weights can be used to represent the time-dependence of Gauss coefficients in internal field models. For further details, see Jackson and Finlay (2007)

where $g^m_{n,p}$ are a set of spline coefficients for each Gauss coefficient g^m_n, defined at knots p that span the time interval of interest. The B-spline basis functions $\mathscr{B}_{K,p}$ are piecewise polynomials of order K. Examples of B-spline basis functions of order $K = 4$ (i.e. cubic B-splines), that when combined with appropriate weights can reproduce the time-dependent signal of interest, are shown in Fig. 12.2. In advanced main models including the latest members of the CHAOS model series (Olsen et al. 2014; Finlay et al. 2016), order 6 B-splines are used, so that the resulting models can easily be differentiated twice in time to study the field acceleration.

12.2.5 Representation of the Field Due to External Sources

The basic parameterization of magnetic fields due to sources located external to observation points, within the potential field framework, takes the form

$$V^{\text{ext}}(r, \vartheta, \lambda, t) = a \sum_{n=1}^{N_{\text{ext}}} \sum_{m=0}^{n} \left[q_n^m(t) \cos m\lambda + s_n^m(t) \sin m\lambda \right] \left(\frac{r}{a} \right)^n P_n^m(\cos \vartheta) \tag{12.10}$$

Although mathematically sufficient, this representation does not account for the specific spatial structure and time dependence of the various magnetospheric current systems.

Maus and Lühr (2005) and Olsen et al. (2005) developed more useful parameterizations of the near-Earth external field including (i) a component expressed in the *Solar Magnetic (SM)* coordinate system (with its z axis parallel to the Earth's magnetic dipole axis and its y axis perpendicular to the plane containing the dipole axis and the Earth–Sun line) that represents well the geometry of the magnetospheric ring current, and (ii) a part in the *Geocentric Solar Magnetospheric (GSM)* coordinate system (with its x axis towards the Sun and its z axis being the projection of the Earth's magnetic dipole axis (positive North) on to the plane perpendicular to the x axis) that is more suitable for studying magnetospheric phenomena strongly influenced by the interplanetary magnetic field direction including magnetotail and magnetopause currents. Further details of these and other magnetic coordinate systems, including how to convert between them, are described by Laundal and Richmond (2017).

Taking as a specific example the parameterization of the external field in the latest version of the CHAOS model series (Olsen et al. 2014; Finlay et al. 2016), one may write the external potential as

$$\begin{aligned} V^{\text{ext}}(r, \vartheta, \lambda, t) &= a \sum_{n=1}^{N_{SM}} \sum_{m=0}^{n} \left[q_n^m \cos m T_d(t) + s_n^m \sin m T_d(t) \right] \left(\frac{r}{a} \right)^n P_n^m(\cos \vartheta_d(t)) \\ &+ a \sum_{n=1}^{N_{GSM}} q_n^{0,\text{GSM}} R_n^0(r, \vartheta, \lambda) \end{aligned} \tag{12.11}$$

where $\vartheta_d(t)$ and $T_d(t)$ are dipole co-latitude and dipole longitude, respectively, of the *(SM)* coordinate system. The upper line is the SM dependent part; it is truncated at degree $N_{SM} = 2$, and includes a special treatment of the $n = 1$ terms (see below). The part in GSM coordinates on the lower line is truncated at degree $N_{GSM} = 2$, but is restricted to order $m = 0$). Here the functions $R_n^0(r, \vartheta, \lambda)$ are modifications of the Legendre functions that explicitly account for field contributions induced in the electrically conducting mantle due to the wobble of the GSM z-axis with respect to the Earth's rotation axis. For a non-conducting Earth, these functions would be $R_n^0 = (r/a)^n P_n^0(\cos \vartheta_{GSM})$ where ϑ_{GSM} is co-latitude in the GSM coordinate system;

considering a plausible 1D model of mantle conductivity leads to a representation of R_n^0 similar to the expansion described by Maus and Lühr (2005).

In addition to the time dependence arising for the time variation of *SM* coordinates in an Earth-fixed, Earth-centred frame (which also results in an induced counterpart), the CHAOS model allows for additional time-dependence of the degree one field in SM coordinates, of the form

$$q_1^0(t) = \hat{q}_1^0 \left[\varepsilon(t) + \iota(t) \left(\frac{a}{r} \right)^3 \right] + \Delta q_1^0(t) \tag{12.12a}$$

$$q_1^1(t) = \hat{q}_1^1 \left[\varepsilon(t) + \iota(t) \left(\frac{a}{r} \right)^3 \right] + \Delta q_1^1(t) \tag{12.12b}$$

$$s_1^1(t) = \hat{s}_1^1 \left[\varepsilon(t) + \iota(t) \left(\frac{a}{r} \right)^3 \right] + \Delta s_1^1(t) \tag{12.12c}$$

where the terms in brackets describe field contributions due to the magnetospheric ring-current and their Earth-induced counterparts as given by the RC index (Olsen et al. 2014). RC is a ground-based index similar to D_{st} but with better baseline control and including a separation into internal and external parts $RC(t) = \varepsilon(t) + \iota(t)$ based on an a priori model of mantle conductivity. If RC were a perfect description of the magnetospheric field at satellite altitude then the values of the regression coefficients would be $\hat{q}_1^0 = -1$, $\hat{q}_1^1 = \hat{s}_1^1 = 0$ and the 'RC baseline corrections' Δq_1^0, Δq_1^1 and Δs_1^0 would vanish. The most recent version of the CHAOS model (Finlay et al. 2016) estimated such baseline corrections in bins of 5 days (for Δq_1^0) and 30 days (for Δq_1^1, Δs_1^1).

12.2.6 Using Data in the Magnetometer Frame: Co-estimation of Magnetometer Attitude

An important issue when using satellite magnetic field data is how to relate the measured field components, made in the reference frame of the magnetometer, $\mathbf{B}_{\text{VFM}}$, to the field components in the Earth-centred, Earth-fixed, North-East-Centre (NEC) frame predicted by global field models $\mathbf{B}_{NEC} = (-B_\vartheta, B_\lambda, -B_r)$. The two representations are related through a rotation that may be expressed in the form

$$\mathbf{B}_{\text{VFM}} = \underline{\underline{\mathbf{R}}}_1 \cdot \underline{\underline{\mathbf{R}}}_2 \cdot \underline{\underline{\mathbf{R}}}_3 \cdot \mathbf{B}_{\text{NEC}} \tag{12.13}$$

Here, the rotation matrix $\underline{\underline{\mathbf{R}}}_3$ rotates the magnetic field from the NEC system to the International Celestial Reference Frame (*ICRF*) and is derived from the satellite position and time (Seeber 2003), $\underline{\underline{\mathbf{R}}}_2$ rotates the magnetic field from the (*ICRF*) system to the Common Reference Frame (*CRF*) of the satellite's star tracker and is constructed from the attitude data collected by the star tracker, and finally $\underline{\underline{\mathbf{R}}}_1$ rotates from the star tracker *CRF* to the orthogonal magnetometer (VFM) frame. $\underline{\underline{\mathbf{R}}}_3$ and

$\underline{\underline{\mathbf{R}}}_2$ are typically well known and in many cases it is also assumed that $\underline{\underline{\mathbf{R}}}_1$ is known exactly and in advance of collecting the data. However, in the most advanced main field models including the CHAOS series, rather than being assumed in advance, the Euler angles defining $\underline{\underline{\mathbf{R}}}_1$ are instead co-estimated as part of the modelling procedure (Olsen et al. 2006). In this case, the relation between the measurements in the VFM frame and the model parameters (coefficients of the internal potential from (12.1) and (12.9), of the external potential from (12.11) and (12.12), and the Euler angles defining $\underline{\underline{\mathbf{R}}}_3$ from (12.13), is nonlinear, and model estimation becomes a nonlinear inverse problem.

12.2.7 Model Estimation: Solution of the Inverse Problem

The relationship between the predicted vector magnetic field data in the vector flux-gate magnetometer (VFM) frame, listed as a vector $\mathbf{d}_{\text{mod}}$ and the model parameters $\mathbf{m}$ may compactly be written in the form

$$\mathbf{d}_{\text{mod}} = \mathbf{g}(\mathbf{m})$$

where $\mathbf{g}(\mathbf{m})$ denotes the non-linear dependence on the model parameters $\mathbf{m}$.

Determination of the model parameters from the magnetic observations $\mathbf{d}_{\text{obs}}$ is a non-linear inverse problem. Furthermore, since it involves downward continuation of observations from satellite altitude, it is also an ill-conditioned problem. Moreover, since there are field sources that vary rapidly in space and time that cannot be captured by the model (i.e. the model is incomplete, for example, failing to account for auroral and polar-cap currents) the residuals between model predictions and the data are often long-tailed and not simply Gaussian distributed.

A well-proven technique for finding suitable solutions to this difficult estimation problem is the regularized, robust, iteratively reweighted least squares (IRLS) algorithm. This approach, which is adopted in the construction of the CHAOS field model series, involves iteratively minimizing a cost function of the form

$$\mathscr{J} = \mathbf{e}^T \underline{\underline{\mathbf{W}}}\mathbf{e} + \lambda_3 \mathbf{m}^T \underline{\underline{\Lambda}}_3 \mathbf{m} + \lambda_2 \mathbf{m}^T \underline{\underline{\Lambda}}_2 \mathbf{m}, \tag{12.14}$$

where $\mathbf{e} = \mathbf{d}_{\text{obs}} - \mathbf{d}_{\text{mod}} = \mathbf{d}_{\text{obs}} - \mathbf{g}(\mathbf{m})$ is the vector of residuals (observations minus the model predictions). $\underline{\underline{\mathbf{W}}} = \underline{\underline{\mathbf{C}}}^{-1/2}\,\underline{\underline{\mathbf{H}}}\,\underline{\underline{\mathbf{C}}}^{-1/2}$ is a data weighting matrix, derived from a pre-specified data error covariance matrix $\underline{\underline{\mathbf{C}}}$ and $\underline{\underline{\mathbf{H}}}$, a diagonal matrix of residual-dependent weights, for example, $H_{jj} = \min(1, c_H\sigma_j/|e^i_j|)$ with $c_H = 1.5$, e^i_j the residual of the jth data at the previous (i.e. ith) iteration (e.g. Constable 1988; Olsen 2002) and σ_j is an a priori estimate of data uncertainty for the jth datum. H_{jj} are known as Huber weights, and they permit robust solutions to be obtained even in the presence of long-tailed error distributions.

$\mathbf{m}^T \underline{\underline{\Lambda}}_3 \mathbf{m}$ is a quadratic norm measuring the mean squared magnitude of the third time derivative of the radial field $\left|\frac{\partial^3 B_r}{\partial t^3}\right|$, integrated over the surface of the outer core Ω_c (radius $c = 3485$ km) and time-averaged over the time span of the model between t_{start} and t_{end}

$$\left\langle \left|\frac{\partial^3 B_r}{\partial t^3}\right|^2 \right\rangle = \frac{1}{(t_{start} - t_{end})} \int_{t_{start}}^{t_{end}} \int \left|\frac{\partial^3 B_r}{\partial t^3}\right|^2 d\Omega_c \, dt = \mathbf{m}^T \underline{\underline{\Lambda}}_3 \mathbf{m} \tag{12.15}$$

Finally, $\mathbf{m}^T \underline{\underline{\Lambda}}_2 \mathbf{m}$ is a similar measure of the mean square magnitude of the second time derivative or secular acceleration of the radial field at the core surface $\left|\frac{\partial^2 B_r}{\partial t^2}\right|$, but evaluated only at the model endpoints t_{start} and t_{end}. λ_3 and λ_2 are regularization parameters for these two norms, specifying their relative importance and chosen to ensure a good balance between fitting field time changes (for example, as seen at ground observatories) while at the same time ensuring that spurious temporal variations of the model are suppressed.

Since (12.14) is a (weakly) nonlinear function of the model parameters, an iterative approach is adopted to linearize it, about the present model $\mathbf{m}_i$. The model is iteratively updated until convergence of the model parameters (e.g. Gubbins 2004), such that

$$\mathbf{m}_{i+1} = \mathbf{m}_i + \delta\mathbf{m}_i$$

where

$$\delta\mathbf{m}_i = \left[\underline{\underline{\mathbf{G}}}_i^T \underline{\underline{\mathbf{W}}}\, \underline{\underline{\mathbf{G}}}_i + \lambda_3 \underline{\underline{\Lambda}}_3 + \lambda_2 \underline{\underline{\Lambda}}_2\right]^{-1} \left[\underline{\underline{\mathbf{G}}}_i^T \underline{\underline{\mathbf{W}}}(\mathbf{d}_{\text{obs}} - \mathbf{g}(\mathbf{m}_i)) - \lambda_3 \underline{\underline{\Lambda}}_3 \mathbf{m}_i - \lambda_2 \underline{\underline{\Lambda}}_2 \mathbf{m}_i\right],$$

$$\text{and} \quad \underline{\underline{\mathbf{G}}}_i = \left.\frac{\partial \mathbf{g}(\mathbf{m})}{\partial \mathbf{m}}\right|_{\mathbf{m}=\mathbf{m}_i}. \tag{12.16}$$

Given its importance, it is worth explicitly stating here how elements of $\underline{\underline{\mathbf{G}}}_i$ are computed. As example, constructing three rows of $\underline{\underline{\mathbf{G}}}_i$ associated with three components of a vector magnetic field measurement in the VFM frame involves (i) taking the spherical polar gradients of the expansions for internal and external potentials (12.8) and (12.11), (ii) multiplying these with the Euler rotation matrices $\underline{\underline{\mathbf{R}}}_1 \cdot \underline{\underline{\mathbf{R}}}_2 \cdot \underline{\underline{\mathbf{R}}}_3$ from the previous iteration to rotate to the VFM frame, then (iii) taking the derivatives with respect to each model parameter in turn, such that each column of $\underline{\underline{\mathbf{G}}}_i$ refers to a derivative with respect to a different model parameter), and evaluating the elements using the model parameters from the previous iteration $\mathbf{m}_i$.

12.3 Use of Field Gradients and Multi-satellite Data in Main Field Modelling

Above we have described the standard approach to geomagnetic field modelling, based on the observed vector or scalar field components. The launch of the *Swarm* multi-satellite constellation (Friis-Christensen et al. 2006; Olsen et al. 2016b), has opened exciting new possibilities for using approximate field gradients, estimated via along-track and across-track differences, in main field modelling. In this section, we outline these new techniques, focusing on how to construct suitable estimates of the field gradients and on how to deal with data from multiple satellites within a single inversion.

12.3.1 Estimates of Field Gradients: Approximation by Along-Track and Across Track Differences

The constellation of the *Swarm* trio of satellites, with two satellites (Alpha and Charlie) flying close together and a third (Bravo) flying at higher altitude and drifting in local time with respect to the lower pair, enables estimates both along-track and cross-track gradients of the geomagnetic field to be made from space. In particular, considering differences in the field recorded by the lower pair permits the cross-track field gradient to be estimated for the first time. This is extremely valuable for constraining small-scale east–west structures in the lithospheric field (Maus et al. 2006a; Thébault et al. 2016), although it provides no constraint on zonal ($m = 0$) and little constraint on near zonal (small m) spherical harmonic components of the field.

Although the *Swarm* constellation does not have an along-track satellite pair (which would have an approximately north–south orientation at mid and low latitudes), one can instead use along-track differences from a single satellite to estimate the gradient in this direction, with the assumption the field does not change appreciably over the time taken to move between the locations differenced. A typical time between the locations differenced is 15 s (Olsen et al. 2015), much shorter than the time scale of large-scale magnetospheric field changes. This corresponds to an along-track spatial separation of about 115 km. (Kotsiaros et al. 2014) have explored the impact on lithospheric field models of using different time separation when computing along-track gradient estimates. There is clearly a trade-off between the signal amplitude (smaller for shorter time differences) and noise (larger for larger time-differences due to the breakdown of the stationarity assumption). The optimal time separation will also depend on the target wavelengths, on the geomagnetic latitude and on geomagnetic conditions (quiet or storm-time, dark or sunlit). There is certainly room to better optimize the calculation of field gradient estimates. Nonetheless, use of field gradients has already proven to be of great value in deriving high-resolution models of the core field (Olsen et al. 2015; Finlay et al. 2016) and lithospheric field (Sabaka et al. 2015; Kotsiaros 2016; Olsen et al. 2017).

How can one use field gradient estimates in the construction of main field models? The approach is a straightforward extension of standard procedure of performing least-square fits. One simply minimizes the square of the residuals between the observed and modelled field differences, which may be either differences of vector components or differences of scalar intensity values.

If we represent residuals between the observed and modelled field differences by

$$\Delta\mathbf{e} = \Delta\mathbf{d} - \Delta\mathbf{d}_{\text{mod}} \tag{12.17}$$

where $\Delta\mathbf{d} = \mathbf{d}(\mathbf{r_1}, t_1) - \mathbf{d}(\mathbf{r_2}, t_2)$ and $\Delta\mathbf{d}_{\text{mod}} = \mathbf{d}_{\text{mod}}(\mathbf{r_1}, t_1) - \mathbf{d}_{\text{mod}}(\mathbf{r_2}, t_2)$ where $(\mathbf{r_1}, t_1)$ is the location and time of the first datum contributing to the field difference while $(\mathbf{r_2}, t_2)$ is the location and time of the second datum that is subtracted from the first. Minimizing the sum of the squares of the field difference residuals $\Delta\mathbf{e}$ then results in an additional contribution to the cost function (12.14) from fitting field differences that may be written

$$\Delta\mathbf{e}^T \underline{\underline{W_\Delta}} \Delta\mathbf{e} \tag{12.18}$$

where

$$\Delta\mathbf{d}_{\text{mod}} = \mathbf{g}_\Delta(\mathbf{m}) \quad \text{with} \quad \mathbf{g}_\Delta(\mathbf{m}) = \mathbf{g}|_{(\mathbf{r_1},t_1)}(\mathbf{m}) - \mathbf{g}|_{(\mathbf{r_2},t_2)}(\mathbf{m})$$

and similar to previously $\underline{\underline{W_\Delta}} = \underline{\underline{C_\Delta}}^{-1/2} \underline{\underline{H_\Delta}}\, \underline{\underline{C_\Delta}}^{-1/2}$ is a data weighting matrix, derived from a specified data error covariance matrix for the difference data $\underline{\underline{C_\Delta}}$ and $\underline{\underline{H_\Delta}}$, a diagonal matrix of 'Huber' weights $H_{jj} = \min(1, c(\sigma_\Delta)_j/(\Delta e^i_j))$ for the difference data that depends on the amplitude of the residuals at the previous iteration Δe^i_j, $(\sigma_\Delta)_j$ is the a priori difference data uncertainty estimate and c_H is a constant, usually set to 1.5.

In the field models CHAOS-6 (Finlay et al. 2016), SIFM (Olsen et al. 2015) and SIFM+ (Olsen et al. 2016a), along-track (or approximately north–south at the equator so denoted north–south below) gradients were approximated by the differences $\Delta B_{\text{NS}} = \pm[B_k(t_k, r_k, \theta_k, \phi_k) - B_k(t_k + 15\,\text{s}, r_k + \delta r, \theta_k + \delta\theta, \phi_k + \delta\phi)]$ of subsequent data, measured by the same satellite k, 15 s later, corresponding to an along-track distance of $\approx$115 km ($\approx 1^\circ$ in latitude near the equator) for the *Swarm* satellites. Here B could be the scalar intensity F or any of the geocentric field components (B_r, B_θ, B_ϕ). The sign of the difference was chosen positive if $\delta\theta > 0$, otherwise negative. The choice of 15 s was found to give a reasonable amplitude of internal field signal while being sufficiently short that much of the large-scale external field is unchanged, so therefore removed on taking the difference. 15 s differences have the advantage of involving differences over lengths similar to the East–West spacing between *Swarm*'s lower satellite pair.

To approximate the East–West gradients the above studies used the difference $\delta B_{\text{EW}} = \pm[B_A(t_1, r_1, \theta_1, \phi_1) - B_C(t_2, r_2, \theta_2, \phi_2)]$, between field components, or the scalar field, measured by the two satellites *Swarm* Alpha and *Swarm* Charlie. Here $t_i, r_i, \theta_i, \phi_i, i = 1 - 2$ are time, radius, geographic co-latitude and longitude of the two observations. The sign of the difference was chosen such that $\delta\phi = \phi_1 - \phi_2 > 0$. For each observation B_A (from *Swarm* Alpha) the corresponding value B_C (from *Swarm* Charlie) was chosen to be that closest in co-latitude θ, with the requirement that $|\delta t| = |t_1 - t_2| < 50$ s. Note there is a time delay between the ground-tracks of *Swarm* Alpha and *Swarm* Charlie to avoid collisions at the pole, so simultaneous values from each satellite do not provide an estimate of the east–west gradient. Taking *Swarm* Charlie values slightly delayed, typically by around 10 s, allows differences to be taken that are very close to the desired East–West configuration. Note that this again requires that large-scale field be stationary over approximately 10 s if they are to cancel on taking the difference.

Turning to the estimation of the field model, including field differences in the inversion simply requires that we augment the vectors and matrices used for conventional field modelling, so that these also contain the data and associated entries of the design matrices appropriate for vector and scalar field differences. The algorithm for updating the model using this augmented dataset (containing both field values and field differences) may be written as

$$\delta \mathbf{m}_i = \left[\underline{\underline{\mathbf{G}}}_i^{aug\,T} \, \underline{\underline{\mathbf{W}}} \, \underline{\underline{\mathbf{G}}}_i^{aug} + \lambda_3 \underline{\underline{\Lambda}}_3 + \lambda_2 \underline{\underline{\Lambda}}_2 \right]^{-1} \left[\underline{\underline{\mathbf{G}}}_i^{aug\,T} \, \underline{\underline{\mathbf{W}}} (\mathbf{d}_{\text{obs}}^{aug} - \mathbf{d}_{\text{mod,i}}^{aug}) - \lambda_3 \underline{\underline{\Lambda}}_3 \mathbf{m}_i - \lambda_2 \underline{\underline{\Lambda}}_2 \mathbf{m}_i \right]$$

$$\text{where } \underline{\underline{\mathbf{G}}}_i^{aug} = \begin{bmatrix} \left. \frac{\partial \mathbf{g}(\mathbf{m})}{\partial \mathbf{m}} \right|_{\mathbf{m}=\mathbf{m}_i} \\ \left. \frac{\partial \mathbf{g}_\Delta(\mathbf{m})}{\partial \mathbf{m}} \right|_{\mathbf{m}=\mathbf{m}_i} \end{bmatrix}, \; \mathbf{d}_{\text{obs}}^{aug} = \begin{bmatrix} \mathbf{d}_{\text{obs}} \\ \Delta \mathbf{d}_{\text{obs}} \end{bmatrix}, \text{ and } \mathbf{d}_{\text{mod,i}}^{aug} = \begin{bmatrix} \mathbf{d}_{\text{mod,i}} \\ \Delta \mathbf{d}_{\text{mod,i}} \end{bmatrix} \quad (12.19)$$

12.3.2 Information Content of Field Gradient Estimates

Olsen et al. (2017) have explored in detail how much field models can be improved by using along-track (north–south) and across-track (east–west) gradient estimates, considering both CHAMP and *Swarm* data in the context of studying the lithospheric field. Figure 12.3, reproduced from their study, shows theoretical model variances of spherical harmonic coefficients as a function of degree (y-axis) and order (x-axis), based on the positions of CHAMP and *Swarm* data, and considering the impact of both vector data and vector gradients, i.e. north–south and east–west gradients. Blue colours show well determined coefficients, yellow colours indicate poorly determined model coefficients.

The results in Fig. 12.3 are based on a simplified linearized version of the inverse problem, where the Euler angles are assumed known and only the internal potential is considered. The plots show diagonal entries of the formal model covariance matrix $(\mathbf{G}^T\mathbf{W}\mathbf{G})^{-1}$, where $\mathbf{G}$ here is simply the matrix for the linear forward problem

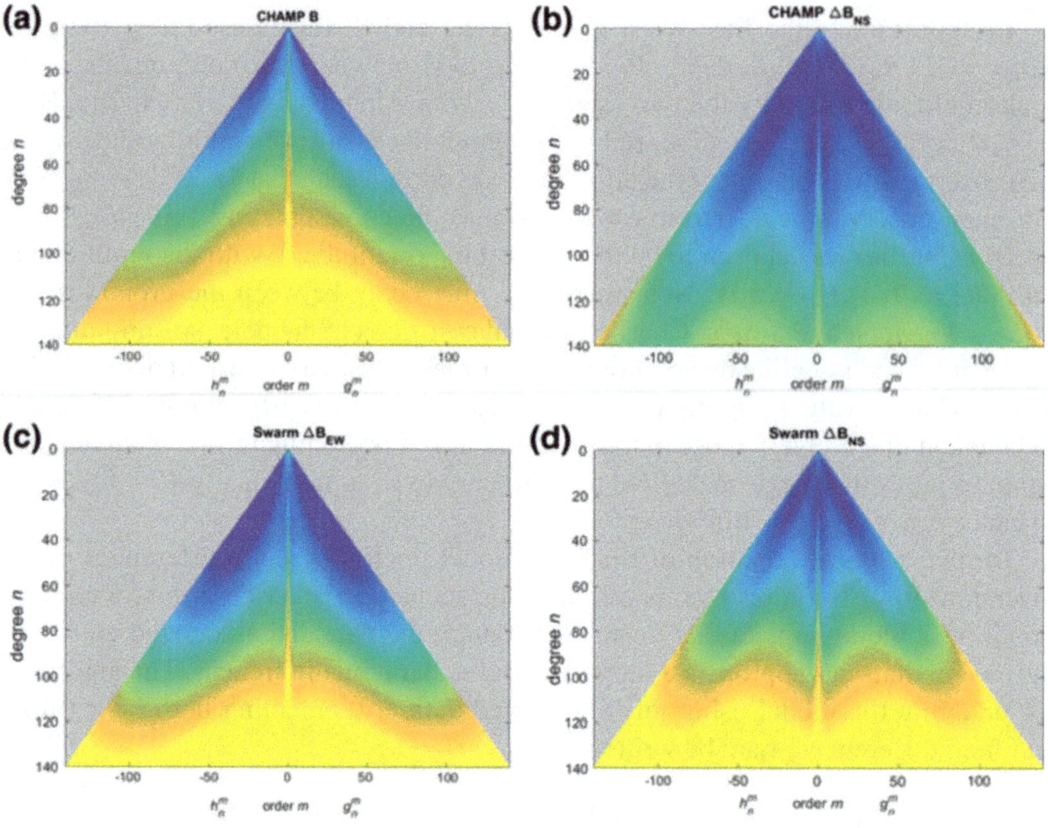

Fig. 12.3 Variance of the spherical harmonic coefficients g_n^m, h_n^m derived with various choices of input data sets. Blue corresponds to well determined coefficients, yellow corresponds to poorly determined coefficients. Reproduced from Olsen et al. (2017)

connecting the satellite data (either vector field or vector field differences from either CHAMP and *Swarm*) to the spherical harmonic model coefficients, and where **W** is a matrix of data weights.

North–south or along-track gradients of CHAMP data clearly provide a much improved resolution (i.e. lower model variances) for the high degree spherical harmonic coefficients, especially at degrees 120–140 which are poorly constrained by field data alone. North–south differences do not, however, constrain well the sectorial and near sectorial spherical harmonics. East–West differences from *Swarm* provide very valuable new constraints on these coefficients. The information provided by *Swarm* on the high degree field will, of course, be enhanced as the satellites descend to lower altitudes (the greater information provided by CHAMP north–south differences compared to *Swarm* north–south differences is purely due to the present higher altitude of the *Swarm* satellites). The poorer determination of high degree near-zonal coefficients is a consequence of the polar gap in the data distribution. Field differences are seen to provide useful information, not just on high degree coefficients associated with the lithospheric field signal, but also on low degree coefficients describing the core field; moreover, unmodelled large-scale magnetospheric fields are effectively suppressed when considering field gradient estimates, which increases the signal to noise ratio.

12.3.3 Examples of Field Gradient Data and Their Interpretation

In order to illustrate the form of signals seen in field gradient (i.e. north–south and east–west difference) data, and how these are related to field data, we present here scalar field and scalar field difference signals collected by the *Swarm* satellites (north–south difference, labelled NS and east–west differences, labelled EW, divided by the separation distance in kilometers between the measurements) on example day-side (Fig. 12.4) and night-side (Fig. 12.5) half orbits. We present the signal remaining after the progressive removal of the estimates of the core field (top), core and lithospheric fields (middle), and core, lithospheric and magnetospheric fields (bottom) as a function of quasi-dipole (QD) latitude. These examples are taken from Olsen and Stolle (2017).

The selected day-side orbit shown in Fig. 12.4 has an equatorial Local Time crossing at 12:12 LT, and is from 2 May 2014, which was a geomagnetically quiet day (Kp $<$ 1+ and Dst $>$-13 nT). The left column presents observations from *Swarm* Alpha; the middle column presents estimates of the East–West field difference measured between *Swarm* Alpha and *Swarm* Charlie, divided by the distance between the two spacecraft; the right column shows an estimate of the North–South gradient, obtained from 15-second north–south differences of *Swarm* Alpha divided by d = 141 km (which is the distance of two satellite measurements taken 15 s apart). For each of the three columns, the blue curves show the difference $F = F_{obs} - F_{mod}$ between the observed magnetic intensity F_{obs} and various model values F_{mod} predicted by CHAOS-6 model, whereas the red curves show predictions of some additional parts of the model. The yellow curve in the bottom left panel shows F_{obs} from *Swarm* Charlie. See figure caption for more details. For this day-side half orbit, the signature of the equatorial electrojet is clearly seen at the magnetic equator, especially in the north–south gradient (right column), resulting from the depression it causes in the field intensity (see left column). The signature of the Sq ionospheric current system is also seen as minima in field intensity close to 30° QD latitude. Large signals remain in the polar regions even after removal of estimates of the core, lithospheric, magnetospheric and Sq fields, especially in sunlit northern hemisphere, where large gradients are seen in both the east–west and north–south differences.

Next, we move to an example night-side half orbit, more typically of the data used in main field modelling. As shown in Fig. 12.5, there is no longer any equatorial electrojet signal at low latitudes or any Sq signal at mid-latitudes. There are, however, some recognizable signals remaining, even after the removal of the core, lithosphere, magnetospheric and Sq-induced signals (note the latter is nonzero, even in dark

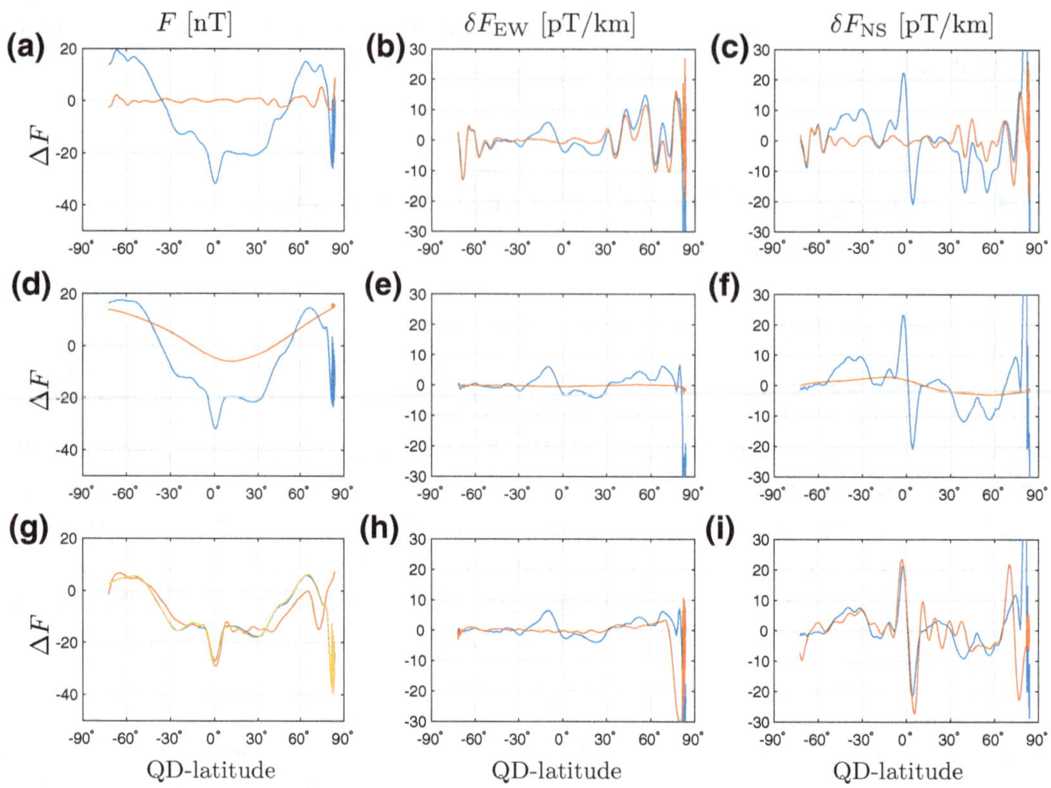

Fig. 12.4 Adapted from Olsen and Stolle (2017). Magnetic field intensity residuals for the day-time part of Swarm orbit number 2464 of 2 May 2014 versus QD latitude. Equator crossing at 18:43:04 UT, corresponding to 12:12 Local Time. **a–c**: The blue curve shows the difference F = $F_{obs} - F_{core}$ between observed magnetic intensity F_{obs} and the core field part F_{core} as given by the CHAOS-6 model. The red curve shows the lithospheric field model predictions. **d–f**: The blue curve presents the difference between the two curves of panels (**a–c**), i.e. the observed values minus model values for core and lithosphere. The red curve shows the modelled contributions of magnetospheric currents. **g–i**: The magnetic field intensity after removal of core, lithospheric and magnetospheric model values (shown by the red curves in panels **d–f**) is shown in blue. The red curves here present ionospheric current contributions as given by the CM5 model. Left panel shows values for Swarm Alpha; middle panel presents East–West gradients based on data from Swarm Charlie minus Swarm Alpha; right panel shows north–south gradients of Swarm Alpha

regions). This is especially evident in the field differences, for example there are characteristic spikes around QD latitude 15° south in the EW differences that are likely due to ionospheric F-region currents related to steep plasma gradients at post-sunset times (these are also obvious in electron density data collected by the *Swarm* satellites, not shown). Ionisation anomalies after sunset are frequently affected by plasma density irregularities (sometimes called 'bubbles') close to ±10° QD latitude that produce magnetic signatures of a few nanotesla; these anomalies have small length scale, so can be different between *Swarm* Alpha and Charlie, that results in clear signatures in the EW differences. Once again the largest unmodelled signals after the removal of the core, lithospheric, magnetospheric and Sq-induced parts, both for field data and field gradient estimates, are found in the polar regions, particularly in

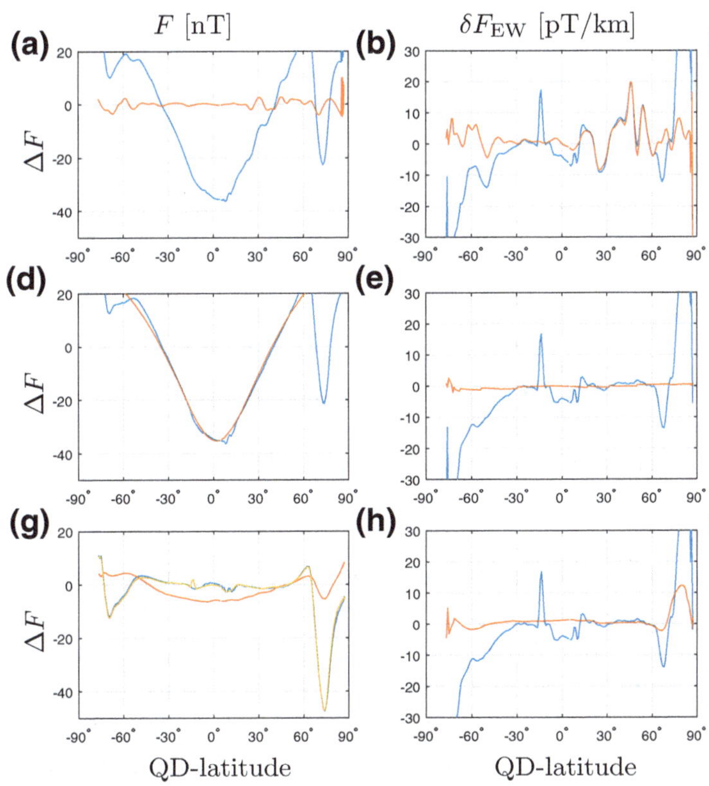

Fig. 12.5 Adapted from Olsen and Stolle (2017). Orbit number 5151 on 25 October 2014. Equator crossing at 01:16:09 UT, corresponding to 20:30 Local Time. Same format as Fig. 12.4

the summer (northern) hemisphere in the example presented here. These unmodelled polar signals are presently the major challenge facing main field modellers (e.g. Finlay et al. 2017).

12.3.4 Simultaneous Inversion of Data from Multiple Satellites

Above we have seen the complexity of the signals contained in low-Earth-orbit satellite data, and how the signal from various sources are either suppressed or enhanced when considering field gradient estimates. A crucial step in using all this information, and in combining such information from multiple satellite missions, is defining suitable data error budgets through the weight matrix **W**. For example, in the recent lithospheric field model LCS-1 (Olsen et al. 2017), an error budget for each component of gradient data for each satellite was developed as a function of quasi-dipole latitude, since many unmodelled near-Earth current systems are organized by the main field geometry. Data uncertainties σ (i.e. the square root of the diagonal entries

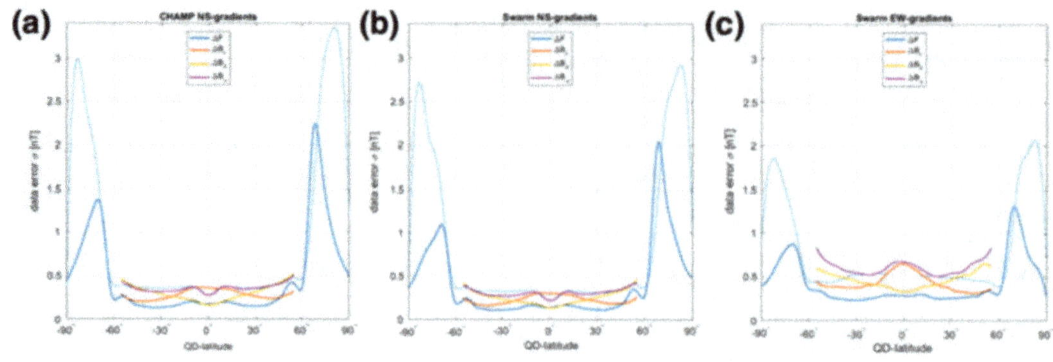

Fig. 12.6 Assigned data uncertainties σ for CHAMP(left, north–south gradients) , and *Swarm* (centre, north–south gradients, right, East–West gradients), versus quasi-dipole latitude. The thin blue line represents scalar gradient data under sunlit conditions. Adapted from (Olsen et al. 2017)

in the data covariance matrix **C** that along with Huber weights H_{jj} make up **W**), as employed in the LCS-1 model, are presented in Fig. 12.6.

Separate uncertainty estimates for CHAMP north–south gradients, *Swarm* north–south gradients and *Swarm* east–west gradients are specified. Uncertainties for the scalar field are shown separately for sunlit and dark regions. These uncertainties were derived by binning the data residuals relative to the CHAOS-6 field model, and estimating the standard deviation σ using a robust (Huber weighting) approach within bins of 5° QD latitude. The largest uncertainties occur in the polar regions, in particular for the sunlit parts, due to the enhanced ionospheric conductivity in these regions, and related magnetosphere-ionosphere coupling. The estimated uncertainties in *Swarm* inter-satellite East–West differences are generally larger than the north–south single satellite differences, and there is also interesting evidence for distinct unmodelled signals in the East–West differences at low quasi-dipole latitudes.

Using these uncertainties within the main or lithospheric field inverse problem essentially downweights data from the polar regions that is more likely to be contaminated by the signature of currents not included in the main or lithospheric field model. Data uncertainties allocated from CHAMP and *Swarm* turn out to be rather similar, showing similar trends as a functions of quasi-dipole latitude. It is also possible to define data and uncertainties from other satellite missions (Ørsted, SAC-C, DMSP etc.), where the uncertainties can be much larger, particularly with regard to how attitude errors influence our determination of the vector field (Olsen et al. 2006).

The error budgets presented here are still rather crude. In reality data errors with respect to internal field modelling are correlated in both space and time due to the structured nature of the unmodelled magnetospheric and ionospheric currents. In particular, the data are correlated along track (Lowes and Olsen 2004) which is why considering along-track or north–south field differences (which decorrelates this error source) is such an advantage. In addition, measurements from similar quasi-dipole latitudes and similar local times are likely to have correlated errors that are not presently taken into account.

12.4 The Internal Field as Seen by the Swarm Multi-satellite Mission

Having now set out the techniques used to construct advanced models of the internal field from multi-satellite data and gradient estimates, we now move on to give a brief summary the latest knowledge from such models.

For many interested in the Earth's magnetic field, either at present its present state or its changes over the past century, the IGRF is a well known and reliable source of information. It is an IAGA/IUGG endorsed model, produced by a international group of scientists every five years from candidate models. It describes the main field up to spherical harmonic degree $n = 13$ and the linear rate of change of the field for the upcoming five years up to degree $n = 8$. The most recent 12th-generation update of IGRF (Thébault et al. 2015b) used data from CHAMP, *Swarm* and ground observatories to provide estimates of the field in 2010, 2015 and a predicted field change for 2015–2020. The advantage of IGRF is that it is an internationally agreed reference. However, it fails to describe the small scale lithospheric field, and it does not catch nonlinear secular variation, including geomagnetic jerk events.

For applications in detailed studies of ionospheric current systems, advantages have been documented in reducing data using more advanced field models that include the small-scale lithospheric field, estimates of the large magnetospheric field, and that follow fast changes in the core field (Stolle et al. 2016; Alken 2016). Such advanced field models include the POMME model developed by Maus and co-workers (Maus et al. 2005, 2006b, 2010), the GRIMM model produced by Lesur and co-workers (Lesur et al. 2008, 2010, 2015a) and the CHAOS model produced by Olsen and colleagues (Olsen et al. 2006, 2009, 2010, 2014; Finlay et al. 2015, 2016). The Comprehensive model series, developed by Sabaka and co-workers (Sabaka et al. 2002, 2004, 2013, 2015), takes an alternative approach and seeks to co-estimate not only the internal field but as far as possible all near-Earth field sources, including ionospheric and oceanic tidal sources. For further details on these models and their differences, the interested reader should consult the above references.

Here, we present the current state of knowledge of the core field, as determined from the latest *Swarm* data in the CHAOS-6 field model (Finlay et al. 2016), and a recent image of the global lithospheric field, from the LCS-1 field model (Olsen et al. 2017), based on the latest data from CHAMP and *Swarm*.

12.4.1 The Core Field

The main part of the geomagnetic field is generated within the liquid iron core of the Earth. Knowledge of the core dynamo process is obtained by downward continuing the internal part of field models to the outer edge of the source region, at the core-mantle boundary (a radius of approx. 3480 km). In doing this, it is assumed that there are no current sources in the mantle on the time scales studied. Maps of the radial component of the magnetic field up to spherical harmonic degree 13 (above which the lithospheric field dominates so we cannot downward continue to the core), as well as of the radial SV and radial SA to degree 16 (the limit of what could be reliably estimated in CHAOS-6) at the core-mantle boundary in 2015 from the CHAOS-6 model are presented in Fig. 12.7. The field in 2015 is determined from vector and scalar field measurements, and also differences of along-track and across-track field measurements made by *Swarm*, as well as ground observatory vector field data (Finlay et al. 2016). Movies showing the time changes of such maps are available at www.spacecenter.dk/files/magnetic-models/CHAOS-6.

The radial field at the core-mantle boundary is characterized by high latitude flux concentrations over Canada and Siberia, and similarly in the Southern hemisphere under the edges of Antarctica towards South America and Australia. It is these features that give rise to the first-order dipolar structure of the geomagnetic field. Other striking features include a train of flux concentrations at low latitude under the Western hemisphere that have been observed to move westwards since the advent of continuous satellite observations in 1999, and the large concentration of reversed flux in the Southern hemisphere.

Turning to the time derivative of the field, known as the secular variation (SV), we find that regions of intense radial SV at the core surface occur close to edges of patches of strong radial field. Intense SV is observed in 2015 to lie in a broad band equatorward of 30° latitude between longitudes 100°E and 90°W and is particularly associated with the westward movement of the intense low latitude flux patches. There is also a well-localized negative–positive–negative series of three patches of radial SV visible under Alaska and Siberia; this appears to be a consequence of the rapid westward movement of intense high latitude radial field concentrations. The SV is also generally high in the Asian longitudinal sector 60°–120°E.

The second time derivative of the radial field, or secular acceleration (SA) at the core-mantle boundary in 2015, displays a prominent positive–negative pair of foci under India–South East Asia, and a series of strong radial SA patches of alternating sign in the region under northern South America, as well as a positive–negative pair at high northern latitudes under Alaska–Siberia. In both the radial SV and SA, there is a striking absence of structure in the southern polar region (Holme et al. 2011; Olsen et al. 2014). Although the Pacific region shows weak radial SV (again see Holme et al. 2011; Olsen et al. 2014), in 2015, there was strong radial SA in the central Pacific, consistent with the aftermath of the jerk observed in 2014 at Hawaii. The SA also changes dramatically on sub-decadal time scales (Chulliat and Maus 2014; Chulliat et al. 2015; Finlay et al. 2015), exhibiting a series of pulses in amplitude. CHAOS-6

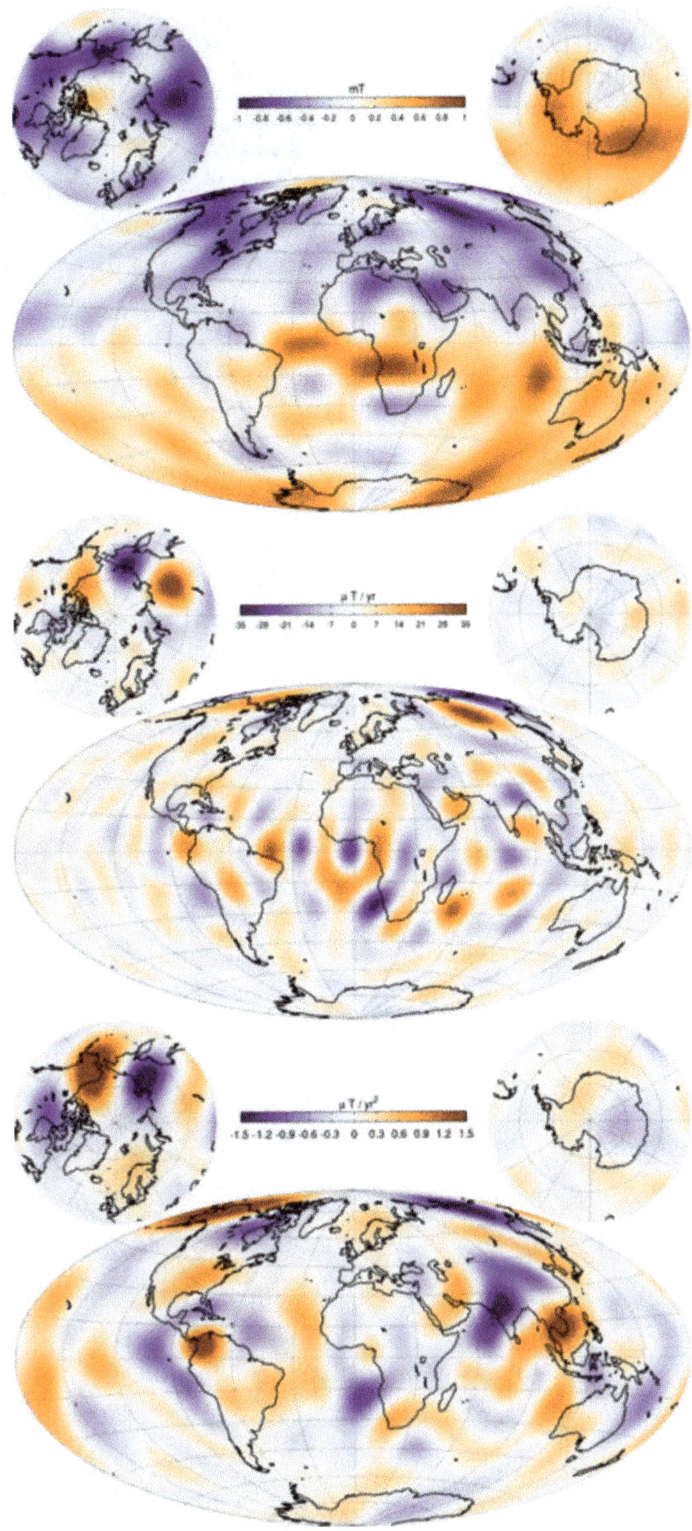

Fig. 12.7 Radial component of the main field at the core-mantle boundary (radius 3480 km) in 2015. Top (radial field up to degree 13) Middle (secular variation of radial field up to degree 16), Bottom (secular acceleration of the radial field up to degree 16). From the CHAOS-6 field model (Finlay et al. 2016)

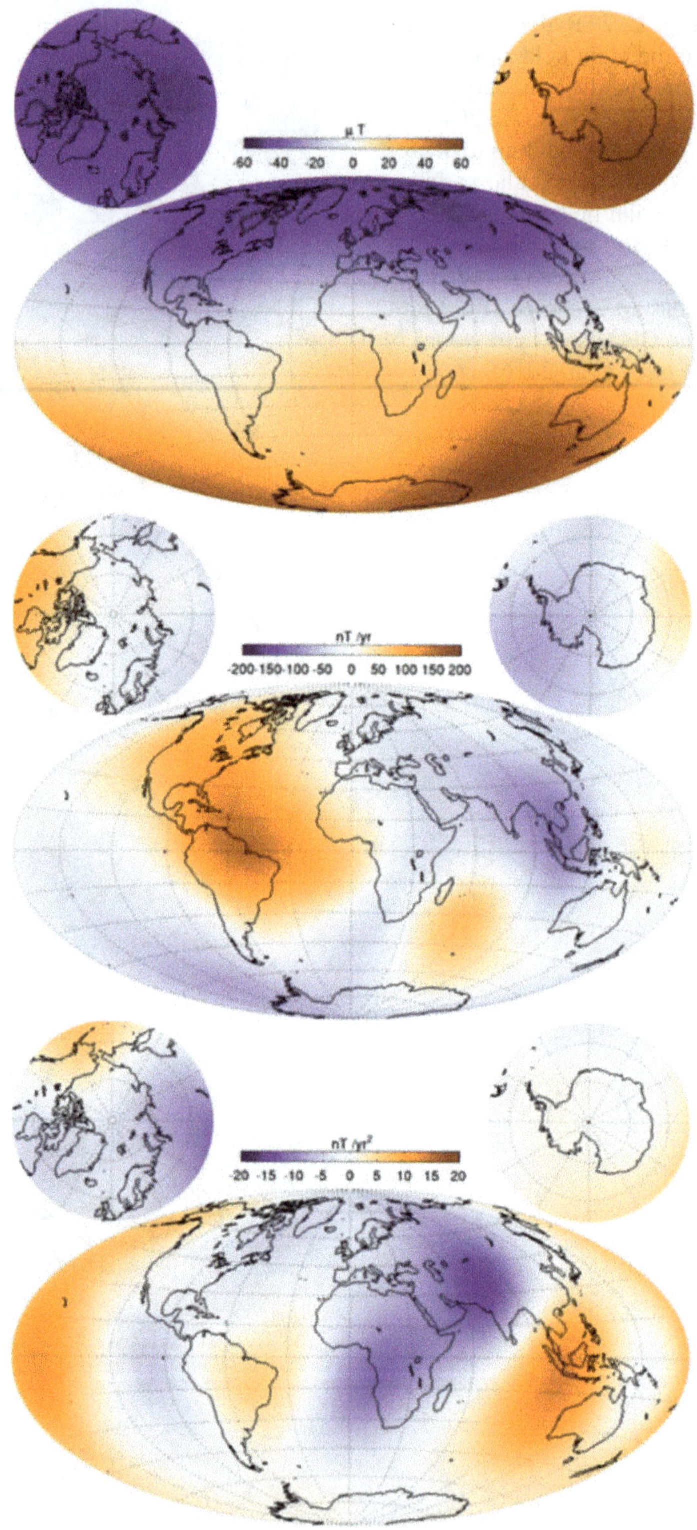

Fig. 12.8 As for Fig. 12.7, but with CHAOS-6 evaluated at satellite altitude of 400 km

shows pulses of SA around 2006, 2009.5 and 2013. Maps and movies of the radial SA at the core surface also show recurring oscillations at particular locations, for example, under northern South America around 40°W close to the equator, and high amplitude SA is often found around longitude 100°E.

Figure 12.8 presents the radial field, its SV and SA, from CHAOS-6, at a typical low-Earth-orbit satellite altitude of 400 km. This is the internal field and its time changes that needs to be accurately accounted for when carrying out data reduction for ionospheric studies. The radial field at 400 km shows clear departures from a tilted dipole, with the high latitudes flux concentrations familiar from the core-mantle boundary again being evident. The radial field in the South Atlantic is distinctly weaker and there is a noticeable kink in the magnetic equator near South America. The radial field at satellite altitude is presently changing most rapidly at low latitudes in the American sector, while there are notable field accelerations taking place close to Indonesia and India–Pakistan, as well as in the mid-Pacific.

12.4.2 The Lithospheric Field

A recent map of the vertical field anomaly at the Earth's surface, due to the magnetized lithosphere is shown in Fig. 12.9 (top). This is derived from the LCS-1 model (Olsen et al. 2017) determined from CHAMP and *Swarm* field differences data and synthesized for spherical harmonic degrees $n = 16 - 185$. The map shows the detailed structure of lithospheric field features throughout the world including the cratonic regions of the continents (especially Archean and Proterozoic domains) that show stronger anomalies, and the long wavelength features associated with and sub-parallel to the oceanic magnetic reversal stripes that are seen consistently on or near widely separated isochrones (green lines). In LCS-1, we see for the first time from the satellite data alone EW oceanic features associated with the reversal stripes formed during the last 50 Ma of separation history of Australia from Antarctica. A number of other features on the ocean crust are evident. For example, there are NS trending lows in the vertical component map associated with the NS trending 85°E ridge in the Bay of Bengal.

The lithospheric power is higher in continental regions compared to oceanic regions, as expected due to the generally thicker continental/cratonic crust. The global average of B_r^2 at Earth's surface is 48.5 nT^2 (for spherical harmonic degrees $n = 16 - 185$), the power in continental regions is 66.1 nT^2 while that of the oceanic regions is only 39.4 nT^2, where the latter numbers are scaled to be whole Earth equivalent values. The LCS-1 model agrees well with the pre-*Swarm* satellite-data-based lithospheric field model MF7 (Maus 2010), up to its truncation degree of 133. If one wishes to study smaller scales of the lithospheric field, satellite data must be combined with near surface Aeromagnetic or marine survey data, for example as collected in the World Digital Magnetic Anomaly Map (WDMAM, currently in its second addition Lesur et al. 2016). Note, however, that the amplitude of small scale lithospheric field signals at satellite altitude is tiny. The bottom part of Fig. 12.9

shows the vertical component of the lithospheric field at satellite altitude (400 km); the scale is then ten times smaller, and the lithospheric signal is on the order of a few to tens of nT, it is for this reason it must be accounted for when studying the magnetic signals due to ionospheric currents, especially when considering weaker current systems.

12.5 Limitations of Present Main Field Models

Although the present generation of main field models are rather impressive and extremely useful for studying ionospheric current systems, it is nonetheless important that users are aware of their limitations, and that space physicists realize that there is a clear opportunity for them to contribute in improving future main field models.

The major factor limiting the accuracy of the internal part of field models is the inability to correctly account for and remove all magnetospheric and ionospheric signals (Finlay et al. 2017). This includes the difficulty in modelling rapid changes in the magnetospheric field. Global coverage (requiring many days with the present satellite missions) are formally required to perform a separation into internal and external field components. Yet the magnetospheric field changes much faster on time scales of minutes to hours; present models try to account for this using activity indices based on ground-based observatory data but there are differences between ground and satellite data that remain poorly understood.

Another source of uncertainty is the internal signal due to currents induced in the electrically conducting mantle by the time-changing external field, which remains difficult to isolate. Present separations often rely on a priori models of the conductivity of the mantle and lithosphere, which although improving (e.g. Kuvshinov 2012) are subject to uncertainties.

Perhaps the most serious issue affecting the accuracy of today's internal field models come from the rapidly changing, high amplitude signals due to polar and auroral current systems that are driven by magnetosphere-ionosphere coupling. Most main field models seek to avoid the impact of strong field-aligned currents at satellite altitude by using only scalar intensity data (field aligned currents have magnetic perturbations that are small in the direction of the main field). However, even scalar data are affected by horizontal currents flowing in the polar ionospheric E-layer that result from the closure of field-aligned currents. These include eastward and westward auroral electrojet currents that can be large even in dark regions, particularly in association with substorm events. At present, no main field model is able to adequately represent these signals, which means (as they are internal to the satellites) their signature can be inadvertently be mapped in the estimated internal field. Attempts have been made to model the time-averaged signature of such horizontal polar ionospheric currents (e.g. Lesur et al. 2008; Olsen et al. 2016a). For example, the latter study included the following extra term in their internal potential model, describing signals arranged by quasi-dipole latitude θ_{QD} and magnetic local time τ

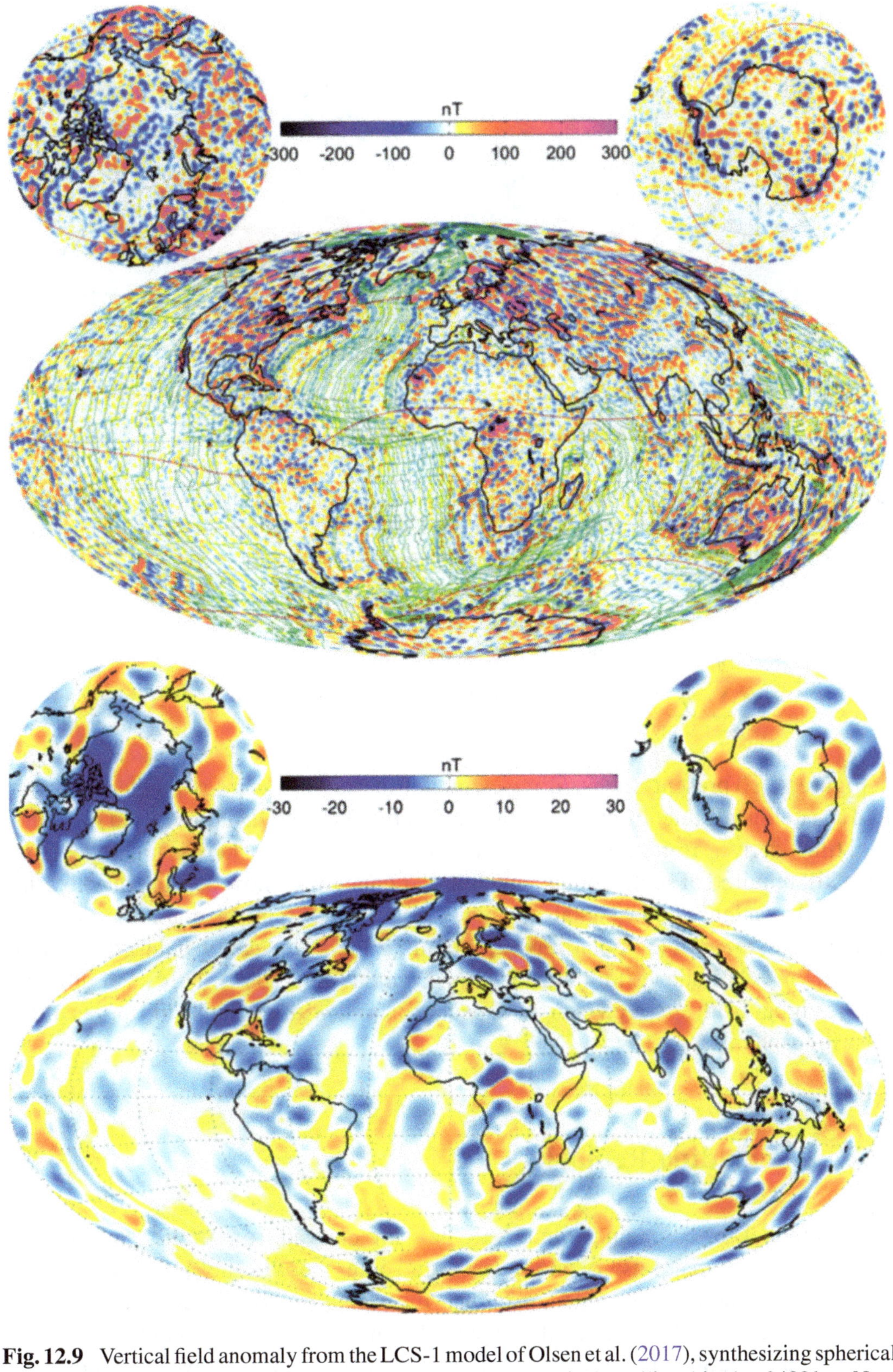

Fig. 12.9 Vertical field anomaly from the LCS-1 model of Olsen et al. (2017), synthesizing spherical harmonic degrees 16–185, at Earth's surface (top) and at a typical satellite altitude of 400 km. Note the change in colour scale from a range of ±300 nT (surface, top) to ±30 nT (400 km altitude, bottom)

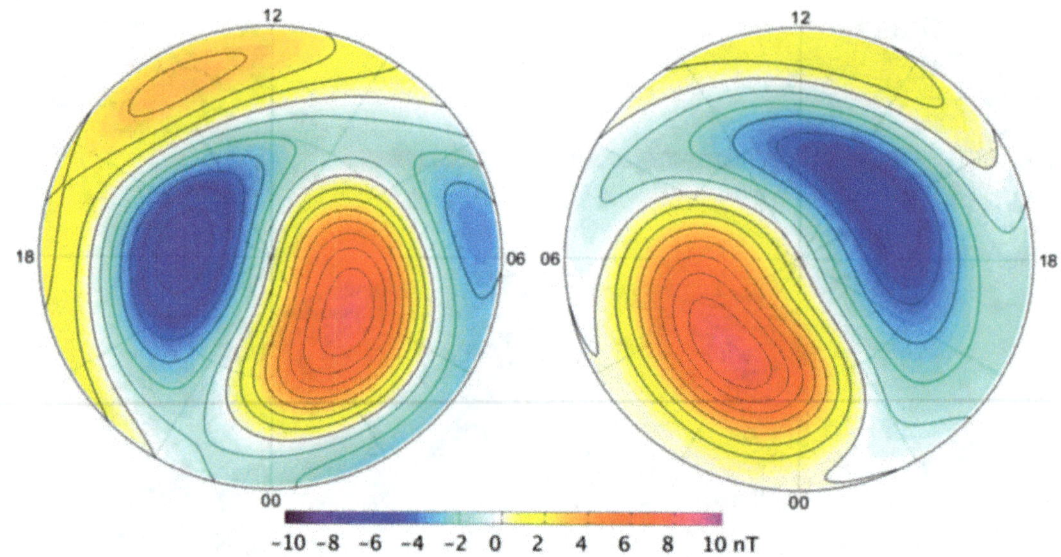

Fig. 12.10 Scalar magnetic field due to ionospheric currents, in dependence on QD latitude and magnetic local time (MLT, in hours) for the Northern (left) and Southern (right) polar regions as given by the potential V_{MLT} of Eq. 12.20. The low-latitude boundary here is at ±60 quasi-dipole latitude. After Olsen et al. (2016a)

$$V^{\text{MLT}} = a\sum_{n=1}^{20}\sum_{m=1}^{n}\left(g_n^{m,\text{MLT}}\cos m\tau + h_n^{m,\text{MLT}}\sin m\tau\right)\left(\frac{a}{r}\right)^{n+1} P_n^m(\cos\theta_{QD}) \tag{12.20}$$

Figure 12.10 presents the resulting estimates of the time-averaged scalar magnetic field at satellite altitude from such sources, as a function of QD latitude and magnetic local time, derived from more than two years of Swarm data. There is much development still required of such models, in particular concerning how best to parameterize the time dependence of the signal, and how to more consistently account for the entire 3D current system in the polar region, including the influence of field-aligned currents on lower latitude vector field data (e.g. Laundal et al. 2016). This work urgently requires input from space physicists, especially those with expertise in studies of the polar ionosphere.

12.6 Concluding Remarks

Modern models of the main geomagnetic field are derived from multi-satellite data, and increasingly make use of along-track and inter-satellite field differences (i.e. approximate gradients) in order to reduce the signatures of large-scale magnetospheric sources and to enhance the signal of small-scale internal fields. Examples from the construction of two such models, CHAOS-6 and LCS-1 have been presented here in detail. Such advanced main field models, including contributions from the small-scale lithospheric field and accounting for the near-Earth signature of magnetospheric sources, now enable more detailed study of relatively weak ionospheric current systems (Stolle et al. 2016).

Field gradient estimates can easily be incorporated within the conventional modelling framework as field differences, while also differencing the rows of the corresponding kernel matrices associated with each data point. It is very important to correctly specify the data uncertainty budget for vector, scalar and field gradient data, treating each satellite separately, in order for these to contribute appropriately during the field estimation procedure. Use of such procedures has enabled new details of the core field (in particular its time-dependent accelerations) and the lithospheric field (anomalies on scales down to 250 km) to be imaged using data from the *Swarm* and CHAMP missions.

Internal field models are nevertheless certainly imperfect, in particular in the polar region. New ideas on how best to parameterize quiet-time field variations in the polar region are much needed; this represents a clear opportunity for the space physics community to apply their expertise in another domain.

Acknowledgements The author is grateful to Nils Olsen and Stavros Kotsiaros for collaboration on the work presented here, and for their help in preparing some of the material presented. Thanks also to Patrick Alken and Clemens Kloss for helpful comments on an early draft. The European Space Agency (ESA) is acknowledged for providing the Swarm data and for financially supporting the work on developing the Swarm Level 1 product 'Magnet'. The CHAMP mission was sponsored by the Space Agency of the German Aerospace Center (DLR) through funds of the Federal Ministry of Economics and Technology. The author thanks the International Space Science Institute in Bern, Switzerland, for supporting the ISSI Working Group: 'Multi-Satellite Analysis Tools, Ionosphere', from which this chapter resulted. The Editors thank Vincent Lesur for his assistance in evaluating this chapter.

References

Alken, P. 2016. Observations and modeling of the ionospheric gravity and diamagnetic current systems from champ and swarm measurements. *Journal of Geophysical Research: Space Physics* 121 (1): 589–601.

Backus, G.E. 1970. Non-uniqueness of the external geomagnetic field determined by surface intensity measurements. *Journal of Geophysical Research* 75 (31): 6339–6341.

Bloxham, J., and A. Jackson. 1992. Time-dependent mapping of the magnetic field at the core-mantle boundary. *Journal of Geophysical Research* 97: 19. https://doi.org/10.1029/92JB01591.

de Boor, C. 1978. A practical guide to splines.

Chulliat, A., and S. Maus. 2014. Geomagnetic secular acceleration, jerks, and a localized standing wave at the core surface from 2000 to 2010. *Journal of Geophysical Research: Solid Earth* 119 (3): 1531–1543.

Chulliat, A., P. Alken, and S. Maus. 2015. Fast equatorial waves propagating at the top of the earth's core. *Geophysical Research Letters* 42 (9): 3321–3329.

Constable, C. 1988. Parameter estimation in non-gaussian noise. *Geophysical Journal International* 94 (1): 131–142.

Finlay, C.C., N. Olsen, and L. Tøffner-Clausen. 2015. Dtu candidate field models for igrf-12 and the chaos-5 geomagnetic field model. *Earth, Planets and Space* 67 (1): 114.

Finlay, C.C., N. Olsen, S. Kotsiaros, N. Gillet, and L. Tøffner-Clausen. 2016. Recent geomagnetic secular variation from swarm. *Earth, Planets and Space* 68 (1): 1–18.

Finlay, C.C., V. Lesur, E. Thébault, F. Vervelidou, A. Morschhauser, and R. Shore. 2017. Challenges handling magnetospheric and ionospheric signals in internal geomagnetic field modelling. *Space Science Reviews* 206 (1–4):157. https://doi.org/10.1007/s11214-016-0285-9.

Friis-Christensen, E., H. Lühr, and G. Hulot. 2006. Swarm: A constellation to study the earth's magnetic field. *Earth, Planets and Space* 58 (4): 351–358.

Gubbins, D. 2004. *Time series analysis and inverse theory for geophysicists*. Cambridge University Press.

Herceg, M., P.S. Jørgensen, and J.L. Jørgensen. 2017. Characterization and compensation of thermo-elastic instability of swarm optical bench on micro advanced stellar compass attitude observations. *Acta Astronautica* 137: 205–213.

Holme, R., N. Olsen, and F. Bairstow. 2011. Mapping geomagnetic secular variation at the core-mantle boundary. *Geophysical Journal International* 186 (2): 521–528.

Jackson, A., and C. Finlay. 2007. Geomagnetic secular variation and its applications to the core. *Treatise on Geophysics* 5: 147–193.

Jørgensen, J., T. Denver, M. Betto, and P. Jorgensen. 2003. Microasc a miniature star tracker, small satellites for earth observations. In: *Fourth International Symposium of the IAA, Berlin*.

Jørgensen, J.L., E. Friis-Christensen, P. Brauer, F. Primdahl, P.S. Jørgensen, T.H. Allin, T. Denver et al. 2008. The swarm magnetometry package. In: *Small Satellites for Earth Observation*, 143–151. Springer.

Kauristie, K., A. Morschhauser, N. Olsen, C. Finlay, R. McPherron, J. Gjerloev, and H.J. Opgenoorth. 2017. On the usage of geomagnetic indices for data selection in internal field modelling. *Space Science Reviews* 206 (1–4): 61–90.

Kotsiaros, S. 2016. Toward more complete magnetic gradiometry with the swarm mission. *Earth, Planets and Space* 68 (1): 130.

Kotsiaros, S., C. Finlay, and N. Olsen. 2014. Use of along-track magnetic field differences in lithospheric field modelling. *Geophysical Journal International* 200 (2): 880–889.

Kuvshinov, A.V. 2012. Deep electromagnetic studies from land, sea, and space: Progress status in the past 10 years. *Surveys in Geophysics* 33(1):169–209. https://doi.org/10.1007/s10712-011-9118-2.

Langel, R.A. 1987. The main field. *Geomagnetism* 1: 249–512.

Laundal, K.M., and A.D. Richmond. 2017. Magnetic coordinate systems. *Space Science Reviews* 206(1–4):27. https://doi.org/10.1007/s11214-016-0275-y.

Laundal, K.M., C.C. Finlay, and N. Olsen. 2016. Sunlight effects on the 3d polar current system determined from low earth orbit measurements. *Earth, Planets and Space* 68(1): 1. https://doi.org/10.1186/s40623-016-0518-x.

Lesur, V., I. Wardinski, M. Rother, and M. Mandea. 2008. Grimm: The GFZ reference internal magnetic model based on vector satellite and observatory data. *Geophysical Journal International* 173 (2): 382–394.

Lesur, V., I. Wardinski, M. Hamoudi, and M. Rother. 2010. The second generation of the GFZ reference internal magnetic model: GRIMM-2. *Earth, Planets and Space* 62 (10): 6.

Lesur, V., M. Rother, I. Wardinski, R. Schachtschneider, M. Hamoudi, and A. Chambodut. 2015a. Parent magnetic field models for the IGRF-12GFZ-candidates. *Earth, Planets and Space* 67 (1): 1–15.

Lesur, V., K. Whaler, and I. Wardinski. 2015b. Are geomagnetic data consistent with stably stratified flow at the core-mantle boundary? *Geophysical Journal International* 201 (2): 929–946.

Lesur, V., M. Hamoudi, Y. Choi, J. Dyment, and E. Thébault. 2016. Building the second version of the world digital magnetic anomaly map (wdmam). *Earth, Planets and Space* 68 (1): 27.

Lowes, F. 1975. Vector errors in spherical harmonic analysis of scalar data. *Geophysical Journal International* 42 (2): 637–651.

Lowes, F., and N. Olsen. 2004. A more realistic estimate of the variances and systematic errors in spherical harmonic geomagnetic field models. *Geophysical Journal International* 157 (3): 1027–1044.

Maus, S. 2010. Magnetic field model MF7.

Maus, S., and H. Lühr. 2005. Signature of the quiet-time magnetospheric magnetic field and its electromagnetic induction in the rotating earth. *Geophysical Journal International* 162 (3): 755–763.

Maus, S., H. Lühr, G. Balasis, M. Rother, and M. Mandea. 2005. Introducing pomme, the potsdam magnetic model of the earth. In: Earth Observation With Champ, Results From Three Years In Orbit, 293–298.

Maus, S., H. Lühr, and M. Purucker. 2006a. Simulation of the high-degree lithospheric field recovery for the swarm constellation of satellites. *Earth, Planets and Space* 58 (4): 397–407.

Maus, S., M. Rother, C. Stolle, W. Mai, S. Choi, H. Lühr, D. Cooke, and C. Roth. 2006b. Third generation of the potsdam magnetic model of the earth (pomme). *Geochemistry, Geophysics, Geosystems* 7 (7).

Maus, S., C. Manoj, J. Rauberg, I. Michaelis, and H. Lühr. 2010. Noaa/ngdc candidate models for the 11th generation international geomagnetic reference field and the concurrent release of the 6th generation pomme magnetic model. *Earth, Planets and Space* 62 (10): 2.

Olsen, N. 2002. A model of the geomagnetic field and its secular variation for epoch 2000 estimated from ørsted data. *Geophysical Journal International* 149 (2): 454–462.

Olsen, N., S. Kotsiaros. 2011. Geomagnetic observations and models. In: *Magnetic Satellite Missions and Data IAGA*, Special Sopron Book Series, vol. 5, 27–44. Dordrecht: Springer.

Olsen, N., and C. Stolle. 2017. *Space Science Reviews* 206: 5. https://doi.org/10.1007/s11214-016-0279-7.

Olsen, N., L. Tøffner-Clausen, T.J. Sabaka, P. Brauer, J.M.G. Merayo, J.L. Jörgensen, J.M. Léger, O.V. Nielsen, F. Primdahl, and T. Risbo. 2003. Calibration of the ørsted vector magnetometer. *Earth, Planets, and Space* 55: 11–18.

Olsen, N., T.J. Sabaka, and F. Lowes. 2005. New parameterization of external and induced fields in geomagnetic field modeling, and a candidate model for IGRF 2005. *Earth, Planets and Space* 57 (12): 1141–1149.

Olsen, N., H. Lühr, T.J. Sabaka, M. Mandea, M. Rother, L. Tøffner-Clausen, and S. Choi. 2006. CHAOS–a model of the earth's magnetic field derived from CHAMP, ørsted, and SAC-C magnetic satellite data. *Geophysical Journal International* 166 (1): 67–75.

Olsen, N., M. Mandea, T.J. Sabaka, and L. Tøffner-Clausen. 2009. CHAOS-2–a geomagnetic field model derived from one decade of continuous satellite data. *Geophysical Journal International* 179 (3): 1477–1487.

Olsen, N., M. Mandea, T.J. Sabaka, and L. Tøffner-Clausen. 2010. The CHAOS-3 geomagnetic field model and candidates for the 11th generation IGRF. *Earth, Planets and Space* 62 (10): 1.

Olsen, N., H. Lühr, C.C. Finlay, T.J. Sabaka, I. Michaelis, J. Rauberg, and L. Tøffner-Clausen. 2014. The CHAOS-4 geomagnetic field model. *Geophysical Journal International* 197 (2): 815–827.

Olsen, N., G. Hulot, V. Lesur, C.C. Finlay, C. Beggan, A. Chulliat, T.J. Sabaka, R. Floberghagen, E. Friis-Christensen, R. Haagmans, et al. 2015. The swarm initial field model for the 2014 geomagnetic field. *Geophysical Research Letters* 42 (4): 1092–1098.

Olsen, N., C.C. Finlay, S. Kotsiaros, and L. Tøffner-Clausen. 2016a. A model of earth's magnetic field derived from 2 years of swarm satellite constellation data. *Earth, Planets and Space* 68 (1): 124.

Olsen, N., C. Stolle, R. Floberghagen, G. Hulot, and A. Kuvshinov. 2016b. Special issue "swarm science results after 2 years in space". *Earth, Planets and Space* 68 (1): 172.

Olsen, N., D. Ravat, C.C. Finlay, and L.K. Kother. 2017. LCS-1: A high resolution global model of the lithospheric magnetic field derived from champ and swarm satellite observations. *Geophysical Journal International*.

Riley, K.F., M.P. Hobson, and S.J. Bence. 2006. *Mathematical methods for physics and engineering: a comprehensive guide*. Cambridge University Press.

Roberts, P.H., and E.M. King. 2013. On the genesis of the earth's magnetism. *Reports on Progress in Physics* 76(9):096801.

Sabaka, T.J., N. Olsen, and R.A. Langel. 2002. A comprehensive model of the quiet-time, near-earth magnetic field: phase 3. *Geophysical Journal International* 151 (1): 32–68.

Sabaka, T.J., N. Olsen, and M.E. Purucker. 2004. Extending comprehensive models of the earth's magnetic field with ørsted and CHAMP data. *Geophysical Journal International* 159 (2): 521–547.

Sabaka, T.J., L. Tøffner-Clausen, and N. Olsen. 2013. Use of the comprehensive inversion method for swarm satellite data analysis. *Earth, Planets and Space* 65 (11): 2.

Sabaka, T.J., N. Olsen, R.H. Tyler, and A. Kuvshinov. 2015. CM5, a pre-swarm comprehensive geomagnetic field model derived from over 12 yr of CHAMP, ørsted, SAC-C and observatory data. *Geophysical Journal International* 200 (3): 1596–1626.

Sabaka, T.J., N. Olsen, R.H. Tyler, and A. Kuvshinov. 2018. A comprehensive model of earth's magnetic field determined from 4 years of swarm satellite observations. *Earth Planets Space* 70: 130. https://doi.org/10.1186/s40623-018-0896-3.

Seeber, G. 2003. Satellite geodesy: foundations, methods, and applications. Walter de gruyter

Stolle, C., I. Michaelis, and J. Rauberg. 2016. The role of high-resolution geomagnetic field models for investigating ionospheric currents at low earth orbit satellites. *Earth, Planets and Space* 68(1):1. https://doi.org/10.1186/s40623-016-0494-1.

Thébault, E., C. Finlay, P. Alken, C. Beggan, E. Canet, A. Chulliat, B. Langlais, V. Lesur, F. Lowes, C. Manoj, M. Rother, and R. Schachtschneider. 2015a. Evaluation of candidate geomagnetic field models for IGRF-12. *Earth, Planets, and Space* 67(112). https://doi.org/10.1186/s40623-015-0273-4.

Thébault, E., C.C. Finlay, C.D. Beggan, P. Alken, J. Aubert, O. Barrois, F. Bertrand, T. Bondar, A. Boness, L. Brocco, et al. 2015b. International geomagnetic reference field: the 12th generation. *Earth, Planets and Space* 67 (1): 79.

Thébault, E., P. Vigneron, B. Langlais, and G. Hulot. 2016. A swarm lithospheric magnetic field model to SH degree 80. *Earth, Planets and Space* 68 (1): 126.

Tøffner-Clausen, L., V. Lesur, N. Olsen, and C.C. Finlay. 2016. In-flight scalar calibration and characterisation of the swarm magnetometry package. *Earth, Planets, and Space* 68: 129. https://doi.org/10.1186/s40623-016-0501-6.

Winch, D., D. Ivers, J. Turner, and R. Stening. 2005. Geomagnetism and schmidt quasi-normalization. *Geophysical Journal International* 160 (2): 487–504.

Yin, F., and H. Lühr. 2011. Recalibration of the CHAMP satellite magnetic field measurements. *Measurement Science and Technology* 22(5):055101.

Correction to: Introduction to Spherical Elementary Current Systems

Heikki Vanhamäki and Liisa Juusola

Correction to:
Chapter 2 in: M. W. Dunlop and H. Lühr (eds.), *Ionospheric Multi-Spacecraft Analysis Tools*, ISSI Scientific Report Series 17, https://doi.org/10.1007/978-3-030-26732-2_2

The original version of this chapter was published without the Electronic Supplementary Material. It has now been included in Chapter 2. The erratum chapter has been updated with the change.

The updated version of this chapter can be found at
https://doi.org/10.1007/978-3-030-26732-2_2

M. W. Dunlop and H. Lühr (eds.), *Ionospheric Multi-Spacecraft Analysis Tools*, ISSI Scientific Report Series 17,
https://doi.org/10.1007/978-3-030-26732-2_13

Index

M. W. Dunlop and H. Lühr (eds.), *Ionospheric Multi-Spacecraft Analysis Tools*, ISSI Scientific Report Series 17,
https://doi.org/10.1007/978-3-030-26732-2

T

V

www.ingramcontent.com/pod-product-compliance
Lightning Source LLC
LaVergne TN
LVHW082118200926
848484LV00008B/1978